Semiconductor Device Modeling with SPICE

Semiconductor Device Modeling with SPICE

Paolo Antognetti   Editor

Professor
Department of Electronics (DIBE)
University of Genoa
Genoa, Italy

Giuseppe Massobrio   Coeditor

Research Associate
Department of Electronics (DIBE)
University of Genoa
Genoa, Italy

McGraw-Hill Book Company

New York   St. Louis   San Francisco   Auckland   Bogotá
Hamburg   London   Madrid   Mexico   Milan
Montreal   New Delhi   Panama   Paris   São Paulo
Singapore   Sydney   Tokyo   Toronto

Library of Congress Cataloging-in-Publication Data

Semiconductor device modeling with SPICE.

Includes index.
1. Electronic circuit design—Mathematical models.
2. Electronic circuit design—Data processing.
I. Antognetti, Paolo.   II. Massobrio, Giuseppe.
TK7867.S46   1987   621.395′0285′53   87-15306
ISBN 0-07-002107-4

234567890   DOC/DOC   89321098

ISBN 0-07-002107-4

*The editors for this book were Daniel A. Gonneau and Nancy Warren,
the designer was Naomi Auerbach, and the production supervisor was
Richard Ausburn. It was set in Century Schoolbook
by University Graphics, Inc.
Printed and bound by R. R. Donnelley & Sons Company.*

Contents

Contributors

Claudio Fasce   *SGS Microelettronica, Milan, Italy*

Giuseppe Massobrio   *Department of Electronics (DIBE), Genoa, Italy*

Ermete Meda   *Department of Electronics (DIBE), Genoa, Italy*

Enrico Profumo   *SGS Microelettronica, Milan, Italy*

Foreword

The SPICE program is used worldwide as an essential computer-aid for circuit design. One can simulate circuits prior to their fabrication and predict detailed performance. However, an essential part of using SPICE is the accurate extraction and understanding of the device model parameters. Namely, the variety of devices used in an integrated circuit includes diodes, FETs, and bipolar transistors. Each device has special features of its operation, and more importantly, the models each have special features and nuances. Moreover, to model integrated circuits over their entire range of operation, effects such as temperature and bias dependence must be considered.

This book can be an invaluable aid in effectively using SPICE and understanding the many details with the device models. Because the emphasis of the book is SPICE-based, the language and examples are well-focused toward that end. The evolving state of device models for scaled-down technology leaves a gap between classical device modeling texts and engineering needs for circuit design. This book helps to bridge this gap by addressing directly the model constraints and assumptions used in SPICE. In many cases there are discrepancies between the physics, models, and measurements. This book does a good job of pointing out many of these pitfalls. There are also examples of both device model limitations and their occurrence in circuit applications where the problems become apparent. Finally, the discussion of the internal structure of SPICE and its execution provide unique documentation to help the SPICE user understand how the models are implemented.

In total this book is a most helpful reference for SPICE models as well as a bridge between device physics and the models needed for circuit design. The material is well-focused toward application and provides quick reference to a host of useful facts. This book will be of great value to both practicing engineers as well as students.

ROBERT W. DUTTON
Stanford University

Preface

Everybody involved in CAD for electronic circuits knows how critical circuit simulation is to the design and optimization of a circuit. Today, SPICE (Simulation Program with Integrated Circuit Emphasis) is the circuit simulator most designers are familiar with; after more than a decade of running on all sorts of computers around the world, after solving critical problems in extremely complex circuits, SPICE can be regarded as the de facto standard in circuit simulation. Designers and engineers consider SPICE the "Berkeley connection," from the university where it originated in the early seventies. Professor D. O. Pederson can be considered the "father" (or "godfather") of SPICE. Engineers and designers are all indebted to Pederson and his students for giving us (free, by the way) such a powerful design aid. All SPICE lovers also know how critical the input model of the transistor is in order to obtain an experiment-matching output of SPICE; this is another way of saying "garbage in, garbage out."

With the explosion of VLSI, designers of integrated circuits are blossoming everywhere, in university courses, research labs, and system houses, as well as, obviously, semiconductor houses. Many new VLSI designers will not need to use circuit simulation, since design methodologies, like gate arrays and cell libraries, do not present the need to simulate the VLSI system at the circuit level. However, for more advanced or full-custom designs, or for improving and optimizing cells, certainly some circuit optimization is needed. It can be obtained from simulators like SPICE.

Today the emphasis is more on logic and system aspects; thus many designers lack the insight into the real transistor, into the physical mechanisms taking place across the junctions, and therefore have some difficulty keeping track of all the parameters which represent the complete model of the transistor (bipolar or MOS). Furthermore, even if one knows the physical meaning of the different parameters, it is often hard to answer the question, "For my transistor, how do I know the right value for this particular model parameter?" In fact, there are so many physical interactions inside the transistor that a correct parameter measurement setup requires a careful examination of the model and of the experimental procedure.

This book has been written as a reference guide for the electronic circuit designer (VLSI or PCB); it covers all the aspects of transistor models that a SPICE user, new or old, may want or need to know, from physics to models to parameter measurements and SPICE use. Certainly there are many books (college-level electronic circuits handbooks or textbooks) that provide details on transistor action and different modeling aspects; some other books may have an appendix describing a circuit simulation program and may mention SPICE, showing an input or output listing. However, for a designer it is certainly not easy to put things together from different sources, with different notations, in order to efficiently use SPICE (or an analogous circuit simulator).

The first part (Chapters 1 to 4) of this book deals with the physical devices (the *pn* junction and the bipolar, JFET, and MOS transistors), starting from the basic aspects and taking the reader step by step through the SPICE model; in fact, the reader does not need to know much about the transistor, since the text is written in a simple and easily understandable way. If some physical background is needed for either bipolar or MOS transistors, it can be found in the appendixes.

Chapters 5 and 6 show how to obtain the numerical values of the parameters of the transistor (bipolar and MOS) model. The parameters are obviously those of the SPICE model, so that SPICE users do not have to do any complicated conversion of parameters.

The final two chapters, Chapters 7 and 8, deal specifically with problems related to the actual use of SPICE; they can also be of great help to those wanting to improve SPICE to adapt it to the particular needs of the designer.

The authors are all graduates of, and work at, the University of Genoa, except two of them, who now work for the SGS Company in Milan.

P. ANTOGNETTI
Genoa, Italy

Semiconductor
Device Modeling
with SPICE

PN-Junction Diode
and Schottky Diode

Giuseppe Massobrio

Department of Electronics (DIBE),
University of Genoa, Genoa, Italy

The analysis and design of integrated circuits depend heavily on the use of suitable models for integrated circuit components. This is true in hand analysis, where simple models are generally used, and in computer analysis, where more complex models are utilized. Since any analysis is only as accurate as the model used, it is essential that the circuit designer know the origin of the models commonly used and their degree of approximation.

Section 1.1 emphasizes the basic aspects of *pn*-junction behavior. The following sections establish dc, large-signal, and small-signal mathematical models for integrated circuit diodes and indicate their equivalent models in SPICE (*s*imulation *p*rogram with *i*ntegrated *c*ircuit *e*mphasis).

SPICE diode behavior is directly regulated by the subroutines *DIODE* (it computes all the required parameters and loads them into the coefficient matrix), *PNJLIM* (it limits, for each iteration, the voltage variation of the *pn*-junction devices), and *TMPUPD* (it updates the temperature-dependent parameters), besides the subroutines concerning each type of analysis.

What will be developed in this chapter is much more important than what diode applications would indicate: it is, in fact, basic to an understanding of the behavior of all junction devices.

1.1   DC Current-Voltage Characteristics

1.1.1   DC characteristics of the ideal diode

a. *PN* diode.   Consider the simple monodimensional *pn* structure shown in Fig. 1-1*a,* whose characteristics at thermodynamic equilibrium are analyzed in Appendix A. Here the following simplifying assumptions are still supposed to be valid.

1. The uniform doping and electrical neutrality in the *p*- and *n*-type regions as shown in Fig. 1-1*b,* where N_A is the acceptor concentration in the *p*-type region and N_D is the donor concentration in the *n*-type region.

2. The abrupt depletion-layer approximation: The built-in potential and

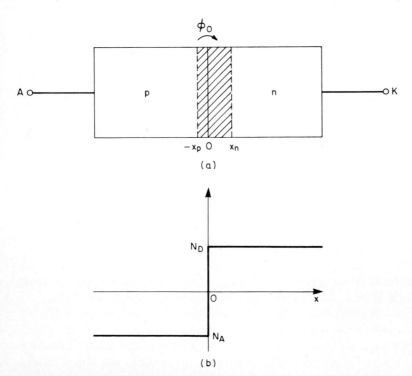

Figure 1-1   (*a*) *PN* junction under equilibrium conditions and (*b*) doping profiles in the *p* and *n* regions.

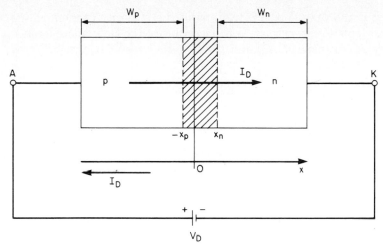

Figure 1-2 *PN* junction under nonequilibrium conditions: forward bias.

applied voltages are supported by a dipole layer with abrupt boundaries, and outside the boundaries the semiconductor is assumed to be neutral.

3. The Boltzmann approximation: Throughout the depletion layer, the Boltzmann relations [see Eqs. (A-45)] are valid.

4. The low-injection assumption: The injected minority-carrier densities are small compared with the majority-carrier densities.

5. No generation current exists in the depletion layer, and the electron and hole currents are constant through the depletion layer.

The analysis of this structure can now be extended to the case in which the junction is directly biased (positive bias on the p-type region with respect to the n-type region) by a constant voltage V_D. The application of the constant voltage source V_D alters the equilibrium conditions existing in the device; thus the circuit and all the transverse sections x of the diode are crossed by a constant current I_D, as shown in Fig. 1-2.

The resulting minority-carrier distribution is shown in Fig. 1-3, where x_p and x_n are the thicknesses of the depletion layer at the two sides of the junction; W_p and W_n are the thicknesses of the p and n neutral regions; and p_{n0}, n_{p0} and p_n, n_p are the minority-carrier concentrations in the neutral regions at equilibrium and at nonequilibrium, respectively.

The applied voltage V_D is sustained entirely across the junction space-charge layer, which is empty of carriers, causing a variation in the contact potential ϕ_0 [see Eqs. (A-25)], which is then given by

$$\phi_J = \phi_0 - V_D \tag{1-1}$$

Consequently, the precise concentrations of the minority carriers at the edges of the depletion layer vary, and, according to the Boltzmann boundary conditions [see Eqs. (A-45) and (A-46)], they result in the following equations:

$$n_p(-x_p) = n_{p0}e^{qV_D/kT} \tag{1-2}$$

$$p_n(x_n) = p_{n0}e^{qV_D/kT} \tag{1-3}$$

where kT/q = thermal voltage = 25.86×10^{-3} V at $T = 300$ K
q = electronic charge = 1.602×10^{-19} C
k = Boltzmann constant = 1.38×10^{-23} J/K

Equations (1-2) and (1-3) characterize the nonequilibrium situation. The minority-carrier concentrations at the edges of the space-charge layer are limited by the applied voltage (or *bias*) V_D. Moreover, the current I_D, being constant, can be calculated for any transverse section x of the diode as the sum of the contribution of the electrons and the holes, on the basis of the general relation

$$I_D = I_n + I_p \tag{1-4}$$

where I_n and I_p are the electron current and the hole current, respectively. They consist of the *drift* component due to field and the *diffusion* component due to carrier concentration gradient according to Eqs. (A-16),

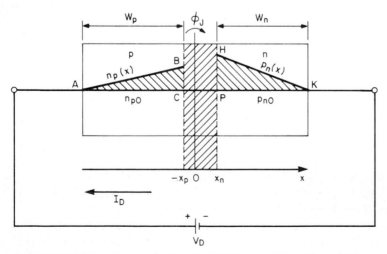

Figure 1-3 *PN* junction under nonequilibrium conditions: forward bias. Distribution of the minority carriers (the vertical axis has a linear scale).

which are given below for convenience in the form

$$I_n = I_{n,\text{diff}} + I_{n,\text{drift}} \tag{1-5}$$

$$I_p = I_{p,\text{diff}} + I_{p,\text{drift}} \tag{1-6}$$

Equations (1-5) and (1-6) represent a special case when they refer to the minority carriers belonging to the neutral regions W_p and W_n (Fig. 1-3). In that case, in fact, the drift currents, $I_{p,\text{drift}}$ in the n-type region and $I_{n,\text{drift}}$ in the p-type region, are quantitatively inferior to the corresponding diffusion currents. Thus it can be written

$$I_n \simeq I_{n,\text{diff}} \qquad \text{in the neutral region } W_p \tag{1-7}$$

$$I_p \simeq I_{p,\text{diff}} \qquad \text{in the neutral region } W_n \tag{1-8}$$

On this basis, the most suitable transverse sections for the evaluation of the total current I_D are those at the edges of the depletion layer, respectively defined at $x = -x_p$ and $x = x_n$.
At $x = -x_p$, from Eq. (1-4) it follows that

$$I_D - I_n(-x_p) + I_p(-x_p) \tag{1-9}$$

Moreover, as the flux of the carriers is constant in the depletion layer (simplifying assumption number 5), it can be written

$$I_p(-x_p) = I_p(x_n) \tag{1-10}$$

and, therefore, combining Eq. (1-9) with Eq. (1-10) yields

$$I_D = I_n(-x_p) + I_p(x_n) \tag{1-11}$$

Equation (1-11) is of fundamental importance in that it makes possible the calculation of the total current I_D as the sum of the single contribution of the diffusion currents of the minority carriers, calculated at $x = -x_p$ and $x = x_n$. Combining Eqs. (1-7) and (1-8) with Eq. (1-11) yields

$$I_D = I_{n,\text{diff}}(-x_p) + I_{p,\text{diff}}(x_n) \tag{1-12}$$

It is important to emphasize the physical significance of Eq. (1-12).

For a forward bias, the holes cross the junction from the p-type into the n-type region, where they become the injected minority current $I_{p,\text{diff}}(x_n)$. Similarly, the electrons cross the junction in the reverse direction (into the p-type region) and become the injected minority current $I_{n,\text{diff}}(-x_p)$.

Holes traveling from left to right constitute a current in the same direction as electrons moving from right to left; hence the resultant current

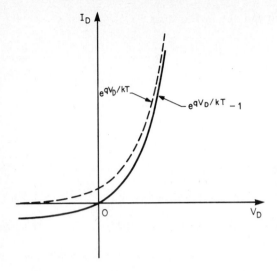

Figure 1-4 Ideal diode characteristic. *(Adapted from Semiconductor Electronics Education Committee [1]. Copyright © 1964, Education Development Center, Inc., Newton, Mass. Used by permission.)*

crossing the junction is the sum of the hole and electron minority currents [see Eq. (1-12)].

The diffusion current is proportional to the gradient of the distribution of the carriers [see Eqs. (A-15)] and is mathematically defined according to the following equations:

$$I_{n,\text{diff}} = qA_JD_n \frac{dn_p(x)}{dx} \qquad \text{for electrons} \qquad (1\text{-}13)$$

$$I_{p,\text{diff}} = -qA_JD_p \frac{dp_n(x)}{dx} \qquad \text{for holes} \qquad (1\text{-}14)$$

where A_J is the area of the transverse section of the diode; D_n and D_p are the electron and hole diffusion coefficients; and $n_p(x)$, $p_n(x)$ are the nonequilibrium distributions of the minority carriers in the neutral regions W_p and W_n. As the distributions of the carriers are linear (see Fig. 1-3), the gradient is obtained from the following simple relations:

$$\frac{dn_p}{dx} = \frac{n_p(-x_p) - n_p(A)}{W_p} = \frac{n_{p0}(e^{qV_D/kT} - 1)}{W_p} = \frac{n_i^2}{N_AW_p}(e^{qV_D/kT} - 1)$$

$$(1\text{-}15)$$

$$\frac{dp_n}{dx} = \frac{p_n(K) - p(x_n)}{W_n} = -\frac{p_{n0}(e^{qV_D/kT} - 1)}{W_n} = -\frac{n_i^2}{N_DW_n}(e^{qV_D/kT} - 1)$$

$$(1\text{-}16)$$

From Eqs. (1-12) to (1-16), it follows that

$$I_D = I_S(e^{qV_D/kT} - 1) \qquad (1\text{-}17)$$

where the quantity

$$I_S = qA_J n_i^2 \left(\frac{D_n}{N_A W_p} + \frac{D_p}{N_D W_n} \right) \tag{1-18}$$

has been defined as the diode *reverse saturation current*.

Equation (1-17) is the well-known *Shockley equation* and represents the *ideal diode* law. The ideal current-voltage relation is shown in Fig. 1-4 in a linear plot. Equations (1-17) and (1-18) "define" what is usually called the *short-base diode* (see Sec. A.2.3d).

In the reverse direction the current saturates at $-I_S$. The term *saturation* indicates that I_D approaches an asymptote and becomes independent of the voltage V_D.

Last, the electrical symbol for the ideal diode is shown in Fig. 1-5.

b. P^+N junction: Asymmetrical diode.   Equation (1-18) allows a much simpler relationship in the case of the one-sided p^+n abrupt junction. In fact, Eq. (1-18) becomes simply

$$I_S \simeq \frac{qA_J n_i^2 D_p}{W_n N_D} \tag{1-19}$$

being, in a p^+n junction, $N_A \gg N_D$.

The physical justification for this simplification has considerable importance, as it is the basic physical characteristic of the bipolar transistor. In fact if $N_A \gg N_D$, it follows from Eqs. (1-15) and (1-16) that $|dp_n/dx| \gg |dn_p/dx|$, so that Eq. (1-12) becomes

$$I_D \simeq I_{p,\text{diff}}(x_n) \tag{1-20}$$

Equation (1-20) means that, in the case of greatly asymmetrical doping, the region which is more heavily doped (p region) emits by diffusion a large number of majority carriers (holes), so by itself it will guarantee the supply of carriers to the current of the diode I_D. In that case the n-type region, being less doped, is poor in majority carriers; therefore, its emission to the adjacent p^+ region is negligible, and it has no relevant influence on the total current I_D.

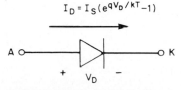

$$I_D = I_S(e^{qV_D/kT} - 1)$$

A ○———▷|———○ K

$+$ V_D $-$

Figure 1-5 Circuit representation of the ideal diode.

1.1.2   Limitations of the ideal diode model

The experimental results for most junction diodes show a range of the forward and reverse static characteristic for which there is good agreement with the idealized pn-junction diode. There are, however, differences over a significant range of useful biases for which the ideal diode equation becomes inaccurate, whether for forward bias ($V_D > \phi_0$) or for reverse bias ($V_D < \phi_0$), as is shown in Figs. 1-6 and 1-7.

In this section several mechanisms that cause the performance of diodes to differ from the predictions of idealized analysis are presented, including the following:

1. Carrier generation-recombination in the space-charge layer (region 1 of Fig. 1-6).

2. High-level injection (region 2 of Fig. 1-6).

3. Voltage drop associated with the electric field in the neutral regions (region 2 of Fig. 1-6).

4. Internal breakdown associated with high reverse voltage (region 3 of Fig. 1-7).

a. Carrier generation-recombination in the space-charge layer. One assumption that defines the idealized pn-junction diode model is that the hole and electron current components are constant throughout the space-charge layer. In reality, under forward-bias conditions, the injected carriers (holes from the p-type region and electrons from the n-type region)

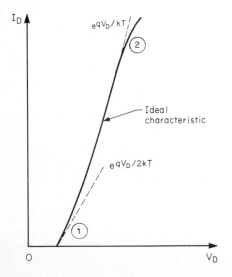

Figure 1-6 Real diode: forward characteristic.

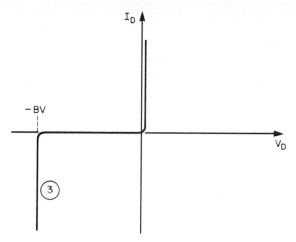

Figure 1-7   Real diode: reverse characteristic.

must cross this region, so that some carriers may be lost in this region by recombination.

To find an expression for the recombination current in the space-charge layer, we use the *Shockley-Read* theory (see Sec. A.1.3*b*). Then, for forward bias, the recombination current may be expressed by

$$I_D = I_S(e^{qV_D/nkT} - 1) \qquad (1\text{-}21)$$

where n, the emission coefficient, is usually about 2.

The measured I_D vs. V_D characteristic of a silicon diode showing the effect of the space-charge layer recombination current at low levels is presented in Fig. 1-6 (region 1).

b. High-level injection. It is known that for *low-level injection* the change in the majority-carrier concentration due to the injected minority carriers is so small in the regions outside the depletion region that its effect can be neglected. Thus, for a uniformly doped p^+n diode there is no electric field in the n region if current flows, and injected carriers move only by diffusion.

At *high-level injection,* on the contrary, there is a significant alteration of the majority-carrier concentration outside the depletion region, which gives rise to an electric field. Thus carrier motion is influenced by both diffusion and drift in this region. The presence of the electric field results in a voltage drop across this region. Hence only a fraction of the applied voltage appears across the junction.

As before, consider an abrupt p^+n junction with a wide, uniformly doped n region. The minority- and majority-carrier concentrations in this region are given in Fig. 1-8 for high-level injection conditions [2].

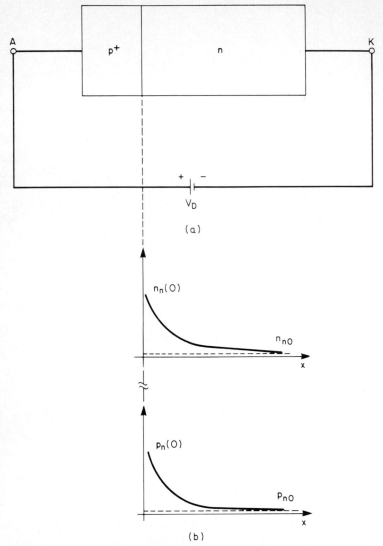

Figure 1-8 (a) p^+n diode structure and (b) minority- and majority-carrier concentrations for high-level injection. *(Adapted from S. K. Ghandhi, Semiconductor Power Devices, copyright © 1977 by John Wiley & Sons, Inc. Used by permission.)*

For charge neutrality, it is assumed that

$$n_n = p_n + N_D \simeq p_n \tag{1-22}$$

$$\frac{dn_n}{dx} \simeq \frac{dp_n}{dx} \tag{1-23}$$

Furthermore, in the p^+n junction, as noted in Sec. 1.1.1b, the electron current I_n is negligible with respect to the hole current I_p. Therefore

$$I_n = qA_J \left(\mu_n n_n E + D_n \frac{dn_n}{dx} \right) \simeq 0 \tag{1-24}$$

The electric field is then obtained from Eq. (1-24) as

$$E = - \frac{D_n}{\mu_n} \frac{1}{n_n} \frac{dn_n}{dx} = - \frac{D_p}{\mu_p} \frac{1}{p_n} \frac{dp_n}{dx} \tag{1-25}$$

when taking into account Eqs. (1-22) and (1-23) and Einstein's relations [see Eqs. (A-17)].

In reality, at high-level injection, the condition of neutrality is not rigorously observed; there exists a space charge that creates the electric field given in Eq. (1-25). However, since n_n is very high, a very small electric field is sufficient to balance the diffusion current and satisfy Eq. (1-24). The departure from the neutrality condition is therefore very negligible, and the assumption of neutrality is acceptable. Knowing the expression for the electric field allows us to determine the hole current

$$I_p = qA_J \left(\mu_p p_n E - D_p \frac{dp_n}{dx} \right) = -2qA_J D_p \frac{dp_n}{dx} \tag{1-26}$$

Regarding the transport of carriers under high-level injection conditions, it can be concluded that the contribution to the current of the electric field is equal to that of diffusion; therefore, it is as if the current were due only to diffusion, with doubled diffusion constant.

Under the assumption of linear distribution of the carriers, it follows that

$$\frac{dp_n}{dx} = \text{const.} = - \frac{p_n(x_n)}{W_n} \tag{1-27}$$

and the profile of the mobile carriers (minority and majority) is linear, being the doping profile constant, provided that the condition of high-level injection is verified.

Last, under quasi-equilibrium [see Eq. (A-59)] and charge-neutrality conditions at high-level injection [see Eq. (1-25)], it follows that

$$p_n(x_n) = n_n(x_n) = n_i e^{qV_D/2kT} \tag{1-28}$$

Taking into account Eqs. (1-26) and (1-27), the expression for the diode current can be obtained from Eq. (1-28). Then

$$I_D = I_p = \frac{2qA_J D_p n_i}{W_n} e^{qV_D/2kT} \tag{1-29}$$

In Eq. (1-29) the current depends again exponentially on the applied voltage V_D, but with the coefficient $q/2kT$ instead of q/kT (see Fig. 1-6, region 2).

c. Voltage drop associated with the electric field in the neutral regions.† The injection or extraction of excess minority carriers across the space-charge layer is invariably accompanied by an electric field in the neutral regions. Consequently, the voltage drops in the neutral regions which are associated with this electric field are linearly dependent on the total current. These voltage drops can be evaluated by integrating the electric field E over the neutral regions. Then, if V_D' is the voltage applied to the diode and V_D is the drop across the pn junction, it can be written that

$$V_D' = V_D + \int_{\text{neutral regions}} (-E)\, dx \qquad (1\text{-}30)$$

Because V_D is related to the total current I_D by Eq. (1-17), which is the result of the idealized model, Eq. (1-30) may be written as

$$V_D' = \underbrace{\frac{kT}{q} \ln\left(1 + \frac{I_D}{I_S}\right)}_{V_D} + \underbrace{r_S I_D}_{V_{rS}} \qquad (1\text{-}31)$$

where r_S is the series resistance given by

$$r_S = \frac{1}{I_D} \int_{\text{neutral regions}} (-E)\, dx \qquad (1\text{-}32)$$

Equation (1-31) suggests that, for small forward currents, the total diode voltage V_D' should vary logarithmically with I_D, in accordance with the idealized model. For large forward currents, on the contrary, the voltage should increase linearly with I_D because $I_D r_S$ increases faster than $(kT/q) \ln (1 + I_D/I_S)$, and thus it dominates V_D', as shown in Fig. 1-6, region 2.

d. Junction breakdown.‡ A more significant deviation of the reverse characteristic from the theoretical one may be seen when the reverse voltage applied range is greatly expanded, as in Fig. 1-7, region 3. As the reverse-bias voltage is increased, there comes a point, called the *breakdown voltage (BV)*, beyond which the current increases rapidly.

---

† The material in this section is taken from Semiconductor Electronics Education Committee [1]. Copyright © 1964, Education Development Center, Inc., Newton, Mass. Used by permission.

‡ The material in this section is derived from Millman [3]. Used by permission.

Two mechanisms of diode breakdown for increasing reverse voltage can be recognized. Consider the following situation. A thermally generated carrier (part of the reverse saturation current) falls down the junction barrier and acquires energy from the applied potential. This carrier collides with a crystal ion and imparts sufficient energy to disrupt a covalent bond. In addition to the original carrier, a new electron-hole pair has now been generated. These carriers may also pick up sufficient energy from the applied field, collide with another crystal ion, and create still another electron-hole pair. Thus each new carrier may, in turn, produce additional carriers through collision and the action of disrupting bonds, a process referred to as *avalanche multiplication*. Large reverse currents result, and the diode is said to be in the region of *avalanche breakdown* [3].

Even if the initially available carriers do not acquire sufficient energy to disrupt bonds, it is possible to initiate breakdown through a direct rupture of the bonds. Because of the existence of the electric field at the junction, a sufficiently strong force may be exerted on a bound electron by the field to tear it out of its covalent bond. The newly created electron-hole pair increases the reverse current. Note that this process, called *Zener breakdown*, does not involve collisions of carriers with the crystal ions (as does avalanche multiplication) [3].

1.2   Static Model

1.2.1   Static model of the ideal diode and its implementation in SPICE

a. DC forward characteristic.   The *dc forward characteristic* of the ideal diode is modeled by the nonlinear current source I_D, whose value is determined by Eq. (1-17), rewritten for convenience in the SPICE implemented form [4]

$$I_D = I_S(e^{qV_D/kT} - 1) + V_D\text{GMIN} \qquad \text{for } V_D \geq -5\frac{kT}{q} \qquad (1\text{-}33)$$

The model derived from Eq. (1-33) is outlined in Fig. 1-9.

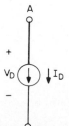

Figure 1-9   SPICE ideal diode dc model.

The forward dc characteristic is determined by only one parameter: the *saturation current* I_S, which is estimated from dc measurements of the forward-biased diode characteristic (see Chap. 5).

A graph of $\ln I_D$ vs. V_D is shown in Fig. 1-10. Then, the diode characteristic is determined by the equation

$$\ln I_D = \ln I_S + \frac{2.3q}{kT} V_D \qquad (1\text{-}34)$$

The ideal saturation current I_S can be determined by extrapolating the vertical axis intercept.

It is important to note that to aid convergence a small conductance GMIN is added by default in SPICE in parallel with every *pn* junction. The value of this conductance is a program parameter that can be set by the user with the *.OPTIONS* card. The default value for GMIN is 10^{-12} mho; moreover, the user *cannot* set GMIN to zero.

b. DC reverse characteristic.   The *dc reverse characteristic* of the ideal diode is modeled in SPICE by a nonlinear current source I_D, as shown in the following equations:

$$I_D = f(V_D)$$

$$= \begin{cases} I_S(e^{qV_D/kT} - 1) + V_D\text{GMIN} & \text{for } -5\dfrac{kT}{q} \le V_D \le 0 \\[2ex] -I_S + V_D\text{GMIN} & \text{for } V_D < -5\dfrac{kT}{q} \end{cases} \qquad (1\text{-}35)$$

Figure 1-11 shows the plot of the dc forward and reverse characteristics of an ideal diode using SPICE default parameter values (see Table 1-1).

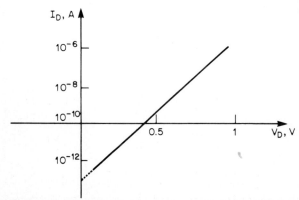

Figure 1-10   Graph of $\ln I_D$ vs. V_D in the forward-bias region.

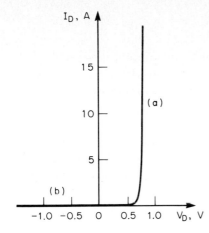

Figure 1-11 SPICE output plot using default parameter values for (a) ideal diode dc forward characteristic and (b) ideal diode dc reverse characteristic.

TABLE 1-1 SPICE Diode Model

Symbol	SPICE 2G keyword	Parameter name	Default value	Unit
I_S	IS	Saturation current	10^{-14}	A
r_S	RS	Ohmic resistance	0	Ω
n	N	Emission coefficient	1	
τ_D	TT	Transit time	0	s
$C_D(0)$	CJO	Zero-bias junction capacitance	0	F
ϕ_0	VJ†	Junction potential	1	V
m	M	Grading coefficient	0.5	
E_g	EG	Energy gap: 1.11 for Si 0.69 for SBD 0.67 for Ge	1.11	eV
p_t	XTI†	Saturation current temperature exponent: 3.0 for pn-junction diode 2.0 for SBD	3.0	
FC	FC	Coefficient for forward-bias depletion capacitance formula	0.5	
BV	BV	Reverse breakdown voltage (positive number)	∞	V
IBV	IBV	Reverse breakdown current (positive number)	10^{-3}	A
k_f	KF	Flicker-noise coefficient	0	
a_f	AF	Flicker-noise exponent	1	
T	‡	Nominal temperature for simulation and at which all input data is assumed to have been measured	27	°C

† In some SPICE versions, VJ is replaced by PB and XTI by PT.
‡ Remember, the temperature T is regarded as an operating condition and not as a model parameter.

15

1.2.2 Static model of the real diode
and its implementation in SPICE

In this section the secondary effects present in the real diode and their implementation in the SPICE diode model are considered. As in the preceding section the effects present in the forward and reverse characteristics are considered separately.

a. Small forward bias: Low-level injection. In some situations, the diode current is determined not by the so-called ideal diffusion component [see Eq. (1-33)] but by a nonideal component resulting from recombination in the space-charge layer. In this case, the diode current is described in SPICE by (see Sec. 1.1.2a)

$$I_D = I_S(e^{qV_D/nkT} - 1) + V_D\text{GMIN} \qquad \text{for } V_D \geq -5\,\frac{nkT}{q} \qquad (1\text{-}36)$$

where n is the emission coefficient ($1 \leq n \leq 2$). Also in this case, the dc characteristic of the diode is modeled by the nonlinear current source I_D, as shown in Fig. 1-9, whose value is now given by Eq. (1-36).

The parameters I_S and n are estimated from dc measurements of the forward-biased diode characteristics, at low-current levels.

It is important to note that the *dc* reverse characteristic also is a function of the parameters I_S and n through the equations

$$I = f(V_D)$$

$$= \begin{cases} I_S(e^{qV_D/nkT} - 1) + V_D\text{GMIN} & \text{for } -5\,\dfrac{nkT}{q} \leq V_D \leq 0 \\[4mm] -I_S + V_D\text{GMIN} & \text{for } V_D < -5\,\dfrac{nkT}{q} \end{cases}$$

$$(1\text{-}37)$$

Note also that n is a constant for the whole static characteristic computation.

Figure 1-12 shows the plot of the dc forward and reverse characteristics of a real diode under low-level bias conditions (determined by the n-parameter value). All the parameters used for the simulation are the same as those used in the ideal diode case, except for the value of the n parameter, which is different from unity.

A more sophisticated diode model, simulating a characteristic that at the same time allows $n = 1$ and $n \neq 1$, can be obtained by starting from the *Gummel-Poon* model of the transistor and short-circuiting one of the two junctions (see Sec. 2.9).

b. Large forward bias: Effect of the ohmic resistance.   At higher levels of bias, the diode current deviates from the ideal exponential characteristic given in Eq. (1-33) (see Sec. 1.1.2b). This deviation is due to the presence of an ohmic resistance in the diode as well as the effects of high-level injection. High-level injection is not included in the SPICE diode model. Therefore, the effects of both ohmic resistance and high-level injection are modeled by the ohmic resistance r_S. The value of r_S is determined by the amount the actual diode voltage deviates from the ideal exponential characteristic at a specific current. In practice, r_S is estimated at several values of I_D and averaged, since the values of r_S depend upon the diode current. In this case, the dc characteristic of the diode is modeled by the nonlinear current source I_D and the linear resistor r_S, so that

$$V_D' = r_S I_D + V_D \tag{1-38}$$

where I_D is given by Eq. (1-36). This model is outlined in Fig. 1-13.

Figure 1-14 shows the plot of the dc forward characteristic of a real diode under high-level bias conditions (determined by the r_S series-ohmic resistance). All the parameters used for the simulation are the same as those used for the ideal diode case except for the r_S value, which is different from zero.

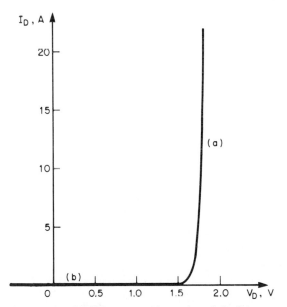

Figure 1-12 SPICE output plot using default parameter values and $n = 2$ for (a) real diode dc forward characteristic and (b) real diode dc reverse characteristic.

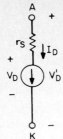

Figure 1-13 SPICE diode dc model at high-level injection.

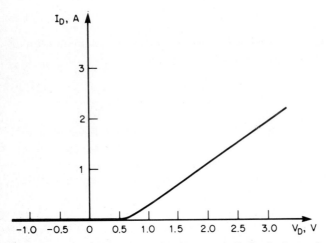

Figure 1-14 SPICE output plot for a real diode dc forward characteristic under high-level injection conditions with $RS = 1\ \Omega$. The other parameters have SPICE default values.

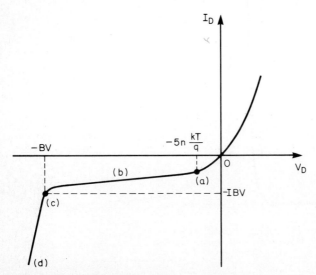

Figure 1-15 Reverse characteristic of the real diode.

c. Large reverse bias: Internal breakdown.   The SPICE reverse characteristic of the real diode has been divided into four distinct regions, (a) to (d), according to the value of the applied voltage V_D, as shown in Fig. 1-15.

The diode is modeled by a nonlinear current source I_D dependent on the applied voltage V_D ($V_D < 0$), according to the following equations:

$$I_D = f(V_D)$$

$$= \begin{cases} I_S(e^{qV_D/nkT} - 1) + V_D\text{GMIN} & \text{for } -5\,\dfrac{nkT}{q} \le V_D \le 0 \\[2ex] -I_S + V_D\text{GMIN} & \text{for } -BV < V_D < -5\,\dfrac{nkT}{q} \\[2ex] -IBV & \text{for } V_D = -BV \\[2ex] -I_S\left(e^{-q(BV+V_D)/kT} - 1 + \dfrac{qBV}{kT}\right) & \text{for } V_D < -BV \end{cases}$$

(1-39)

It must be noted that since (c) and (d) of Fig. 1-15 are obtained by different analytical expressions, problems of convergence for voltage values V_D close to both regions can arise. In fact, since from Eq. (1-39) it follows that

$$I_D(-BV) = IBV \tag{1-40}$$

Eq. (1-39), too, must converge to

$$\lim_{V_D \to -BV} I_D = -I_S\,\frac{qBV}{kT} = -IBV \tag{1-41}$$

The numerical value for IBV and BV is therefore of extreme importance, as it characterizes point (c) in Fig. 1-15.

The forward and reverse characteristics are modeled in SPICE by the following parameters, which the user can specify in the SPICE .MODEL card [4]:

IS = saturation current I_S
RS = ohmic resistance r_S
N = emission coefficient n
BV = breakdown voltage BV (it is a positive number)
IBV = breakdown current IBV (it is a positive number)

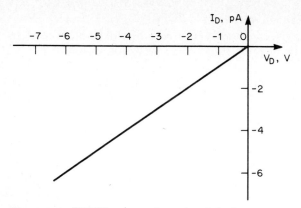

Figure 1-16 SPICE output plot using default parameter values for an ideal diode dc reverse characteristic.

Figure 1-16 shows the dc reverse characteristic of the ideal diode using SPICE default parameter values.

Figure 1-17 shows the dc reverse characteristic of the real diode with a specified BV parameter value, while the other parameter values are the same as those used for the ideal diode case.

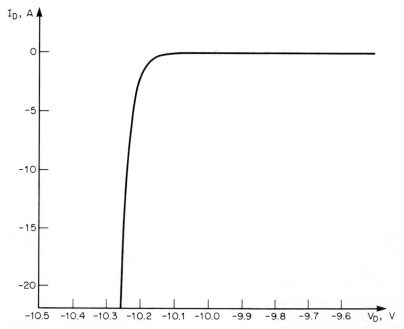

Figure 1-17 SPICE output plot for a real diode dc reverse characteristic with $BV = 10$ V. The other parameter values are SPICE default values.

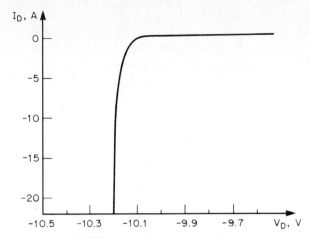

Figure 1-18 SPICE output plot for a real diode dc reverse characteristic with $BV = 10$ V and $IBV = 10$ mV. The other parameter values are SPICE default values.

Figure 1-18 shows the dc reverse characteristic of the real diode with specified BV and IBV parameter values, while the other parameter values are the same as those used for the ideal diode case.

In order to simulate a *Zener* diode, the macromodel shown in Fig. 1-19 can be used. Diode D_1 and the threshold voltage source V_0 represent the normal forward and reverse behavior [see region (*b*) in Fig. 1-15]. Diode D_2, the voltage source BV, and the resistance R_B define the breakdown

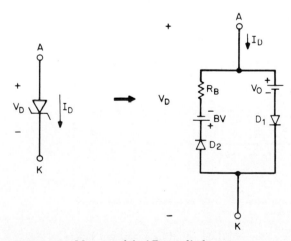

Figure 1-19 Macromodel of Zener diode.

region [region (d) in Fig. 1-15]. Diode D_2 does not conduct until $V_D = -BV$; if the reverse voltage is also increased, this diode becomes forward-biased and the reverse current crosses the resistor R_B.

A Zener diode model has been proposed and implemented in SPICE by Laha and Smart [5]. In this model, breakdown is modeled as a sum of one or two exponential terms; if breakdown is neglected, the model reduces to an ordinary diode model. A new limiting algorithm is also presented to ensure numerical stability during the iterative solution of the circuit equations.

1.3   Large-Signal Model

1.3.1   Large-signal model of the diode

Having defined the complete set of dc transport relationships for the *pn*-junction diode, let us consider the charge-storage effects of the device. The importance of these effects is clear. If there were no charge storage, the device would be infinitely fast; that is, there would be no charge inertia, and currents could be changed in zero time. The two forms of charge storage are the minority-carrier injection Q_s and the space charge Q_d.

a. Stored charge Q_s.   The electrons injected into the p side and the holes injected into the n side of the diode (see Fig. 1-20) not only generate the current I_D but also represent the charge Q_{sn} and Q_{sp} stored in the diode. These charges are described by the dashed areas of the triangles ABC and KHP.

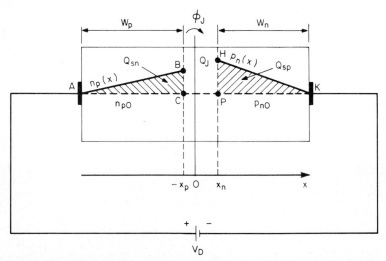

Figure 1-20 PN junction under nonequilibrium conditions: forward bias. Distribution of minority carriers (the vertical axis has a linear scale).

For the asymmetric p^+n diode $Q_{sn} \ll Q_{sp}$, so

$$Q_s \simeq Q_{sp} = \frac{qA_JW_np_{n0}(e^{qV_D/kT} - 1)}{2} = \frac{qA_JW_nn_i^2}{2N_D}(e^{qV_D/kT} - 1) \quad (1\text{-}42)$$

Charge storage results from excess minority carriers injected across the junction in forward bias. This charge is proportional to the total current injected across the junction, and hence it is exponential in voltage.

Comparing Eq. (1-42) with Eqs. (1-17) and (1-19), we can write

$$Q_s = \tau_D I_D(V_D) \quad (1\text{-}43)$$

where

$$\tau_D \equiv \frac{W_n^2}{2D_p} \quad (1\text{-}44)$$

The proportionality constant τ_D between the current and the charge is a time constant which represents the minimum time required to either store or remove the charge; it is called the *transit time* of the diode. To gain an intuitive appreciation of the concept of transit time, consider that the time required for the holes to travel across the n-type region of length W_n of a p^+n diode can be calculated as

$$\tau_D = \int_0^{W_n} \frac{dx}{v_p(x)} \quad (1\text{-}45)$$

where $v_p(x)$ is the hole speed, which is related to the hole current by

$$I_p = A_Jqv_p(x)[p_n(x) - p_{n0}] \quad (1\text{-}46)$$

Furthermore, if we remember that

$$I_p = -A_JqD_p\frac{dp}{dx} \quad (1\text{-}47)$$

the following expression for the velocity $v_p(x)$ of the holes injected into the less doped n-type region can be derived:

$$\frac{1}{v_p(x)} = \frac{1}{D_p}(W_n - x) \quad (1\text{-}48)$$

Last, substituting Eq. (1-48) into Eq. (1-45) and integrating across the edges of the n-type neutral region, it follows that

$$\tau_D = \int_0^{W_n} \frac{W_n - x}{D_p}dx = \frac{W_n^2}{2D_p} \quad (1\text{-}49)$$

which is Eq. (1-44) once more.

For the *short-base diode* with uniform doping concentrations, the transit time, as defined by Eq. (1-44), is equal to the time it takes for the minority carriers (holes) to travel across the *n*-type region of a p^+n structure.

b. Space charge Q_d.   In the ideal model, the magnitude of the total charge on either side of the metallurgical junction may be written [6] as

$$Q_d = A_J q x_p N_A = A_J q x_n N_D \tag{1-50}$$

In addition, from Eq. (A-36) it follows that

$$\phi_0 - V_D = \frac{q}{2\epsilon_s} (N_A x_p^2 + N_D x_n^2) \tag{1-51}$$

A change in V_D requires a variation in x_p and x_n (see Fig. 1-21); a change in $-x_p$ and x_n implies a change in the charge associated with the charge layer, from Eq. (1-50). Thus, this charge must be moved into or out of the space-charge region to balance a change in the junction voltage. This means that a capacitance must be associated with the space-charge region.

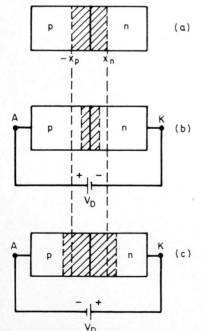

(a)

(b)

(c)

Figure 1-21 Behavior of *pn*-junction depletion-layer capacitance C_d as a function of the bias voltage V_D: (*a*) equilibrium condition; (*b*) forward-bias condition; (*c*) reverse-bias condition.

The equivalent capacitance associated with a change dV_D of the applied voltage V_D is then

$$C_d = \frac{dQ_d}{dV_D} \qquad (1\text{-}52)$$

The charge on one side of the capacitor is

$$Q_d = A_J q x_p N_A = A_J q x_n N_D \qquad (1\text{-}53)$$

Besides, if $N_A \gg N_D$, then x_n may be written as follows [see Eq. (A-36)]:

$$x_n = \sqrt{\frac{2\epsilon_s(\phi_0 - V_D)}{qN_D}} \qquad (1\text{-}54)$$

Hence

$$Q_d = A_J \sqrt{2\epsilon_s q N_D} \sqrt{\phi_0 - V_D} \qquad (1\text{-}55)$$

and

$$C_d = \frac{dQ_d}{dV_D} = \frac{A_J}{2} \frac{\sqrt{2\epsilon_s q N_D}}{\sqrt{\phi_0 - V_D}} \qquad (1\text{-}56)$$

Equation (1-56) can be written in the following two useful ways:

$$C_d = \frac{C_d(0)}{\sqrt{1 - V_D/\phi_0}} \qquad (1\text{-}57)$$

$$C_d = \frac{\epsilon_s A_J}{x_n} \qquad (1\text{-}58)$$

Equation (1-57) gives C_d, at any given voltage V_D, in terms of the zero-bias capacitance $C_d(0)$ and built-in voltage ϕ_0; Eq. (1-58) shows that the capacitance can be calculated by the same expression as a parallel-plate capacitor of area A_J spacing x_n.

To obtain the space charge Q_d it is necessary to integrate the space-charge capacitance with respect to voltage V; that is,

$$Q_d = \int_0^{V_D} C_d \, dV = 2\phi_0 C_d(0) \sqrt{1 - \frac{V_D}{\phi_0}} \qquad (1\text{-}59)$$

Now two important observations can be made. For reverse bias and small forward bias, Q_d is the dominant charge storage. For moderate forward bias and beyond, the injected charge Q_s dominates; hence the effects of Q_d become negligible.

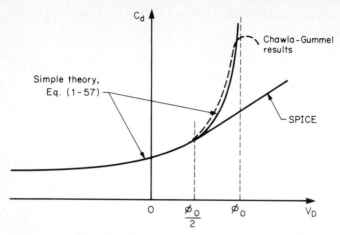

Figure 1-22 Behavior of pn-junction depletion-layer capacitance C_d as a function of the bias voltage V_D. *(From P. R. Gray and R. G. Meyer, Analysis and Design of Analog Integrated Circuits, copyright © 1977 by John Wiley & Sons, Inc. Used by permission.)*

This second observation is extremely important, since as V_D approaches ϕ_0, the space-charge capacitance [see Eq. (1-57)] becomes infinite, whereas charge Q_d becomes zero [see Eq. (1-55)]. Of course, in this case, Eq. (1-57) is no longer valid.

A more exact analysis of the behavior of C_d as a function of V_D gives the result shown in Fig. 1-22 [7].

For forward-bias voltages up to about $\phi_0/2$, the values of C_d predicted by Eq. (1-57) are very close to the more accurate value. Some computer programs approximate C_d for $V_D > \phi_0/2$ by a linear extrapolation of Eq. (1-57) [8].

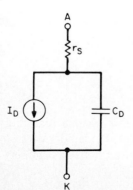

Figure 1-23 SPICE large-signal diode model.

1.3.2   Large-signal model of the diode and its implementation in SPICE

Having established the basic relationships of charge-storage effects given by Eqs. (1-43) and (1-59), we can now discuss in a straightforward way the details of the large-signal model implemented in SPICE.

The charge-storage element $Q_D = Q_s + Q_d$ that is shown in Fig. 1-23 models the charge storage of the diode in SPICE. It is determined by the following relations:

$$
Q_D = \begin{cases}
\underbrace{\tau_D I_D}_{Q_s} + \underbrace{C_d(0) \int_0^{V_D} \left(1 - \frac{V}{\phi_0}\right)^{-m} dV}_{Q_d} & \text{for } V_D < FC \times \phi_0 \\[3em]
\underbrace{\tau_D I_D}_{Q_s} + \underbrace{C_d(0)F_1 + \frac{C_d(0)}{F_2} \int_{FC \times \phi_0}^{V_D} \left(F_3 + \frac{mV}{\phi_0}\right) dV}_{Q_d} & \text{for } V_D \geq FC \times \phi_0
\end{cases}
$$

$$(1\text{-}60)$$

These elements can be defined equivalently by the capacitance relations

$$
C_D = \frac{dQ_D}{dV_D} = \begin{cases}
\underbrace{\tau_D \frac{dI_D}{dV_D}}_{C_s} + \underbrace{C_d(0) \left(1 - \frac{V_D}{\phi_0}\right)^{-m}}_{C_d} & \text{for } V_D < FC \times \phi_0 \\[3em]
\underbrace{\tau_D \frac{dI_D}{dV_D}}_{C_s} + \underbrace{\frac{C_d(0)}{F_2} \left(F_3 + \frac{mV_D}{\phi_0}\right)}_{C_d} & \text{for } V_D \geq FC \times \phi_0
\end{cases}
$$

$$(1\text{-}61)$$

where F_1, F_2, and F_3 are SPICE constants whose values are

$$
F_1 = \frac{\phi_0}{1 - m} [1 - (1 - FC)^{1-m}]
$$

$$
F_2 = (1 - FC)^{1+m}
$$

$$(1\text{-}62)$$

$$
F_3 = 1 - FC(1 + m)
$$

Then, the charge-storage element Q_D models two distinct charge-storage mechanisms as shown in Secs. 1.3.1a and b.

a. Stored charge Q_s. The stored charge Q_s due to injected minority carriers is modeled by the exponential term in Eq. (1-36) using Eq. (1-60), where I_S is the saturation current, n is the emission coefficient ($n = 1$ for the ideal diode), and τ_D is the transit time.

It is important to note that Q_s is proportional to I_D so that it is modeled by a highly nonlinear capacitor.

b. Space charge Q_d. The space charge Q_d is modeled by the parameter $C_D(0)$, the diode junction capacitance at zero bias ($V_D = 0$); m, the junction grading coefficient; ϕ_0, the junction potential (built-in voltage); and FC, the forward-bias junction capacitance coefficient.

Typically, ϕ_0 may range from 0.2 to 1 V, and m is 0.33 for a linearly graded junction or 0.5 for an abrupt junction.

FC determines how the depletion capacitance is calculated when the junction is forward-biased (by default, $FC = 0.5$); thus, when $V_D \geq FC \times \phi_0$, SPICE approximates C_d [see Eq. (1-57)] by a linear extrapolation (see Fig. 1-22).

The SPICE *large-signal* model for the diode is shown in Fig. 1-23.

1.4   Small-Signal Model

The models of the diode discussed so far maintain the strictly nonlinear nature of the physics of the device, and therefore are able to represent, both statically and dynamically, the behavior of the diode for both types of bias V_D. In some circuit situations, however, the characteristics of the device must be represented only in a restricted range of currents and voltages.

In particular, for small variations around the operating point which is fixed by a constant source, the nonlinear characteristic of the diode can be *linearized* so that the incremental current of the diode becomes proportional to the incremental voltage, if the variations are sufficiently small.

The development of an incremental linear model is not necessarily linked to a graphic interpretation of the relationships between the variables; the mathematical development can be based on the linearization of a nonlinear functional relationship, by means of the expansion in *Taylor series* truncated after the first-order term.

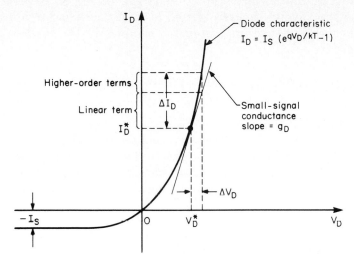

Figure 1-24 Relationship between diode static I_D vs. V_D characteristic and the small-signal conductance. *(From Semiconductor Electronics Education Committee [1]. Copyright © 1964, Education Development Center, Inc., Newton, Mass. Used by permission.)*

1.4.1 Small-signal model of the diode and its implementation in SPICE†

The *small-signal* model can be found by considering small-voltage variations of V_D around a given dc voltage V_D^* and writing $V_D = V_D^* + \Delta V_D$, as indicated in Fig. 1.24.

The interest is in the corresponding small change of current, which is proportional to the change in voltage if the voltage change is small enough. This linear relationship can be used to define a *small-signal conductance* $g_D = dI_D/dV_D$.

As the conductance is a dc or *low-frequency* concept, it can be evaluated by using the large-signal dc current-voltage relationship of Eq. (1-36), which may be expanded in a Taylor series about V_D^*:

$$I_D = I_S(e^{qV_D^*/nkT} - 1)$$
$$+ I_S e^{qV_D^*/nkT} \left[\frac{q\,\Delta V_D}{nkT} + \frac{1}{2!}\left(\frac{q\,\Delta V_D}{nkT}\right)^2 + \cdots \right] \qquad (1\text{-}63)$$

The first term on the right-hand side of Eq. (1-63) is the operating-point current I_D^*; the second term is I_D, the small-signal component of the

---

† The material in this section is taken in part from Semiconductor Electronics Education Committee [1]. Copyright © 1964, Education Development Center, Inc., Newton, Mass. Used by permission.

current, which is approximately linearly related to V_D if the second term in the brackets is much less than the first one. Consequently, the *small-signal conductance* is

$$g_D = \left.\frac{dI_D}{dV_D}\right|_{op} = \frac{qI_S}{nkT}\, e^{qV_D^*/nkT} \tag{1-64}$$

Figure 1-24 shows that g_D is equal to the slope of the static characteristic at the operating point. With moderate reverse bias, the current I_D is negative and approaches the saturation current, so that the conductance g_D is zero. On the other hand, for moderate forward bias, the current I_D is much larger than the saturation current, and the conductance is approximately proportional to I_D.

Further, the charge Q_D stored in the diode changes an amount dQ_D for a small voltage V_D; so if $Q_D = Q_s + Q_d$, it can be written

$$
\begin{aligned}
C_s &= \left.\frac{dQ_s}{dV_D}\right|_{op} \\[2mm]
C_d &= \left.\frac{dQ_d}{dV_D}\right|_{op}
\end{aligned}
\tag{1-65}
$$

From Eqs. (1-60) and (1-64), it follows that

$$
C_d = \left.\frac{dQ_D}{dV_D}\right|_{op} =
\begin{cases}
\underbrace{\tau_D g_D}_{C_s} + \underbrace{C_d(0)\left(1 - \frac{V_D}{\phi_0}\right)^{-m}}_{C_d} & \text{for } V_D < FC \times \phi_0 \\[6mm]
\underbrace{\tau_D g_D}_{C_s} + \underbrace{\frac{C_d(0)}{F_2}\left(F_3 + \frac{mV_D}{\phi_0}\right)}_{C_d} & \text{for } V_D \geq FC \times \phi_0
\end{cases}
\tag{1-66}
$$

where F_2 and F_3 are expressed by Eq. (1-62).

Of course, g_D and C_D are evaluated at the dc operating point (I_D, V_D). The SPICE *small-signal* linearized model for the diode is shown in Fig. 1-25.

1.5 Schottky Diode and Its Implementation in SPICE

A rectifying barrier is usually formed when a metal is in contact with a semiconductor. These structures are electrically similar to abrupt one-sided p^+n junctions, but several important differences make these diodes, called *Schottky barrier diodes (SBDs),* attractive and useful.

First, the diode operates as a majority-carrier device under low-level injection conditions. As a consequence, storage time (due to storage of minority carriers) is eliminated and a fast response is obtained [9, 10].

Second, for a fixed voltage bias, the current through the device is typically more than two orders of magnitude greater than it is for the junction device, as shown in Fig. 1-26.

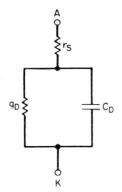

Figure 1-25 SPICE linearized, small-signal diode model.

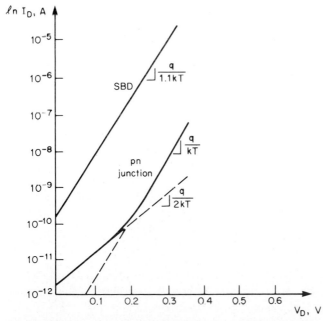

Figure 1-26 Current-voltage characteristic for Schottky barrier diode (SBD) and junction diode (*pn*).

A detailed understanding of the Schottky diode involves energy-band theory, which is beyond the scope of this book. In this case, it is sufficient to say that the SBD current I_D is given by

$$I_D = I_S(e^{qV_D/nkT} - 1) \qquad (1\text{-}67)$$

where
$$I_S = RT^2 e^{-q\phi_B/kT} \qquad (1\text{-}68)$$

in which R is a constant and ϕ_B is the Schottky barrier voltage, depending only on the material (metal-semiconductor). It is not a function of semiconductor doping, and it is only a weak function of temperature.

Equation (1-67) is similar to Shockley's equation for the *pn* junction [see Eq. (1-21)]. However, Eq. (1-68), representing the saturation current, is quite different from Eq. (1-18) and is a function of the barrier voltage.

Not all metal-semiconductor contacts have diode characteristics. When the semiconductor is heavily doped (N_A or $N_D \gg 10^{17}\ \text{cm}^{-3}$), the depletion layer becomes so narrow that carriers can travel through the potential barrier in either direction by a quantum-mechanical carrier transport mechanism known as *tunneling*. Under these conditions, current flows equally well in either direction, resulting in what is usually described as an *ohmic* contact. In fact, this is the physical explanation for almost all ohmic contacts to semiconductor devices [11].

At sufficiently high reverse voltages, avalanche breakdown occurs, similar to that observed in *pn*-junction diodes. The previous *pn*-junction diode models implemented in SPICE (dc, large-signal, and small-signal) apply just as well to the Schottky barrier diode, but with I_S given by Eq. (1-68) and, of course, $\tau_D = 0$.

1.6   Temperature and Area Effects on the Diode Model

This section analyzes the effects that temperature and area factors have on the diode model implemented in SPICE. However, in order to more easily understand the relations introduced, there follows a brief description of the physical quantities that have a marked dependence on temperature.

1.6.1   Temperature dependence of the diode characteristic†

In many applications of junction diodes, the temperature dependence of the diode characteristic is important. Although the diode equation con-

---

† The material in this section is taken in part from the Semiconductor Electronics Education Committee [1]. Copyright © 1964, Education Development Center, Inc., Newton, Mass. Used by permission.

tains the absolute temperature explicitly in the exponent qV_D/kT, the principal temperature dependence of the characteristic results from the extremely strong implicit temperature dependence of the saturation current I_S [12].

The saturation current contains several parameters which depend on temperature. However, the temperature dependence of I_S is dominated by the strong dependence on temperature of the equilibrium minority-carrier concentrations. These equilibrium concentrations are related to the intrinsic carrier concentration and to the doping according to Eqs. (A-7) and (A-9). Consequently, the saturation current is [see Eq. (1-18)]

$$I_S = qA_J n_i^2 \left(\frac{D_n}{N_A W_p} + \frac{D_p}{N_D W_n} \right)$$
(1-69)

The factor in parentheses is not strongly temperature-dependent. Therefore, the principal temperature dependence of the saturation current is that of n_i^2, which is

$$n_i^2 = BT^3 e^{-E_g(0)/kT}$$
(1-70)

where B is a constant and $E_g(0)$ is the width of the energy gap extrapolated to absolute zero (0 K).

In fact, E_g is also a function of temperature, according to the general relation [13]

$$E_g(T) = E_g(0) - \frac{\alpha T^2}{\beta + T}$$
(1-71)

For Si experimental results give $\alpha = 7.02 \times 10^{-4}$, $\beta = 1108$, and $E_g(0) = 1.16$ eV.

Last, it is necessary to emphasize that even the built-in voltage ϕ_0 [see Eqs. (A-27)], rewritten here for convenience,

$$\phi_0 = \frac{kT}{q} \ln \frac{N_A N_D}{n_i^2}$$
(1-72)

is dependent on temperature both explicitly and through n_i [see Eq. (1-70)].

1.6.2 Temperature dependence of the diode model parameters

All input data for SPICE are assumed to have been measured at 27°C (300 K). The simulation also assumes a *nominal temperature (TNOM)* of 27°C that can be changed with the TNOM option of the .OPTIONS control card.

The circuits can be simulated at temperatures other than TNOM by using a *.TEMP* control card. Temperature appears explicitly in the exponential term of the diode model equation.

In addition, four diode parameters are modified in SPICE to reflect changes in the temperature of *pn* and SBD diodes. They are I_S, the saturation current; ϕ_0, the junction potential; $C_D(0)$, the zero-bias junction capacitance; and FC, the coefficient for forward-bias depletion capacitance formula.

The temperature dependence of I_S in the SPICE diode model is determined by

$$I_S(T_2) = I_S(T_1) \left(\frac{T_2}{T_1}\right)^{p_t/n} \exp\left[-\frac{qE_g(300)}{kT_2}\left(1 - \frac{T_2}{T_1}\right)\right] \quad (1\text{-}73)$$

where two new model parameters are introduced: p_t, the saturation current temperature exponent, and E_g, the energy gap.

It is important to note that these parameters assume different values in the simulation of *pn* diodes or SBDs; that is, $p_t = 3$ for a *pn* diode and $p_t = 2$ for an SBD; E_g (300 K) = 1.11 eV for a *pn* diode (Si) and E_g (300 K) = 0.69 for an SBD.

The effect of temperature on ϕ_0 is modeled from Eqs. (1-70) and (1-72) by

$$\phi_0(T_2) = \frac{T_2}{T_1} \phi_0(T_1) - 2\frac{kT_2}{q} \ln\left(\frac{T_2}{T_1}\right)^{1.5} - \left[\frac{T_2}{T_1} E_g(T_1) - E_g(T_2)\right] \quad (1\text{-}74)$$

where $E_g(T_1)$ and $E_g(T_2)$ are described by the general Eq. (1-71).

The temperature dependence of $C_D(0)$ is determined by

$$C_D(T_2) = C_D(T_2)\left\{1 + m\left[400 \times 10^{-6}(T_2 - T_1) - \frac{\phi_0(T_2) - \phi_0(T_1)}{\phi_0(T_1)}\right]\right\} \quad (1\text{-}75)$$

The temperature dependence of the coefficient for forward-bias depletion capacitance formula is determined by

$$FCPB(T_2) = FC \times \phi_0 = FCPB(T_1)\frac{\phi_0(T_2)}{\phi_0(T_1)}$$

$$F_1(T_2) = F_1(T_1)\frac{\phi_0(T_2)}{\phi_0(T_1)} \quad (1\text{-}76)$$

In the preceding equations, T_1 and T_2 must be considered as follows:

- If TNOM in the .OPTIONS card is not specified, then T_1 = TNOM = TREF = 27°C (300 K); this is the default value for TNOM.

■ If a TNOM$_{new}$ value is specified, then T_1 = TNOM = 27°C; T_2 = TNOM$_{new}$.

These two cases are valid only if one temperature is requested.

If more than one temperature is requested (using the .TEMP card), then T_1 = TNOM = 27°C and T_2 is the first temperature specified in the .TEMP card. Afterward, T_1 is the last working temperature (T_2) and T_2 is the next temperature in the .TEMP card to be calculated.

It is very important to note that the temperature T is regarded as an operating condition and *not* as a model parameter, so that it is not possible to simulate the behavior of devices in the same circuit at different temperatures at the same time. In other words, the *same* temperature T is used in the *whole* circuit to be simulated.

1.6.3   Area dependence of the diode model parameters

The AREA factor used in SPICE for the diode model determines the number of equivalent parallel devices of a specified model. The diode model parameters affected by the AREA factor are I_S, r_S, and $C_D(0)$; that is,

$$I_S = I_S \times \text{AREA}$$

$$C_D(0) = C_D(0) \times \text{AREA} \tag{1-77}$$

$$r_S = r_S/\text{AREA}$$

The default value for the AREA parameter is 1.

REFERENCES

1. P. E. Gray, D. DeWitt, A. R. Boothroyd, and J. F. Gibbons, *Physical Electronics and Circuit Models of Transistors,* vol. 2, Semiconductor Electronics Education Committee, Wiley, New York, 1964.
2. S. K. Ghandhi, *Semiconductor Power Devices,* Wiley, New York, 1977.
3. J. Millman, *Microelectronics,* McGraw-Hill, New York, 1979.
4. L. W. Nagel, SPICE2: A Computer Program to Simulate Semiconductor Circuits, Electronics Research Laboratory Rep. No. ERL-M520, University of California, Berkeley, 1975.
5. A. Laha and D. Smart, A Zener Diode Model with Application to SPICE2, *IEEE J. Solid-State Circuits,* **16,** 1981.
6. J. F. Gibbons, *Semiconductor Electronics,* McGraw-Hill, New York, 1966.
7. B. R. Chawla and H. K. Gummel, Transition Region Capacitance of Diffused *pn* Junctions, *IEEE Trans. Electron Devices,* **ED18,** 1971.
8. P. R. Gray and R. G. Meyer, *Analysis and Design of Analog Integrated Circuits,* Wiley, New York, 1977.
9. A. S. Grove, *Physics and Technology of Semiconductor Devices,* Wiley, New York, 1967.
10. A. B. Glaser and G. Subak-Sharpe, *Integrated Circuit Engineering,* Addison-Wesley, Reading, Mass., 1979.

11. D. A. Hodges and H. G. Jackson, *Analysis and Design of Digital Integrated Circuits,* McGraw-Hill, New York, 1983.
12. P. E. Gray and C. L. Searle, *Electronic Principles: Physics, Models, and Circuits,* Wiley, New York, 1969.
13. S. M. Sze, *Physics of Semiconductor Devices,* Wiley, New York, 1969.
14. R. S. Muller and T. I. Kamins, *Device Electronics for Integrated Circuits,* Wiley, New York, 1977.
15. *SPICE Version 2G User's Manual.*
16. E. Cohen, Program Reference for SPICE2, Electronics Research Laboratory Rep. No. ERL-M592, University of California, Berkeley, 1976.
17. A. G. Milnes, *Semiconductor Devices and Integrated Electronics,* Van Nostrand Reinhold, New York, 1980.
18. A. A. Barna and D. Horelick, A Simple Diode Model Including Conductivity Modulation, *IEEE Trans. Circuit Theory,* **CT-18,** 1971.
19. S. C. Choo, Analytical Approximations for an Abrupt *pn* Junction under High-Level Conditions, *Solid-State Electron.,* **16,** 1973.
20. J. Cornu, Analytical Model for *pn* Junction under High-Injection Conditions, *Electron. Lett.,* **7,** 1971.
21. P. R. Wilson, Avalanche Breakdown Voltage of Diffused Junctions in Silicon, *Solid-State Electron.,* **16,** 1973.
22. W. Shockley, The Theory of *pn* Junctions in Semiconductors and *pn* Junction Transistors, *Bell Syst. Technol.,* **28,** 1949.
23. H. K. Gummel and D. L. Scharfetter, Depletion-Layer Capacitance of p^+n Step Junction, *J. Appl. Phys.,* **38,** 1967.

Bipolar Junction Transistor (BJT)

Giuseppe Massobrio

Department of Electronics (DIBE),
University of Genoa, Genoa, Italy

In Chap. 1, it was shown that a *pn* junction under forward bias conducts current because holes are injected from the *p* region and electrons are injected from the *n* region. These are majority carriers in each region, and their supply is not limited near the junction; consequently, current rises rapidly as voltage is increased. Current is much smaller under the reverse-bias condition because it is carried only by minority carriers that are generated either in the junction space-charge region or in other nearby regions. The current flowing by a reverse-biased junction will, however, be increased if the number of minority carriers in the vicinity of the junction is increased. One means of enhancing the minority-carrier concentration in the vicinity of a reverse-biased *pn* junction is simply to locate a forward-biased *pn* junction very close to it. This method is especially advan-

Note: The material contained in this chapter is based in part upon the book *Modeling the Bipolar Transistor,* by Ian E. Getreu, copyright © Tektronix, Inc., 1976. All rights reserved. Reproduced with permission. The material in the first paragraph is taken from R. S. Muller and T. Kamins, *Device Electronics for Integrated Circuits,* copyright © 1977 by John Wiley & Sons, Inc. Reprinted by permission.

tageous because it brings the minority-carrier concentration under electrical control, that is, under the control of the bias applied to the nearby forward-biased junction. Modulating the current-flow properties of one *pn* junction by changing the bias on a nearby junction is called *bipolar transistor action* [1].

Transistors may be divided into two classes: *bipolar transistors* and *unipolar transistors* (see Chap. 3). In bipolar transistors, both positive and negative free carriers take part in the device operation; hence the term *bipolar*. In unipolar transistors, the current is carried only by the free majority carriers in the conducting channel; hence the term *unipolar*.

Thus the bipolar junction transistor (BJT) has the physical configuration of two *pn* junctions back to back with a thin *p*-type or *n*-type region between them. In accordance with the order of the layers of the two junctions, we have two typical structures: *npn* or *pnp* configuration. The middle layer, through which the minority carriers pass, is called the *base* of the BJT; one outer layer in which the minority carriers originate is called the *emitter*; and the other outer layer is called the *collector*.

In our analysis of BJT operation, we will use the one-dimensional model of an *npn* transistor. The operation of a *pnp* device is similar in every respect if the roles of holes and electrons are interchanged and the polarities of the terminal currents and voltages are reversed.

This chapter describes the BJT models implemented in version 2G of SPICE. The available models, their equations as calculated by SPICE, their parameters, the effects of these parameters on the characteristics of the models, and their physical derivation are described.

The BJT model in SPICE is a more comprehensive version of the classic *Gummel-Poon* model [2]. The model is implemented as an extension of the *Ebers-Moll* model [3] so that the model defaults to the Ebers-Moll model by specifying the appropriate parameters.

The original Ebers-Moll model (a nonlinear dc model only) has been modified by many people to include effects such as charge storage, β variation with current, base-width modulation, and so forth. Different levels of complexity and notation can be found in the literature.

The Gummel-Poon model is originally based on the classic charge-control Gummel-Poon model. However, the model can be treated as an extension of the Ebers-Moll model. Several secondary effects are implemented in the model, and each modification can be introduced independently to obtain greater accuracy over the Ebers-Moll model.

Here the BJT model analysis follows the SPICE notation, i.e., (1) the *Ebers-Moll* model and (2) the *Gummel-Poon* model, even if item 1 contains something more than the original Ebers-Moll model [3] and item 2 contains something different from the classic integral charge-control model by Gummel and Poon [2].

SPICE BJT behavior is directly regulated by the subroutines *BJT* (it computes all the required parameters and loads them into the coefficient matrix), *PNJLIM* (it limits, for each iteration, the voltage variation of the *pn*-junction devices) and *TMPUPD* (it updates the temperature-dependent parameters), besides the subroutines of each type of analysis.

2.1 Transistor Conventions and Symbols

The circuit symbols for *npn* and *pnp* BJTs are shown in Fig. 2-1. The emitter shows an arrow indicating the actual current direction under normal operation, in which the emitter region injects minority carriers into the base region.

To define the circuit currents and voltages of a BJT, the emitter, base, and collector currents, as well as the emitter-base, emitter-collector, and base-collector voltages, must be specified. Independent of the actual direction of current flow, all the terminal currents are conventionally defined as *positive* when *directed inward,* as illustrated in Fig. 2-2*a*, while voltages between the terminals are defined according to the notation of Fig. 2-2*b* [4].

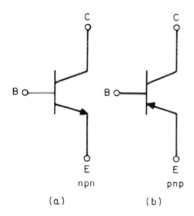

Figure 2-1 Circuit symbols for the BJT: (*a*) *npn* type and (*b*) *pnp* type.

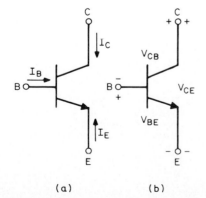

Figure 2-2 Terminal variables for the BJT: (*a*) terminal currents and (*b*) interterminal voltages.

Currents and voltages are related through the basic circuit laws, valid for both *npn* and *pnp* BJTs. From *Kirchhoff's current law (KCL)*

$$I_E + I_B + I_C = 0 \qquad (2\text{-}1a)$$

and from *Kirchhoff's voltage law (KVL)*

$$V_{EB} + V_{BC} + V_{CE} = 0 \qquad (2\text{-}1b)$$

Figure 2-3 is a useful aid for understanding transistor behavior, because it defines the regions of device operation in terms of the applied junction voltages.

2.2   Ebers-Moll Static Model

The bipolar junction transistor (BJT) can be considered to be an *interacting* pair of *pn* junctions, and the approach to the problem is basically the same as that used for the diode. Two diode solutions are needed, and the boundary conditions for one diode are the same as the boundary conditions for the other. The Ebers-Moll equations express this formally. They provide the emitter and collector currents I_E and I_C of the transistor in terms of the emitter and collector diode voltages V_{BE} and V_{BC}.

2.2.1   Static model and its implementation in SPICE

a. Equation formulation.   Currently there are two versions of the Ebers-Moll model: the *injection version* (Fig. 2-4) and the *transport version*

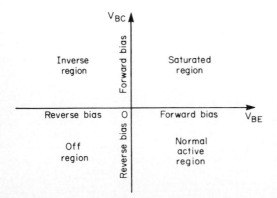

Figure 2-3 Regions of operation for a BJT as defined by base-emitter and base-collector bias polarities.

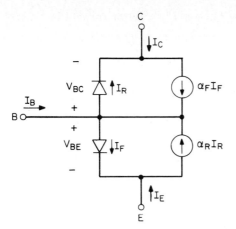

Figure 2-4 Ebers-Moll static model for an *npn* ideal transistor: injection version.

(Fig. 2-5). These versions are mathematically the same; however, since the second version is preferred for computer simulation, it will be the base for the analysis of the BJT behavior.

In the injection version (Fig. 2-4), the reference currents are I_F and I_R, the currents through the diodes. Thus it can be written

$$I_F = I_{ES}(e^{qV_{BE}/kT} - 1) \tag{2-2a}$$

$$I_R = I_{CS}(e^{qV_{BC}/kT} - 1) \tag{2-2b}$$

where I_{ES} and I_{CS} are the base-emitter and base-collector saturation currents, respectively.

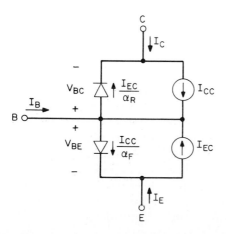

Figure 2-5 Ebers-Moll static model for an *npn* ideal transistor: transport version.

The terminal currents can now be expressed in terms of I_F and I_R as follows:

$$I_C = \alpha_F I_F - I_R$$

$$I_E = -I_F + \alpha_R I_R \qquad (2\text{-}3)$$

$$I_B = (1 - \alpha_F)I_F + (1 - \alpha_R)I_R$$

where α_F and α_R are, respectively, the large-signal forward and reverse current gains of a common-base BJT.

A simple intuitive method that justifies this model can be obtained by inspection. The diodes represent the BJT's base-emitter and base-collector junctions. I_F (I_R) is the current that would flow across the base-emitter (base-collector) junction for a given V_{BE} (V_{BC}) if the collector (emitter) region were replaced by an ohmic contact without disturbing the base. The two current-dependent current sources model the coupling between the two junctions, physically represented by the base region.

Consider now the BJT to be biased in the forward active region. In this situation the base-collector diode can be approximated by an open circuit, and thus the model reduces to the $\alpha_F I_F$ current source and the base-emitter diode. I_F represents the total current through the base-emitter junction, while α_F is the fraction of that current that is collected at the base-collector junction.

A dual consideration is valid for a BJT operating in the reverse active region; it is sufficient to consider I_R and α_R instead of I_F and α_F and to change the role of the two junctions. From Eqs. (2-1) to (2-3), it follows that four parameters are needed to describe the model of Fig. 2-4, which, however, can be reduced by one when the reciprocity property is applied. This property gives

$$\alpha_F I_{ES} = \alpha_R I_{CS} \equiv I_S \qquad (2\text{-}4)$$

where I_S is the BJT saturation current.

The physical meaning of I_S is the following. As explained in Appendix A, the saturation current in a pn junction is due to two components. As far as the BJT is concerned, the $\alpha_F I_{ES}$ term represents the portion of the emitter-base saturation current (I_{ES}) coming from the analysis of the base region. Similarly, $\alpha_R I_{CS}$ is the portion of the collector-base saturation current (I_{CS}) coming from the analysis of the base region, too. Equation (2-4) means that this analysis of the base region is the same for both I_{ES} and I_{CS} and that I_S is the common portion of both saturation currents [5].

Finally, the following familiar relations must be considered.

$$\beta_F = \frac{\alpha_F}{1 - \alpha_F}$$

$$\beta_R = \frac{\alpha_R}{1 - \alpha_R} \tag{2-5}$$

where β_F and β_R are, respectively, the large-signal forward and reverse current gain of a common-emitter BJT.

In the transport version (Fig. 2-5), the reference currents I_{CC} and I_{EC} are those flowing through the model's current sources and are given by

$$I_{CC} = I_S(e^{qV_{BE}/kT} - 1) \tag{2-6}$$

$$I_{EC} = I_S(e^{qV_{BC}/kT} - 1) \tag{2-7}$$

The BJT's terminal currents are then given by

$$I_C = I_{CC} - \frac{I_{EC}}{\alpha_R}$$

$$I_E = -\frac{I_{CC}}{\alpha_F} + I_{EC} \tag{2-8}$$

$$I_B = \left(\frac{1}{\alpha_F} - 1\right) I_{CC} + \left(\frac{1}{\alpha_R} - 1\right) I_{EC}$$

A change in the model topology of Fig. 2-5 can be made. The change, as shown in Fig. 2-6, consists of replacing the two reference current

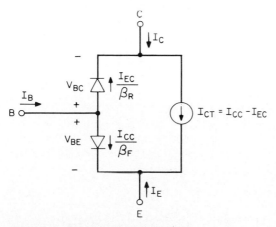

Figure 2-6 Ebers-Moll static model.

sources with a single current source (I_{CT}) between the emitter and the collector, defined by the expression

$$I_{CT} = I_{CC} - I_{EC} = I_S(e^{qV_{BE}/kT} - e^{qV_{BC}/kT}) \tag{2-9}$$

Equation (2-9) causes a change in the equations for the diode saturation currents. Thus the diode currents become

$$\frac{I_{CC}}{\beta_F} = \frac{I_S}{\beta_F}(e^{qV_{BE}/kT} - 1) \tag{2-10}$$

$$\frac{I_{EC}}{\beta_R} = \frac{I_S}{\beta_R}(e^{qV_{BC}/kT} - 1) \tag{2-11}$$

The BJT's terminal currents are then given by

$$I_C = I_{CT} - \frac{I_{EC}}{\beta_R}$$

$$I_E = -\frac{I_{CC}}{\beta_F} - I_{CT} \tag{2-12}$$

$$I_B = \frac{I_{CC}}{\beta_F} + \frac{I_{EC}}{\beta_R}$$

b. Implementation in SPICE.   The model's terminal currents for the four regions of Fig. 2-3 can now be written as they are implemented in SPICE.

As outlined in Sec. 1.2.1a, to aid convergence, a small conductance GMIN (its default value is 10^{-12} mho) is added automatically by SPICE [7] and put in parallel with every *pn* junction. This yields the following equations.

Normal active region

For $V_{BE} > -5\dfrac{kT}{q}$ and $V_{BC} \leq -5\dfrac{kT}{q}$:

$$I_C = I_S\left(e^{qV_{BE}/kT} + \frac{1}{\beta_R}\right) + \left[V_{BE} - \left(1 + \frac{1}{\beta_R}\right)V_{BC}\right]\text{GMIN}$$

$$I_B = I_S\left[\frac{1}{\beta_F}(e^{qV_{BE}/kT} - 1) - \frac{1}{\beta_R}\right] + \left(\frac{V_{BE}}{\beta_F} + \frac{V_{BC}}{\beta_R}\right)\text{GMIN} \tag{2-13}$$

Inverse region

For $V_{BE} \leq -5\dfrac{kT}{q}$ and $V_{BC} > -5\dfrac{kT}{q}$:

$$I_C = -I_S \left[e^{qV_{BC}/kT} + \frac{1}{\beta_R} (e^{qV_{BC}/kT} - 1) \right]$$

$$+ \left[V_{BE} - \left(1 + \frac{1}{\beta_R} \right) V_{BC} \right] GMIN$$

$$I_B = -I_S \left[\frac{1}{\beta_F} - \frac{1}{\beta_R} (e^{qV_{BC}/kT} - 1) \right] + \left(\frac{V_{BE}}{\beta_F} + \frac{V_{BC}}{\beta_R} \right) GMIN$$

$$(2\text{-}14)$$

Saturated region

For $V_{BE} > -5\dfrac{kT}{q}$ and $V_{BC} > -5\dfrac{kT}{q}$:

$$I_C = I_S \left[(e^{qV_{BE}/kT} - e^{qV_{BC}/kT}) - \frac{1}{\beta_R} (e^{qV_{BC}/kT} - 1) \right]$$

$$+ \left[V_{BE} - \left(1 + \frac{1}{\beta_R} \right) V_{BC} \right] GMIN$$

$$I_B = I_S \left[\frac{1}{\beta_F} (e^{qV_{BE}/kT} - 1) + \frac{1}{\beta_R} (e^{qV_{BC}/kT} - 1) \right]$$

$$+ \left(\frac{V_{BE}}{\beta_F} + \frac{V_{BC}}{\beta_R} \right) GMIN$$

$$(2\text{-}15)$$

Off region

For $V_{BE} \leq -5\dfrac{kT}{q}$ and $V_{PC} \leq -5\dfrac{kT}{q}$:

$$I_C = \frac{I_S}{\beta_R} + \left[V_{BE} - \left(1 + \frac{1}{\beta_R} \right) V_{BC} \right] GMIN$$

$$I_B = -I_S \left(\frac{\beta_F + \beta_R}{\beta_F \beta_R} \right) + \left(\frac{V_{BE}}{\beta_F} + \frac{V_{BC}}{\beta_R} \right) GMIN$$

$$(2\text{-}16)$$

For a *pnp* transistor, the voltage and current polarities must be changed appropriately. In SPICE, the parameter values of the model are always considered to be positive, and the appropriate sign changes are implemented internally by the program; the user need only specify the transistor type *npn* or *pnp* in the .MODEL card.

Although it is very simple in form and requires, at most, three parameters—β_F, β_R, and I_S—the Ebers-Moll model just derived is accurate and is recommended when the BJT is working as a dc switch or in a specific and narrow bias range. In fact, this principle must be taken into account: Use the simplest model that will do the job. This saves modeling effort and computer time, and the results are easier to understand.

Figure 2-7 shows the I_C vs. V_{CE} characteristics of the Ebers-Moll model for the ideal BJT as SPICE outputs when the values of the default parameters (see Table 2-1) are used for $I_B = 0.1$ A and $I_B = 0.5$ A.

The three parameters I_S, β_F, and β_R are user-specified parameters. The saturation current I_S is the extrapolated intercept current of ln I_C vs. V_{BE} in the forward region and ln I_E vs. V_{BC} in the reverse region, as shown in Fig. 2-8. The parameters β_F and β_R are the *maximum* forward and reverse short-circuit current gains, respectively, which are assumed not to vary with the operating point.

The limitation of this simple model lies mainly in its neglect of transistor charge storage (i.e., no diffusion or junction capacitances) and ohmic resistances to the terminals. These effects are included in the model described in the next section.

To summarize, the ideal model just described is characterized in SPICE

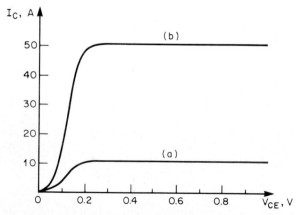

Figure 2-7 SPICE I_C vs. V_{CE} characteristics of the Ebers-Moll model of the ideal BJT when default parameter values are used for $I_B = 0.1$ A and $I_B = 0.5$ A.

TABLE 2-1 SPICE BJT Model Parameters

Symbol	SPICE 2G keyword	SPICE 2E/F keyword	Parameter name	Default value	Unit
I_S	IS	IS/JS	Saturation current	10^{-16}	A
β_F	BF	BF/BF	Ideal maximum forward current gain	100	
β_R	BR	BR/BR	Ideal maximum reverse current gain	1	
n_F	NF	.... /NF	Forward current emission coefficient	1	
n_R	NR	.... /NR	Reverse current emission coefficient	1	
C_2	ISE $= C_2 I_S$	C2/JLE	Base-emitter leakage saturation current	0	A
C_4	ISC $= C_4 I_S$	C4/JLC	Base-collector leakage saturation current	0	A
I_{KF}	IKF	IK/JBF	Corner for forward β high-current roll-off	∞	A
I_{KR}	IKR	IKR/JBR	Corner for reverse β high-current roll-off	∞	A
n_{EL}	NE	NE/NLE	Base-emitter leakage emission coefficient	1.5	
n_{CL}	NC	NC/NLC	Base-collector leakage emission coefficient	2	
V_A	VAF	VA/VBF	Forward Early voltage	∞	V
V_D	VAR	VB/VBR	Reverse Early voltage	∞	V
r_C	RC	RC/RC	Collector resistance	0	Ω
r_E	RE	RE/RE	Emitter resistance	0	Ω
r_B	RB	RB/RB	Zero-bias base resistance	0	Ω
r_{BM}	RBM	.... /RBM	Minimum base resistance at high currents	RB	Ω
I_{rB}	IRB	.... /JRB	Current where base resistance falls halfway to its minimum value	∞	A
τ_F	TF	TF/TF	Ideal forward transit time	0	s
τ_R	TR	TR/TR	Ideal reverse transit time	0	s
$X_{\tau F}$	XTF	.... /XTF	Coefficient for bias dependence of TF	0	
$V_{\tau F}$	VTF	.... /VTF	Voltage describing V_{BC} dependence of TF	∞	V
$I_{\tau F}$	ITF	.... /JTF	High-current parameter for effect on TF	0	A
$P_{\tau F}$	PTF	.... /PTF	Excess phase at $f = 1/2\pi\tau_F$	0	°
C_{JE}	CJE	CJE/CJE	Zero-bias base-emitter depletion capacitance	0	F
ϕ_E	VJE	PE/VJE	Base-emitter built-in potential	0.75	V
m_E	MJE	ME/MJE	Base-emitter junction grading coefficient	0.33	
C_{JC}	CJC	CJC/CJC	Zero-bias base-collector depletion capacitance	0	F
ϕ_C	VJC	PC/VJC	Base-collector built-in potential	0.75	V
m_C	MJC	MC/MJC	Base-collector junction grading coefficient	0.33	

TABLE 2-1 SPICE BJT Model Parameters (*Continued*)

Symbol	SPICE 2G keyword	SPICE 2E/F keyword	Parameter name	Default value	Unit
C_{JS}	CJS	CCS/CJS	Zero-bias collector-substrate capacitance	0	F
ϕ_S	VJS	.... /VJS	Substrate-junction built-in potential	0.75	V
m_S	MJS	.... /MJS	Substrate-junction exponential factor	0	
X_{CJC}	XCJC	.... /CDIS	Fraction of base-collector depletion capacitance connected to internal base node	1	
FC	FC	FC	Coefficient for forward-bias depletion capacitance formula	0.5	
$X_{T\beta}$	XTB	.... /TB	Forward and reverse β temperature coefficient	0	
X_{TI}	XTI	PT/PT	Saturation current temperature exponent	3	
E_g	EG	EG/EG	Energy gap for temperature effect on I_S	1.11	eV
k_f	KF	KF/KF	Flicker-noise coefficient	0	
a_f	AF	AF/AF	Flicker-noise exponent	1	
T	T†	T/T	Nominal temperature for simulation and at which all input data is assumed to have been measured	27	°C

†Remember, the temperature T is regarded as an operating condition and not as a model parameter.

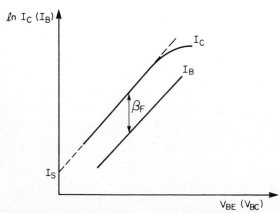

Figure 2-8 Typical current vs. voltage plots to show the meaning of I_S and β_F.

by the following three parameters the user can specify in the .MODEL card:

IS   Saturation current (I_S)

BF   Ideal maximum forward current gain (β_F)

BR   Ideal maximum reverse current gain (β_R)

2.2.2   Static model: Second-order effects and their implementation in SPICE

The ideal BJT model described in the preceding section is very simple because of the following reasons:

1. It neglects parasitic base, collector, and emitter resistances.

2. It neglects the dependence of I_S on V_{BC} (Early effect, base-width modulation).

3. It assumes *ideal* base current components (Boltzmann-type) or β_F and β_R current-independent.

4. It neglects high-level effects in the base and collector regions.

In this section, an improved and more accurate dc model of the BJT is developed. We introduce four new parameters, r_E, r_B, r_C, and V_A, which take into account the base, collector, and emitter resistances and the dependence of I_S on V_{BC}. We illustrate them with the SPICE Ebers-Moll model.

In Sec. 2.5, we discuss the nonideal base current components and high-level effects in the base and collector regions using the SPICE Gummel-Poon model.

a. Ohmic resistances.   The inclusion of three constant resistors (r_C, r_E, and r_B) improves the dc characterization of the model. They represent the transistor's ohmic resistances from its active region to its collector, emitter, and base terminals, respectively. These resistors are included in the model as shown in Fig. 2-9. The internal nodes of these resistances are denoted by the letters C, E, and B in the model circuit.The voltages used in describing the two ideal diodes and the current source are the *internal voltages*. In the following, the effects of ohmic resistances on the static model are described [5].

Collector resistance r_C.   The effect of the *collector resistance r_C* is shown in Fig. 2-10, in which collector characteristics of the actual model (solid lines) are compared to collector characteristics of the ideal model (dashed lines). Resistance r_C decreases the slope of the curves in the saturated region for low collector-emitter voltages.

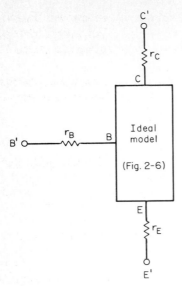

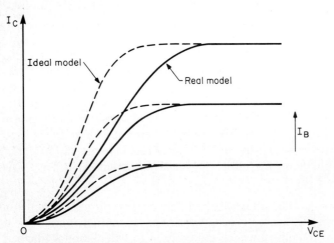

Figure 2-9 Effect of the ohmic resistances [5].

In the real model, r_C is assumed to be constant. In an actual device, however, it will be a function of collector current and base-collector voltage. Therefore, the biggest problem in obtaining r_C is not how to measure it but which value to use. A more detailed description of this problem is given in the measurement section on r_C (see Chap. 5). The collector resistance can limit the current-handling capability of a BJT, and it also affects the maximum operating frequency at high currents.

Figure 2-10 Effect of the collector resistance on the I_C vs. V_{CE} characteristics. Dashed lines represent $r_C = 0$. Solid lines represent $r_C \neq 0$.

Emitter resistance r_E. The emitter is the most heavily doped region in most present-day transistors in order to produce a high emitter injection efficiency and therefore a high β_F. For this reason, the dominant component of *emitter resistance r_E* is normally the contact resistance (usually on the order of 1 Ω). r_E, which is often neglected, can normally be assumed to have a small, constant value.

Its main effect is a reduction in the voltage seen by the emitter-base junction by a factor of $r_E I_E$. In this effect on V_{BE}, r_E is equivalent to a base resistance of $(1 + \beta_F)r_E$. Therefore r_E affects the collector current as well as the base current, as shown in Fig. 2-11. This effect can be significant, and r_E can cause substantial errors in the determination of r_B. Resistance r_E can also affect the collector characteristics in the saturation region if the transistor has a low r_C value.

Base resistance r_B. *Base resistance r_B* is an important model parameter. Its greatest impact is its effect on the small-signal and transient responses. It is also one of the most difficult parameters to measure accurately, partly because of its strong dependence on operating point (due to crowding) and partly because of the error introduced by the small but finite value of r_E.

In the Ebers-Moll model, r_B is assumed to be constant. The dc effect of r_B is seen on the ln I_C and ln I_B vs. V_{BE} curve, as illustrated in Fig. 2-11.

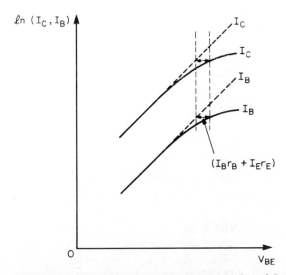

Figure 2-11 The effect of r_B and r_E on the ln I_C and I_B vs. V_{BE} characteristics of the real model [5].

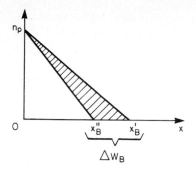

Figure 2-12 Effect of increases in V_{CE} on the collector-base space-charge layer and base width of an *npn* BJT.

b. Base-width modulation (Early effect). Base-width modulation (the so-called *Early effect*) is the change in base width, W_B, that results from a change in the collector-base junction voltage. In the normal active region, the emitter-base junction is forward-biased and the collector-base junction is reverse-biased. The width of the space-charge layer of a *pn* junction is a strong function of the applied potential. Large variations in V_{BC}, for example, may cause the collector-base space-charge layer to vary significantly. This, in turn, changes the normally thin base width, as shown in Fig. 2-12.

The total effect of the base-width modulation is a modification (as a function of V_{BC}) of I_S and thereby of the collector current, β_F, and τ_F through the base transit time τ_{BF} (see Sec. 2.3.1).

Only one extra parameter, the *Early voltage* V_A, is used to model base-width modulation in the forward active region. Figure 2-13 shows the

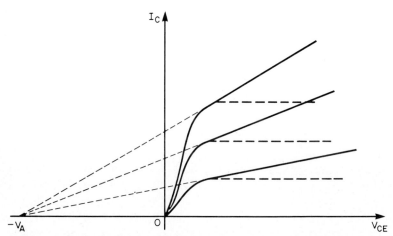

Figure 2-13 The effect of base-width modulation on the I_C vs. V_{CE} characteristics. The dashed lines illustrate the zero slope obtained with the ideal model, where base-width modulation is not included [5].

effect of base-width modulation on the variation of collector current I_C with collector-emitter voltage (I_C vs. V_{CE})—a nonzero slope in the normal active region. (The dashed lines illustrate the zero slope obtained with the ideal model, where base-width modulation is not included.)

The analysis, which assumes that the transistor is operated in the linear region, first determines the effect of base-width modulation on the base width and then on the three base-width-related parameters. The results of the analysis are as follows [5]:

$$W_B(V_{BC}) = W_B(0) \left(1 + \frac{V_{BC}}{V_A} \right) \tag{2-17}$$

$$I_S(V_{BC}) = \frac{I_S(0)}{1 + V_{BC}/V_A} \simeq I_S(0) \left(1 - \frac{V_{BC}}{V_A} \right) \tag{2-18}$$

$$\beta_F(V_{BC}) = \frac{\beta_F(0)}{1 + V_{BC}/V_A} \simeq \beta_F(0) \left(1 - \frac{V_{BC}}{V_A} \right) \tag{2-19}$$

$$\tau_{BF}(V_{BC}) = \tau_{BF}(0) \left(1 + \frac{V_{BC}}{V_A} \right)^2 \tag{2-20}$$

where V_A is defined as

$$V_A = \begin{cases} \left[\dfrac{1}{W_B(0)} \dfrac{dW_B}{dV_{BC}} \bigg|_{V_{BC}=0} \right]^{-1} & \text{for } npn \text{ transistor} \quad (2\text{-}21a) \\[4ex] \left[-\dfrac{1}{W_B(0)} \dfrac{dW_B}{dV_{BC}} \bigg|_{V_{BC}=0} \right]^{1} & \text{for } pnp \text{ transistor} \quad (2\text{-}21b) \end{cases}$$

and $I_S(0)$, $\beta_F(0)$, $\tau_{BF}(0)$ are the values of the parameters at $V_{BC} = 0$.

The difference between the definitions for the npn and pnp transistors lies only in the sign of V_{BC}. For an npn transistor in the normal active region, V_{BC} is negative (i.e., reverse bias). An increase in V_{BC} (a decrease in the reverse bias) results in an increase in the base width, and the derivative in Eq. (2-21a) is positive. The minus sign in Eq. (2-21b) preserves the positive nature of V_A for a pnp transistor.

Equation (2-17) describes the (assumed linear) variation of the base width with V_{BC}. Equations (2-18) to (2-20), which give the variation of the three model parameters with V_{BC}, follow directly from Eq. (2-17) and the assumptions that the constant-base doping relationships $I_S \propto 1/W_B$, $\beta_F \propto 1/W_B$, and $\tau_{BF} \propto W_B^2$ are also approximately valid in general.

The second forms of Eqs. (2-18) and (2-19) (obtained using the binomial expansion assuming $|V_{BC}| \ll V_A$) are preferred computationally, since the first forms become infinite at $V_{BC} = V_A$ ($V_{BC} = V_A$ may not necessarily be the correct value but could be a temporary value while the computer is iterating to the solution).

V_A has no physical counterpart in the circuit model, only a mathematical effect whereby existing equations are modified. (This process of altering equations or parameters without altering the *form* of the equivalent circuit will be observed for other effects in the section on the Gummel-Poon model.)

With the exception of τ_{BF}, which is described later, the total effect of base-width modulation is accounted for if I_S and β_F are modified as in Eqs. (2-18) and (2-19). The expressions for I_{CT} and I_B [Eqs. (2-9) and (2-12), respectively] then become

$$I_{CT} = \frac{I_S(0)}{1 + V_{BC}/V_A} \left(e^{qV_{BE}/kT} - e^{qV_{BC}/kT} \right) \tag{2-22}$$

$$I_B = \frac{I_S(0)}{\beta_F(0)} \left(e^{qV_{BE}/kT} - 1 \right) + \frac{I_S(0)}{\beta_R} \left(e^{qV_{BC}/kT} - 1 \right) \tag{2-23}$$

In the first term of Eq. (2-23), the similar dependence of I_S and β_F on V_{BC} (since both are assumed to be proportional to $1/W_B$) results in a cancellation. Therefore, in the normal active region where the second term is negligible, I_B is independent of V_{BC}. The V_{BC} variation in β_F is achieved by keeping I_B constant and modifying I_C. This introduces a *very* important concept: The correct variation of β_F (with V_{BC}, as discussed here, or with I_C, as discussed later) is *not* obtained by varying β_F but by modeling correctly the expressions for I_C and/or I_B. A similar analysis for base-width modulation when the device is operated in the inverse mode is included in the discussion of the Gummel-Poon model (see Sec. 2.5.4).

The Early voltage V_A can be obtained directly from the I_C vs. V_{CE} characteristics. This can be shown mathematically [5]. The slope of these characteristics in the normal active region, g_o, is obtained from Eq. (2-22) by first dropping the (negligible) second term and then differentiating with respect to V_{BC} (V_{BE} is assumed constant). The result is

$$g_o = \left. \frac{dI_C}{dV_{CE}} \right|_{V_{BE}=\text{const.}} \simeq \frac{I_C(0)}{V_A} \tag{2-24}$$

The geometrical interpretation of Eq. (2-24) shows that V_A is obtained from the intercept of the extrapolated slope on the V_{CE} axis (as shown in the curve in Fig. 2-13). For example, a slope of $(50 \text{ k}\Omega)^{-1}$ at $I_C(0) = 1$ mA gives, from Eq. (2-24), $V_A = 50$ V. A more detailed description of the geometrical interpretation is given in Ref. [5].

c. Implementation in SPICE.   The physics-based static Ebers-Moll model for the *real* BJT requires, besides the parameters affecting the *ideal* model, only four new model parameters that the user can specify in the .MODEL card. These parameters are as follows:

RC   Collector resistance (r_C)

RE   Emitter resistance (r_E)

RB   Zero-bias base resistance (r_B); it can be dependent on high current (see Sec. 2.5.5)

VAF   Forward Early voltage (V_A)

The Early voltage V_A affects the basic model equations for I_C and I_B [see Eqs. (2-13) to (2-16)].

To show how these equations are modified in SPICE, the more general relationships of Eqs. (2-15) are taken into account; the other equations follow immediately.

$$
\begin{aligned}
I_C &= I_S\left[(e^{qV_{BE}/kT} - e^{qV_{BC}/kT})\left(1 - \frac{V_{BC}}{V_A}\right) - \frac{1}{\beta_R}(e^{qV_{BC}/kT} - 1)\right] \\
&\quad + \left[V_{BE} - \left(1 + \frac{1}{\beta_R}\right) V_{BC}\right] \text{GMIN} \\
I_B &= I_S\left[\frac{1}{\beta_F}(e^{qV_{BE}/kT} - 1) + \frac{1}{\beta_R}(e^{qV_{BC}/kT} - 1)\right] \\
&\quad + \left(\frac{V_{BE}}{\beta_F} + \frac{V_{BC}}{\beta_R}\right) \text{GMIN}
\end{aligned}
\tag{2-25}
$$

Obviously the models of Figs. 2-6 and 2-9 are still valid if the terminal currents are changed appropriately in order to model the above-described effects.

Figure 2-14 shows the SPICE-generated I_C vs. V_{CE} characteristics that plot the effects of parasitic base, collector, and emitter resistances and the Early effect on the Ebers-Moll model.

2.3   Ebers-Moll Large-Signal Model

Having defined the complete set of dc transport relationships for the BJT (or better, the Ebers-Moll static model), we will now consider the charge-storage effects of the device. The importance of these effects has been outlined in Sec. 1.3.1, which you may refer to in order to understand the physics presiding over the large-signal behavior of the device (keep in mind that the BJT can be considered to be an interacting pair of two *pn* junctions).

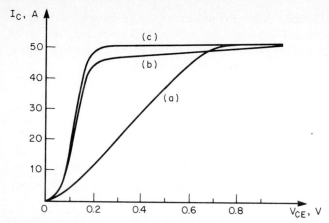

Figure 2-14 SPICE I_C vs. V_{CE} characteristics showing the second-order effects on the Ebers-Moll model: (*a*) $r_C = 10$ mΩ, (*b*) $V_A = 10$ V, and (*c*) default values, with $I_B = 0.5$ A. The other parameters have default values.

2.3.1 Large-signal model

Charge storage in the BJT is modeled by the introduction of three types of capacitors: two nonlinear junction capacitors, two nonlinear diffusion capacitors, and a constant substrate capacitor.

a. Stored charge Q_{DE} and Q_{DC}.   The charge associated with the *mobile carriers* in a BJT is modeled by the *diffusion capacitances*. This charge is divided into two components: one associated with the reference collector current source, I_{CC}, and the other with the reference emitter current source, I_{EC}. Each component is represented by a capacitor.

To evaluate the diffusion capacitance associated with I_{CC}, the total mobile charge associated with this current must be considered. Therefore, the base-emitter junction is assumed to be forward-biased and $V_{BC} = 0$.

For the simplified one-dimensional case of constant base doping, negligible base recombination, and low-level injection of the BJT of Fig. 2-15, the total mobile charge associated with I_{CC}, Q_{DE}, can be written as the sum of the individual minority charges:

$$Q_{DE} = Q_E + Q_{JE} + Q_{BF} + Q_{JC} \tag{2-26}$$

where Q_E is the mobile minority charge stored in the neutral emitter region, Q_{JE} is the mobile minority charge in the emitter-base space-charge region associated with I_{CC} (normally considered to be zero), Q_{BF} is the minority mobile charge stored in the neutral base region, and Q_{JC} is the mobile minority charge in the collector-base space-charge region associated with I_{CC}. Because of charge neutrality, there will be identical majority charges stored in the neutral regions. However, to determine diffusion capacitance, only one (minority or majority) needs to be considered.

From Eq. (2-26) the total mobile charge associated with I_{CC} can also be expressed as [5]

$$Q_{DE} = (\tau_E + \tau_{EB} + \tau_{BF} + \tau_{CB})I_{CC} \equiv \tau_F I_{CC} \qquad (2\text{-}27)$$

where τ_E is the emitter delay; τ_{EB} is the emitter-base space-charge layer transit time; τ_{BF} is the base transit time; τ_{CB} is the base-collector space-charge layer transit time; and τ_F is the total forward transit time (assumed here to be a constant), which represents the mean time for the minority carriers to cross (diffuse) the neutral base region from the emitter to the collector.

A similar analysis of the total mobile charge associated with I_{EC} results in

$$Q_{DC} = Q_C + Q_{JC} + Q_{BR} + Q_{JE} \qquad (2\text{-}28)$$

where Q_C is the mobile minority charge stored in the neutral collector region, Q_{JC} is the mobile minority charge in the collector-base space-

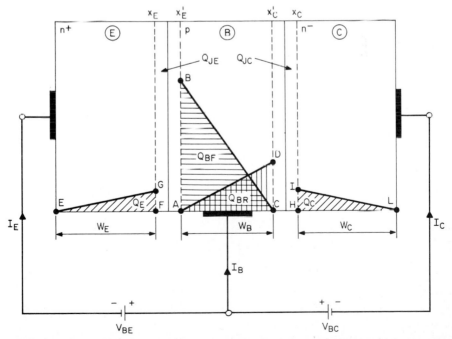

Figure 2-15 Cross section of an n^+pn^- transistor, indicating the location of the charge components.

charge region associated with I_{EC}, Q_{BR} is the minority mobile charge stored in the neutral base region, and Q_{JE} is the mobile minority charge in the emitter-base space-charge region associated with I_{EC}. If charge Q_{JC} is assumed to be zero, then it follows from Eq. (2-28) that [5]

$$Q_{DC} = (\tau_C + \tau_{CB} + \tau_{BR} + \tau_{EB})I_{EC} \equiv \tau_R I_{EC} \qquad (2\text{-}29)$$

where τ_C is the collector delay, τ_{BR} is the reverse base transit time, and τ_R is the total reverse transit time (assumed to be a constant here also).

For the saturated mode (that is, V_{BE} and V_{BC} are both forward-biased), both Q_{DE} and Q_{DC} are assumed to occur independently and the total minority charge stored in the BJT is the sum of all the components.

The two charges Q_{DE} and Q_{DC} are modeled by two *nonlinear* capacitors

$$
\begin{aligned}
C_{DE} &\equiv \frac{dQ_{DE}}{dV_{BE}} = \frac{d(\tau_F I_{CC})}{dV_{BE}} \\[2mm]
C_{DC} &\equiv \frac{dQ_{DC}}{dV_{BC}} = \frac{d(\tau_R I_{EC})}{dV_{BC}}
\end{aligned}
\qquad (2\text{-}30)
$$

as shown in Fig. 2-16.

b. Space charges Q_{JE} and Q_{JC}. The incremental *fixed charges* Q_{JE} and Q_{JC} stored in the BJT's space-charge layers for incremental changes in the associated junction voltages can be modeled by two *junction capac-*

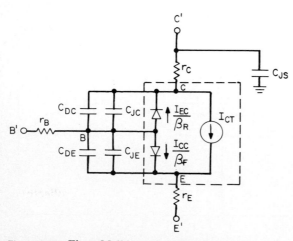

Figure 2-16   Ebers-Moll large-signal model.

itances (sometimes called *depletion capacitances*). These capacitances, denoted by C_{JE} for the base-emitter junction and C_{JC} for the base-collector junction, are included in the model as shown in Fig. 2-16. Each junction capacitance is a nonlinear function of the voltage across the junction with which it is associated.

A simple analysis of C_{JC} and C_{JE} makes use of the depletion approximation (which assumes that at the junction of interest the space-charge layer is depleted of carriers). The depletion capacitance for a *pn* junction is calculated by solving the one-dimensional *Poisson* equation at the junction [see Eq. (A-29)]. This equation is solved by assuming that the change in concentration from one region to another takes place abruptly *(step junction)*. The junction capacitance obtained through the above equation is given by Eq. (A-42*a*).

The same approach can be used for a linearly graded junction where the net dopant concentration varies linearly from the *p*-type to the *n*-type material [see Eq. (A-42*b*)].

It can be seen that both these capacitances have the following form:

$$C = \frac{C(0)}{(1 - V/\phi)^m} \qquad (2\text{-}31)$$

Since the actual junctions are usually somehow between these two cases, the capacitance model assumes the parameter m as an exponent, which should usually be between 0.5 and 0.33. The doping profile at emitter-base junctions is usually *Gaussian* and closer to an abrupt junction, and m should be close to 0.5. The collector-base junctions, on the other hand, are more like linearly graded junctions, and m is close to 0.33. Then, for a step (or abrupt) junction and for a linear (or graded) junction, the variation of the emitter and collector junction capacitances with base-emitter and base-collector junction voltages can be written for an *npn* transistor as follows:

$$C_{JE}(V_{BE}) = \frac{C_{JE}(0)}{(1 - V_{BE}/\phi_E)^{m_E}}$$

$$C_{JC}(V_{BC}) = \frac{C_{JC}(0)}{(1 - V_{BC}/\phi_C)^{m_C}} \qquad (2\text{-}32)$$

where $C_{JE}(0)$ and $C_{JC}(0)$ are the values of the emitter-base and collector-base junction capacitances at $V_{BE} = 0$ or $V_{BC} = 0$, ϕ_E and ϕ_C are the emitter-base and collector-base barrier potentials, and m_E and m_C are the emitter-base and collector-base capacitance gradient factors.

To obtain the space charges Q_{JE} and Q_{JC}, it is necessary to integrate the space-charge capacitances with respect to their voltages, i.e., for the Ebers-Moll model.

$$Q_{JE} = \int_0^{V_{BE}} C_{JE}\, dV = \frac{C_{JE}(0)}{1 - m_E}\left(1 - \frac{V_{BE}}{\phi_E}\right)^{1 - m_E}$$

$$Q_{JC} = \int_0^{V_{BC}} C_{JC}\, dV = \frac{C_{JC}(0)}{1 - m_C}\left(1 - \frac{V_{BC}}{\phi_C}\right)^{1 - m_C}$$

(2-33)

It is important to note that the physical considerations that have led to the space-charge capacitance in the diode analysis are still valid for obtaining C_{JE} and C_{JC} for the BJT.

Figure 2-17 shows three plots of the variation of the junction capacitance as a function of voltage. The solid line represents Eqs. (2-32). Under forward bias these equations predict infinite capacitance when the internal junction voltage equals the built-in voltage. It has been shown by Chawla and Gummel [6] that under forward bias, the depletion approximation is no longer valid and that Eqs. (2-32) no longer apply. The dashed line in Fig. 2-17 shows the (noninfinite) variation of the junction capacitance obtained by Chawla and Gummel. An expression requiring four parameters has been fitted to this curve by Gummel and Poon [9]. The third curve in Fig. 2-17 represents the straight-line approximation made by usual computer programs for $V > \phi/2$. The equation for this straight line, obtained by matching slopes at $\phi/2$, is given by

$$C_J = 2^m C_J(0)\left[2m\,\frac{V}{\phi} + (1 - m)\right] \qquad \text{for } V \geq \frac{\phi}{2} \qquad (2\text{-}34)$$

where the junction subscripts have been omitted. This approximation, while avoiding the infinite capacitance, is not as accurate as the Chawla-Gummel curve [6]. However, it is acceptable because under forward bias, the diffusion capacitances, described next, are dominant and inherently include the effect of the mobile charges in the space-charge layers [5].

SPICE [7–8] uses a straight-line approximation for C_J similar to the line c of Fig. 2-17; Eq. (2-34) is replaced by the following general relationships:

$$C_J = C_J(0)\left(1 + m\,\frac{V}{\phi}\right) \qquad \text{for } V \geq 0 \qquad (2\text{-}35)$$

Besides C_{JE} and C_{JC}, another capacitance must be taken into account in the design of integrated circuits: the *substrate capacitance* C_{JS}. Although it is actually a junction capacitance in the way it varies with the epitaxial-layer–substrate potential, it is modeled here as a constant-value capacitor. This representation is adequate for most cases, since the epitaxial-layer–substrate junction is reverse-biased for isolation purposes.

Theoretically, C_{JS} for an *npn* transistor is connected to the collector (see Fig. 2-16). However, for a *pnp* transistor, C_{JS} may not be connected to the collector. Instead, for a lateral *pnp* device, C_{JS} is connected between the base and the substrate; while for a substrate *pnp* device, C_{JS} is set to zero, since it is already modeled in the C_{JC} capacitance [5]. In SPICE, however, C_{JS} is connected to the collector for both *npn* and *pnp* transistors.

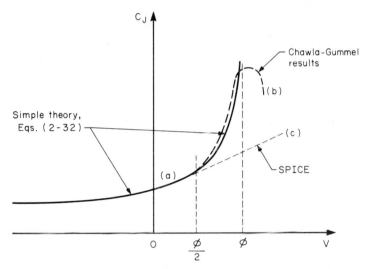

Figure 2-17 Plot of the variation of junction capacitances with voltage [5].

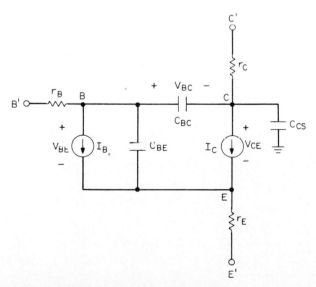

Figure 2-18 SPICE BJT large-signal model.

2.3.2 Large-signal model and its implementation in SPICE

Having established the basic relationships of the charge-storage effects, we can discuss the details of the large-signal model implemented in SPICE in a straightforward way.

The charge-storage elements $Q_{BE} = Q_{DE} + Q_{JE}$ and $Q_{BC} = Q_{DC} + Q_{JC}$, or, better, the capacitors C_{BE} and C_{BC}, shown in Fig. 2-18 define the SPICE charge-storage model of the BJT.

The charges Q_{BE} and Q_{BC} are determined by the following relations:

$$
Q_{BE} =
\begin{cases}
\underbrace{\tau_F I_{CC}}_{Q_{DE}} + \underbrace{C_{JE}(0) \int_0^{V_{BE}} \left(1 - \frac{V}{\phi_E}\right)^{-m_E} dV}_{Q_{JE}} & \text{for } V_{BE} < FC \times \phi_E \\[4ex]
\underbrace{\tau_F I_{CC}}_{Q_{DE}} + \underbrace{C_{JE}(0)F_1 + \frac{C_{JE}(0)}{F_2} \int_{FC \times \phi_E}^{V_{BE}} \left(F_3 + \frac{m_E V}{\phi_E}\right) dV}_{Q_{JE}} & \text{for } V_{BE} \geq FC \times \phi_E
\end{cases}
\tag{2-36}
$$

$$
Q_{BC} =
\begin{cases}
\underbrace{\tau_R I_{EC}}_{Q_{DC}} + \underbrace{C_{JC}(0) \int_0^{V_{BC}} \left(1 - \frac{V}{\phi_C}\right)^{-m_C} dV}_{Q_{JC}} & \text{for } V_{BC} < FC \times \phi_C \\[4ex]
\underbrace{\tau_R I_{EC}}_{Q_{DC}} + \underbrace{C_{JC}(0)F_1 + \frac{C_{JC}(0)}{F_2} \int_{FC \times \phi_C}^{V_{BC}} \left(F_3 + \frac{m_C V}{\phi_C}\right) dV}_{Q_{JC}} & \text{for } V_{BC} \geq FC \times \phi_C
\end{cases}
\tag{2-37}
$$

$$
Q_{CS} =
\begin{cases}
C_{JS}(0) \int_0^{V_{CS}} \left(1 - \frac{V}{\phi_S}\right)^{-m_S} dV & \text{for } V_{CS} < 0 \\[4ex]
C_{JS}(0) \int_0^{V_{CS}} \left(1 + \frac{m_S V}{\phi_S}\right) dV & \text{for } V_{CS} \geq 0
\end{cases}
\tag{2-38}
$$

These charge-storage elements can be equivalently represented by the following SPICE voltage-dependent capacitance equations.

$$
C_{BE} = \frac{dQ_{BE}}{dV_{BE}} = \begin{cases} \underbrace{\tau_F \frac{dI_{CC}}{dV_{BE}}}_{C_{DE}} + \underbrace{C_{JE}(0) \left(1 - \frac{V_{BE}}{\phi_E}\right)^{-m_E}}_{C_{JE}} & \text{for } V_{BE} < FC \times \phi_E \\[4mm] \underbrace{\tau_F \frac{dI_{CC}}{dV_{BE}}}_{C_{DE}} + \underbrace{\frac{C_{JE}(0)}{F_2} \left(F_3 + \frac{m_E V_{BE}}{\phi_E}\right)}_{C_{JE}} & \text{for } V_{BE} \geq FC \times \phi_E \end{cases}
$$

(2-39)

$$
C_{BC} = \frac{dQ_{BC}}{dV_{BC}} = \begin{cases} \underbrace{\tau_R \frac{dI_{EC}}{dV_{BC}}}_{C_{DC}} + \underbrace{C_{JC}(0) \left(1 - \frac{V_{BC}}{\phi_C}\right)^{-m_C}}_{C_{JC}} & \text{for } V_{BC} < FC \times \phi_C \\[4mm] \underbrace{\tau_R \frac{dI_{EC}}{dV_{BC}}}_{C_{DC}} + \underbrace{\frac{C_{JC}(0)}{F_2} \left(F_3 + \frac{m_C V_{BC}}{\phi_C}\right)}_{C_{JC}} & \text{for } V_{BC} \geq FC \times \phi_C \end{cases}
$$

(2-40)

$$
C_{CS} = \begin{cases} C_{JS}(0) \left(1 - \frac{V_{CS}}{\phi_S}\right)^{-m_S} & \text{for } V_{CS} < 0 \\[4mm] C_{JS}(0) \left(1 + \frac{m_S V_{CS}}{\phi_S}\right) & \text{for } V_{CS} > 0 \end{cases}
$$

(2-41)

where, for the base-emitter junction,

$$
F_1 = \frac{\phi_E}{1 - m_E} [1 - (1 - FC)^{1-m_E}]
$$

$$
F_2 = (1 - FC)^{1+m_E}
$$

(2-42)

$$
F_3 = 1 - FC(1 + m_E)
$$

and for the base-collector junction

$$F_1 = \frac{\phi_C}{1 - m_C} [1 - (1 - FC)^{1-m_C}]$$

$$F_2 = (1 - FC)^{1+m_C} \tag{2-43}$$

$$F_3 = 1 - FC(1 + m_C)$$

FC is a factor between 0 and 1, which is used to compute the voltage ($FC \times \phi_E$ and $FC \times \phi_C$) in the forward-bias region, beyond which the capacitance is modeled by a linear extrapolation. This has been done to prevent infinite capacitances at $V = \phi_E$ and at $V = \phi_C$ and also to guarantee a continuous function for the capacitances and derivatives. The default value for FC is set by SPICE to 0.5.

The physics-based Ebers-Moll model for the real BJT requires new model parameters that the user can specify in the .MODEL card. These parameters, which take into account the large-signal effects, are as follows [7–8]:

CJE   Base-emitter zero-bias depletion capacitance (C_{JE})

CJC   Base-collector zero-bias depletion capacitance (C_{JC})

CJS   Collector-substrate zero-bias capacitance (C_{JS}); it is assumed constant

VJE   Base-emitter built-in potential (ϕ_E)

VJC   Base-collector built-in potential (ϕ_C)

VJS   Substrate junction built-in potential (ϕ_S)

TF   Ideal total forward transit time (τ_F); it is assumed constant

TR   Ideal total reverse transit time (τ_R); it is assumed constant

FC   Coefficient for forward-bias depletion capacitance (FC)

The emitter-base and the collector-base capacitance gradient factors, m_E and m_C, even if included in Eqs. (2-36) to (2-41), have not been included in the parameters of the actual Ebers-Moll model, since they affect only the Gummel-Poon model (see Sec. 2.6.1). For the Ebers-Moll model these parameters are fixed at a value of 0.33.

2.4   Ebers-Moll Small-Signal Model

The models of the BJT discussed so far maintain the strictly nonlinear nature of the physics of the device and therefore are able to represent, both statically and dynamically, the behavior of the BJT for both types of bias.

In some circuit situations, however, the characteristics of the device must be represented only in a restricted range of currents and voltages. In particular, for small variations around the operating point, which is fixed by a constant source, the nonlinear characteristic of the BJT can be *linearized,* so that the incremental current becomes proportional to the incremental voltage if the variations are sufficiently small.

The development of an incremental linear model is not necessarily linked to a graphic interpretation of the relationships between the variables; the mathematical development can be based on the linearization of a nonlinear functional relationship by means of the expansion in a Taylor series truncated after the first-order terms, as obtained for the diode small-signal model.

2.4.1 Small-signal model and its implementation in SPICE

a. Parameter definition.   When transistors are biased in the active region and used for amplification, it is often worthwhile to approximate their behavior under conditions of small voltage variations at the base-emitter junction. If these variations are smaller than the thermal voltage kT/q, it is possible to represent the transistor by a linear equivalent circuit. This representation can be of great aid in the design of amplifying circuits. It is called the *small-signal transistor model.*

When a transistor is biased in the active mode, collector current is related to base-emitter voltage by Eq. (2-9), which is repeated here for convenient reference (assuming $V_{BE} \gg kT/q$ and $V_{BC} \leqslant 0$):

$$I_C = I_S e^{qV_{BE}/kT} \tag{2-44}$$

Hence, if V_{BE} varies incrementally, I_C also varies according to

$$\frac{dI_C}{dV_{BE}} = \frac{qI_S}{kT} e^{qV_{BE}/kT} = \frac{qI_C}{kT} \equiv g_m \tag{2-45}$$

This derivative is recognized as the *transconductance* and is given the usual symbol g_m.

The variation of base current with base-emitter voltage can be found directly from Eq. (2-44), that is,

$$\frac{dI_B}{dV_{BE}} = \frac{d(I_C/\beta_F)}{dV_{BE}} = \frac{1}{\beta_F} \frac{I_S q}{kT} e^{qV_{BE}/kT} = \frac{g_m}{\beta_F} \equiv g_\pi \tag{2-46}$$

It must be noted that β_F is here a constant (variations of β_F are described in Sec. 2.5.1).

From the analysis of Sec. 2.2.2b, it is known that the voltage across the collector-base junction influences collector current chiefly as a result of the Early effect. The variation of I_C with V_{BC} is shown in Eq. (2-24) to be the ratio of the collector current I_C to the Early voltage V_A. In terms of the small-signal parameters,

$$\frac{dI_C}{dV_{BC}} = \frac{I_C}{|V_A|} = \frac{g_m kT}{q|V_A|} \equiv g_o \tag{2-47}$$

Any change in base minority charge results in a change in base current as well as in collector current. Thus, the variation in V_{BC} results in a change in I_B; this effect can be modeled by inclusion of a resistor, r_μ, from collector to base. If V_{BE} is assumed constant, we can write

$$\frac{dI_B}{dV_{BC}} = \frac{g_o}{\beta_F} \equiv g_\mu \tag{2-48}$$

An additional effect has to be considered: the base resistance. Current-crowding effects cause an overall dc base resistance that is a function of collector current. In the Ebers-Moll model, however, this variation is not taken into account; thus

$$g_x = \frac{1}{r_B} \tag{2-49}$$

Expressing the charge-storage elements as voltage-dependent small-signal capacitances (linearized capacitances†) yields

$$\frac{dQ_{BE}}{dV_{BE}} = \underbrace{\frac{d(Q_{DE} + Q_{JE})}{dV_{BE}}}_{C_\pi}$$

$$= \underbrace{\tau_F \frac{qI_S}{kT} e^{qV_{BE}/kT}}_{\tau_F g_{mF}} + \underbrace{C_{JE}(0)\left(1 - \frac{V_{BE}}{\phi_E}\right)^{-m_E}}_{C_{JE}(V_{BE})} \tag{2-50}$$

---

† This linearization is performed inside the computer program so the user is required only to specify τ_F. By specifying τ_F, the user models Q_{DE} in either a nonlinear mode (for *transient* analyses) or a linear mode (for ac analyses) since τ_F is assumed to be constant.

$$\frac{dQ_{BC}}{dV_{BC}} = \underbrace{\frac{d(Q_{DC} + Q_{JC})}{dV_{BC}}}_{C_{\mu}} = \underbrace{\tau_R \frac{qI_S}{kT} e^{qV_{BC}/kT}}_{\tau_R g_{mR}} + \underbrace{C_{JC}(0)\left(1 - \frac{V_{BC}}{\phi_C}\right)^{-m_C}}_{C_{JC}(V_{BC})} \qquad (2\text{-}51)$$

Combination of the above small-signal circuit elements yields the small-signal model of the bipolar transistor shown in Fig. 2-19. This is valid for both *npn* and *pnp* devices and is called the *hybrid-π model*.

The elements for this linear hybrid-π model, shown in Fig. 2-19, are summarized as follows:

$$g_{mF} = \frac{q}{kT} I_C$$

$$r_{\pi} = \frac{\beta_F}{g_{mF}}$$

$$r_o = \frac{q}{kT g_{mF}} V_A$$

$$r_{\mu} = \frac{\beta_R}{g_{mR}} \qquad (2\text{-}52)$$

$$r_x = r_B$$

$$C_{\pi} = g_{mF}\tau_F + C_{JE}(V_{BE})$$

$$C_{\mu} = g_{mR}\tau_R + C_{JC}(V_{BC})$$

$$C_{CS} = C_{JS}(V_{CS})$$

In the normal region of operation, reverse transconductance g_{mR} is essentially zero, so that resistance r_{μ} can be regarded as infinite and capacitance $C_{\mu} = C_{JC}(V_{BC})$.

b. Implementation in SPICE. Equations (2-52) are expressed in SPICE by the following relations, which define the BJT SPICE Ebers-Moll small-signal model:

$$g_m = \begin{cases} \dfrac{qI_S}{kT} e^{qV_{BE}/kT} + \text{GMIN} - g_o & \text{for } V_{BE} > -5\dfrac{kT}{q} \\[4mm] -\dfrac{I_S}{V_{BE}} + \text{GMIN} - g_o & \text{for } V_{BE} \leq -5\dfrac{kT}{q} \end{cases} \qquad (2\text{-}53)$$

$$g_\pi \equiv \frac{1}{r_\pi} = \begin{cases} \dfrac{I_S}{\beta_F}\dfrac{q}{kT}\,e^{qV_{BE}/kT} + \dfrac{\text{GMIN}}{\beta_F} & \text{for } V_{BE} > -5\,\dfrac{kT}{q} \\[3ex] -\dfrac{I_S}{\beta_F V_{BE}} + \dfrac{\text{GMIN}}{\beta_F} & \text{for } V_{BE} \le -5\,\dfrac{kT}{q} \end{cases} \qquad (2\text{-}54)$$

$$g_o \equiv \frac{1}{r_o} = \begin{cases} \dfrac{qI_S}{kT}\,e^{qV_{BC}/kT} + \text{GMIN} & \text{for } V_{BC} > -5\,\dfrac{kT}{q} \\[3ex] -\dfrac{I_S}{V_{BC}} + \text{GMIN} & \text{for } V_{BC} \le -5\,\dfrac{kT}{q} \end{cases} \qquad (2\text{-}55)$$

$$g_\mu \equiv \frac{1}{r_\mu} = \begin{cases} \dfrac{I_S}{\beta_R}\dfrac{q}{kT}\,e^{qV_{BC}/kT} + \dfrac{\text{GMIN}}{\beta_R} & \text{for } V_{BC} > -5\,\dfrac{kT}{q} \\[3ex] -\dfrac{I_S}{\beta_R V_{BC}} + \dfrac{\text{GMIN}}{\beta_R} & \text{for } V_{BC} \le -5\,\dfrac{kT}{q} \end{cases} \qquad (2\text{-}56)$$

$$g_x = \frac{1}{r_B} \qquad (2\text{-}57)$$

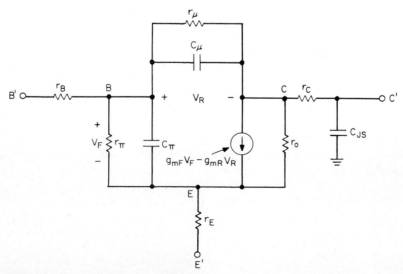

Figure 2-19 Linear hybrid-π model.

$$C_\pi \equiv C_{BE} = \begin{cases} \tau_F \dfrac{qI_S}{kT} e^{qV_{BE}/kT} + C_{JE}(0)\left(1 - \dfrac{V_{BE}}{\phi_E}\right)^{-m_E} \\[2em] \qquad\qquad\qquad\qquad \text{for } V_{BE} < FC \times \phi_E \\[1.5em] \tau_F \dfrac{qI_S}{kT} e^{qV_{BE}/kT} + \dfrac{C_{JE}(0)}{F_2}\left(F_3 + \dfrac{m_E V_{BE}}{\phi_E}\right) \\[2em] \qquad\qquad\qquad\qquad \text{for } V_{BE} \geq FC \times \phi_E \end{cases} \quad (2\text{-}58)$$

$$C_\mu \equiv C_{BC} = \begin{cases} \tau_R \dfrac{qI_S}{kT} e^{qV_{BC}/kT} + C_{JC}(0)\left(1 - \dfrac{V_{BC}}{\phi_C}\right)^{-m_C} \\[2em] \qquad\qquad\qquad\qquad \text{for } V_{BC} < FC \times \phi_C \\[1.5em] \tau_R \dfrac{qI_S}{kT} e^{qV_{BC}/kT} + \dfrac{C_{JC}(0)}{F_2}\left(F_3 + \dfrac{m_C V_{BC}}{\phi_C}\right) \\[2em] \qquad\qquad\qquad\qquad \text{for } V_{BC} \geq FC \times \phi_C \end{cases} \quad (2\text{-}59)$$

$$C_{CS} = \begin{cases} C_{JS}(0)\left(1 - \dfrac{V_{CS}}{\phi_S}\right)^{-m_S} & \text{for } V_{CS} < 0 \\[1.5em] C_{JS}(0)\left(1 + \dfrac{m_S V_{CS}}{\phi_S}\right) & \text{for } V_{CS} > 0 \end{cases} \quad (2\text{-}60)$$

where F_2 and F_3 are derived from Eqs. (2-40) and (2-41).

2.5   Gummel-Poon Static Model

The Ebers-Moll model lacks a representation of many important second-order effects present in actual devices; the two most important effects are those of *low current* and *high-level injection*. The low-current effects result from additional base current due to recombination that degrades current gain. The effects of high-level injection also reduce current gain and in addition cause an increase in τ_F and τ_R. To account for these second-order effects, the *Gummel-Poon model* has been implemented in SPICE.

The Gummel-Poon model [2] is based on an extension of the *Moll-Ross* formula. The basic idea underlying the Moll-Ross formula is the integral charge concept Q, which is the charge represented by the total majority carriers per unit area in the neutral emitter, base, or collector regions.

The following derivation of the Gummel-Poon static model proceeds basically in three major steps according to the modeling of the following second-order effects:

1. Low-current effects (low-current drop in β)

2. Complete description of base-width modulation

3. High-level injection

It must be noted that item 1 affects modifications in the expression for I_B, while items 2 and 3 are incorporated by modifying existing equations (for I_{CT} and C_{DE} only).

Involved in the derivation of these two last effects is a new definition of I_S in terms of the internal physics of the BJT.

It should be emphasized that the Gummel-Poon model derivation (like the previous Ebers-Moll model) assumes a one-dimensional transistor structure.

2.5.1 β variation with current

As described in Sec. A.2.5, recombination in the space-charge layer leads to modified diode relationships for the junction currents. These can be modeled by adding four parameters to the Ebers-Moll model in order to define base current in terms of a superposition of ideal diode and nonideal diode components. These extra components of base current affect the low-current drop in the current gain β (discussed next), allowing introduction of the four new parameters.

In general, there are three regions of interest in the variation of β with current. Figure 2-20 shows a typical variation of β_F with I_C. Region 1 is the low-current region in which β_F increases with I_C. Region 2 is the mid-current region in which β_F is constant (β_{FM}). Region 3 is the high-current region in which β_F drops as the current is increased. Before we analyze these regions, several points should be noted about Fig. 2-20.

The curve in Fig. 2-20 is drawn for constant V_{BC}, in this case, $V_{BC} = 0$.

In the following, it will be assumed that all data corresponds to $V_{BC} = 0$. Nonzero V_{BC} data can be reduced appropriately by the application of Eq. (2-19). For simplicity, most of the following analysis considers only the variation of β_F with I_C. A similar analysis can be performed for the variation of β_R with I_E at constant V_{BE}.

It will be shown that region 1 is governed by additional components of I_B, while region 3 results from a change in I_C. This information is not evident from Fig. 2-20. Therefore, an alternative form of presenting the above information is used. This alternative form is a plot of $\ln I_C$ and $\ln I_B$ as a function of V_{BE}, as shown in Fig. 2-21. Because of the logarithmic nature of the vertical axis, β_F is obtained directly from the plot as the distance between the I_C and I_B curves. Not only is it evident from this plot what causes the variation of β_F with I_C, but all the model parameters needed to characterize this variation can be obtained directly from it. It is still assumed that $V_{BC} = 0$ for all points.

For some transistors, there may not appear to be a region in which β_F is constant. For these transistors, regions 1 and 3 have simply overlapped. The analysis and subdivision into the three regions is still valid, since the model parameters can still be obtained from Fig. 2-21 even when region 2 in the β_F vs. I_C curve does not exist.

An analysis of the regions is now performed, after Getreu [5].

a. Region 2: Midcurrents. In this region, the Ebers-Moll model for the ideal BJT holds; the β_F used in this model applies only to region 2 and is now called β_{FM}. The two currents in this region (for $V_{BC} = 0$) are given by

$$I_C = I_S(0)(e^{qV_{BE}/kT} - 1)$$

$$I_B = \frac{I_S(0)}{\beta_{FM}(0)}(e^{qV_{BE}/kT} - 1) \tag{2-61}$$

Values of $\beta_{FM}(0)$ and $I_S(0)$ can be obtained directly from Fig. 2-21.

b. Region 1: Low currents. The drop in β_F at low currents is caused by extra components of I_B that until now have been ignored. For the normal

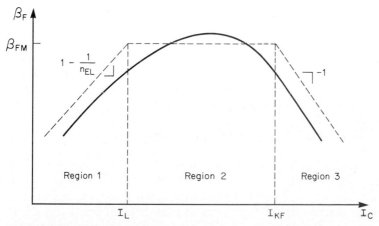

Figure 2-20 Graph of β_F vs. I_C.

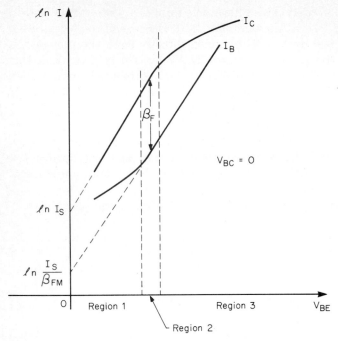

Figure 2-21   Graph of $\ln I_C$ and $\ln I_B$ vs. V_{BE}, with $V_{BC} = 0$ [5].

active region with $V_{BC} = 0$, there are three extra components, which are caused by the following:

1. The recombination of carriers at the surface
2. The recombination of carriers in the emitter-base space-charge layer
3. The formation of emitter-base surface channels

All three components have a similar variation with base-emitter voltage V_{BE}. These three components should be added to the base current of Eq. (2-61). Fortunately, this can be simplified. A composite current can be made of all three extra components. It has the form [5]

$$I_B = I_{S,\text{composite}}(e^{qV_{BE}/n_{EL}kT} - 1) \qquad (2\text{-}62)$$

where n_{EL} is called the *low-current, forward-region emission coefficient* and lies between 1 and 4.

For most cases, a fit to Eq. (2-62) can be made with reasonable accuracy. Since channeling and surface recombination can both be made small with careful processing, the dominant component is normally the recombination in the emitter-base space-charge layer, and n_{EL} is normally close to 2. Therefore, at $V_{BC} = 0$, the base current is approximated by

$$I_B = \frac{I_S(0)}{\beta_{FM}(0)} (e^{qV_{BE}/kT} - 1) + C_2 I_S(0)(e^{qV_{BE}/n_{EL}kT} - 1) \quad (2\text{-}63)$$

where the term $I_{S,\text{composite}}$ in Eq. (2-62) has been replaced by $C_2 I_S(0)$; that is, it has simply been normalized to $I_S(0)$. The two additional model parameters are C_2 and n_{EL}.

When the base-collector junction is forward-biased, there will generally be three similar additional components of I_B at low-current levels: surface recombination, collector-base space-charge layer recombination, and collector-base channeling. In a similar way, they can be lumped together into a composite component that depends on V_{BC}. The expression for I_B in general then becomes

$$I_B = \frac{I_S(0)}{\beta_{FM}(0)} (e^{qV_{BE}/kT} - 1) + C_2 I_S(0)(e^{qV_{BE}/n_{EL}kT} - 1)$$

$$+ \frac{I_S(0)}{\beta_{RM}} (e^{qV_{BC}/kT} - 1) + C_4 I_S(0)(e^{qV_{BC}/n_{CL}kT} - 1) \quad (2\text{-}64)$$

where the two extra model parameters n_{CL} (the low-current, inverse-region emission coefficient) and C_4 have been introduced.

The additional components of base current I_B are included in the circuit model by means of two nonideal diodes, as shown in Fig. 2-22.

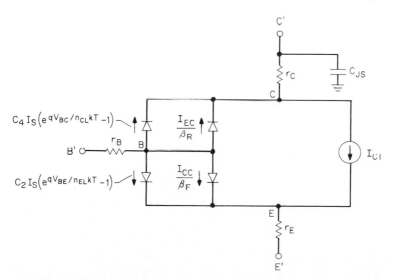

Figure 2-22 Effects of the extra components of I_B.

The plot of $\ln I_B$ vs. qV_{BE}/kT for $V_{BC} = 0$, shown in Fig. 2-23, illustrates the two components of base current I_B: the ideal component with the slope of 1 and the nonideal component with the slope of $1/n_{EL}$. The extrapolation of these straight-line components to the line defined by $V_{BE} = 0$ gives the values of $C_2 I_S(0)$ and $I_S(0)/\beta_{FM}(0)$.

A similar plot of $\ln I_B$ as a function of V_{BC} for inverse operation yields values for the model parameters C_4 and n_{CL}. A typical value for C_2 (and C_4) is 10^3, and a typical value for n_{EL} (and n_{CL}) is 2.

c. Region 3: High currents.   The drop in β at high currents, caused by the effects of high-level injection in the base, will be discussed later.

2.5.2   Physical definition of I_S

In this section, as pointed out previously, a new expression for I_S is obtained from an examination of the current-density equations in the BJT. This new expression inherently incorporates the effects of base-width modulation and high-level injection.

The derivation starts with the one-dimensional, dc equations for the electron current density J_n [see Eq. (A-16b)] and the hole current density J_p [see Eq. (A-16a)] in an npn transistor:

$$J_n = q\mu_n n(x)E(x) + qD_n \frac{dn(x)}{dx}$$

$$J_p = q\mu_p p(x)E(x) - qD_p \frac{dp(x)}{dx}$$

$$(2\text{-}65)$$

where $E(x)$ is the electric field, $n(x)$ is the free electron concentration, and $p(x)$ is the hole concentration. No restriction is placed on the variation of the carrier concentrations, so the following analysis applies for any doping profile. Equations (2-65) apply for both high- and low-level injection. At this point it is assumed that the hole current is zero. This is not exactly the case, but it is normally a reasonable approximation. The approximation is justified by showing that there is no place where a large hole current could go.

The *base-emitter junction* is normally designed for high emitter-injection efficiency (high emitter doping with respect to the base doping). This means that the current injected into the emitter from the base is small even when the emitter-base junction is forward-biased [5].

In the normal active region, the *collector-base junction* is reverse-biased, and therefore no significant current flows across it from the base to the collector. In the inverse and saturated regions of operation, however, the collector-base junction is forward-biased. It is assumed that for most cases of interest, the hole current that flows from the base to the collector is still small.

Obviously, then, the following analysis will be reasonably valid only for the normal active region and when the device is weakly saturated. For cases of strong saturation and inverse operation, it may not hold. However, if the J_p term is retained, it will be multiplied by $n(x)/p(x)$, which reduces its effect even further, since $n(x) \ll p(x)$, except at high-level injection.

Then if the hole current is assumed to be zero, the second expression of Eqs. (2-65) becomes [5]

$$J_p = q\mu_p p(x)E(x) - qD_p \frac{dp(x)}{dx} = 0 \qquad (2\text{-}66)$$

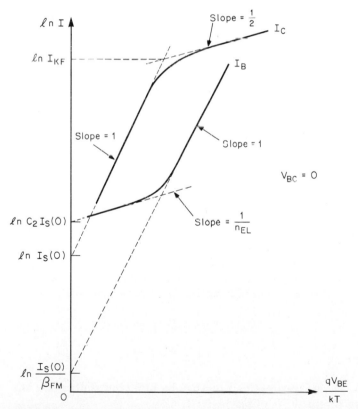

Figure 2-23 Plot of $\ln I_C$ and $\ln I_B$ vs. qV_{BE}/kT for $V_{BC} = 0$ [5].

If Eq. (2-66) is solved for the field E, it yields

$$E(x) = \frac{D_p}{\mu_p} \frac{1}{p(x)} \frac{dp(x)}{dx} \tag{2-67}$$

Substituting Eq. (2-67) in the first expression of Eqs. (2-65) results in

$$J_n = q\mu_n \left[\frac{D_p}{\mu_p} \frac{1}{p(x)} \frac{dp(x)}{dx} \right] n(x) + qD_n \frac{dn(x)}{dx} \tag{2-68}$$

or $$p(x)J_n = qD_n n(x) \frac{dp(x)}{dx} + qD_n p(x) \frac{dn(x)}{dx} \tag{2-69}$$

where the Einstein relationship [see Eqs. (A-17)] has been used. The multiplication of Eq. (2-68) by $p(x)$ is the mathematical trick that, with the following use of the differential product rule, will result in the importance of the majority-carrier concentration, $p(x)$, in the base. The use of the product rule on Eq. (2-69) results in

$$p(x)J_n = qD_n \frac{d}{dx} [n(x)p(x)] \tag{2-70}$$

Both sides of this equation are now integrated from x_E to x_C, where x_E is the position of the emitter side of the emitter-base space-charge layer and x_C is the position of the collector side of the collector-base space-charge layer, as illustrated in Fig. 2-24. Figure 2-24 also defines x'_E and x'_C, the positions of the *base* sides of these space-charge layers [5].

Since current density J_n is constant for dc and is independent of x, assuming negligible recombination in the base region (recombination in the space-charge layers is independently modeled by the two nonideal diodes in Fig. 2-22, so it is assumed to be zero here), it is taken out of the integral. Then

$$J_n \int_{x_E}^{x_C} p(x)\, dx = qD_n \int_{x_E}^{x_C} \frac{d}{dx} [n(x)p(x)]\, dx$$

$$= qD_n [n(x_C)p(x_C) - n(x_E)p(x_E)]$$

$$J_n = \frac{qD_n [n(x_C)p(x_C) - n(x_E)p(x_E)]}{\displaystyle\int_{x_E}^{x_C} p(x)\, dx} \tag{2-71}$$

Equation (2-71) can be simplified further by applying Boltzmann statistics to the pn products [see Eqs. (A-59)].

$$n(x_C)p(x_C) = n_i^2 e^{qV_{BC}/kT}$$

$$n(x_E)p(x_E) = n_i^2 e^{qV_{BE}/kT}$$

(2-72)

Therefore

$$J_n = -\frac{qD_n n_i^2 (e^{qV_{BE}/kT} - e^{qV_{BC}/kT})}{\int_{x_E}^{x_C} p(x)\,dx}$$

(2-73)

At this point, the *depletion approximation* is made [5]. This assumes that there are no (or negligible) mobile carriers in a space-charge layer (see Sec. A.2.1b). That is, the field experienced by the carriers and the thickness of the space-charge layer are such that the carriers are transported instantaneously across it. This approximation, as pointed out in Sec. 2.3.1, is not valid for junctions under forward bias.

The analysis that follows therefore first assumes the depletion approximation to be true and then later fixes up the solution to take the mobile charges in the space-charge layers into account. In effect, the depletion approximation will be applied only for the thermal equilibrium condition.

The application of the depletion approximation results in the limits of the integral in Eq. (2-73) being replaced by x_C' and x_E' and the integration being performed in the neutral base region only. The change in integration limits results, of course, from the assumption that $p(x)$ is approxi-

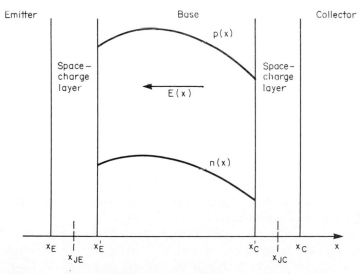

Figure 2-24 Base profile for an *npn* transistor defining the integral limits x_E, x_C, x_E', x_C' [5].

mately zero inside the space-charge layers. Equation (2-73) can therefore be rewritten as

$$I_n \simeq - \frac{qD_n A_J n_i^2 [(e^{qV_{BE}/kT} - 1) - (e^{qV_{BC}/kT} - 1)]}{\displaystyle\int_{x_E'}^{x_C'} p(x)\,dx} \tag{2-74}$$

where A_J is the one-dimensional cross-sectional area, and the -1 term has been introduced for convenience.

I_n represents the total dc minority current in the positive x direction that results from minority carriers injected into the base at the emitter and/or the collector. It is represented in the model of Fig. 2-22 by the current generator I_{CT}. (The other components of the collector current in Fig. 2-22 are components of base current resulting from the injection of holes, for an npn transistor, from the base toward the collector.)

The equation for I_{CT}, previously obtained in the Ebers-Moll model [see Eq. (2-9)], can be written as

$$I_{CT} = I_S [(e^{qV_{BE}/kT} - 1) - (e^{qV_{BC}/kT} - 1)] \tag{2-75}$$

The sign of I_{CT} is opposite to that of I_n [see Eq. (2-74)] since I_n has the opposite direction (out of the collector terminal) to that assumed for I_C (into the collector region from the terminal).

A direct comparison of Eqs. (2-74) and (2-75) yields a physical definition of I_S. In the Ebers-Moll model, I_S has been considered a fundamental constant of the device. Yet the integral in Eq. (2-74) is not constant in that, under high-level injection, $p(x)$, the majority-carrier concentration, is a function of the applied bias. To reconcile this difference, a new symbol, I_{SS}, is used in the Gummel-Poon model and is defined from Eq. (2-74) under low-level injection conditions only [5].

At low-level injection, the combination of Eqs. (2-74) and (2-75) gives

$$I_{CT(low-level)} = \frac{qD_n n_i^2 A_J}{\displaystyle\int_{x_E'}^{x_C'} N_A(x)\,dx} [(e^{qV_{BE}/kT} - 1) - (e^{qV_{BC}/kT} - 1)] \tag{2-76}$$

where $p(x)$ has been replaced by $N_A(x)$, since, at low-current levels in the neutral base region, $p(x)$ approximates $N_A(x)$.

Before the definition of I_{SS} is made, though, more attention must be paid to the limits of the integration, x_E' and x_C'. Because of the variation of the space-charge layer widths with applied voltage, x_E' and x_C' are functions of the appropriate bias voltage (and, in fact, will be seen later to incorporate the effects of base-width modulation).

The fundamental constant I_{SS} is therefore defined at *zero* V_{BE} and V_{BC} as

$$I_{SS} = \frac{qD_n n_i^2 A_J}{\int_{x'_{E0}}^{x'_{C0}} N_A(x)\, dx} \qquad (2\text{-}77)$$

where x'_{E0} and x'_{C0} are the values of x'_E and x'_C when the applied junction voltages are zero. The fundamental nature of I_{SS} is seen immediately from Eq. (2-77), since it is uniquely determined once the base-doping profile is fixed. It must be noted that in SPICE, I_S and I_{SS} coincide, as they are specified in the .MODEL card [8] by the same parameter IS.

In the derivation of Eq. (2-77), the diffusion constant D_n has been assumed to be constant and independent of x. In practice, this assumption is not valid. The diffusion constant should be included in the denominator integral. Instead, D_n is interpreted in Eq. (2-77) as an effective diffusion constant in the base.

At this point in the analysis, the depletion approximation can be removed. The application of the depletion approximation allowed the above derivation to concentrate solely on the majority carriers in the neutral base region. As a result, the mobile carriers in the space-charge layers have been ignored. If these carriers are now taken into account, the definition of I_{SS} has to be updated as [5]

$$I_{SS} = \frac{qD_n n_i^2 A_J}{\int_{x_{E0}}^{x_{C0}} p_0(x)\, dx} \qquad (2\text{-}78)$$

where $p_0(x)$ is the equilibrium hole concentration in the neutral base and space-charge regions.

2.5.3 Q_B concept

The general expression for I_S (which is now a function of bias voltages) can be obtained in terms of the zero-bias constant I_{SS}. However, before this is done, it is worthwhile to make a few definitions and to introduce some new concepts.

When multiplied by q and A_J, the integral in Eq. (2-74) represents the total *majority* charge in the neutral base region and is given the symbol Q_B:

$$Q_B = \int_{x'_E}^{x'_C} qA_J p(x)\, dx \qquad (2\text{-}79)$$

The *zero-bias* majority base charge, Q_{B0}, is defined by

$$Q_{B0} = \int_{x'_{E0}}^{x'_{C0}} qA_J N_A(x)\, dx \qquad (2\text{-}80)$$

Finally, the *normalized* majority base charge, q_b, is defined as

$$q_b = \frac{Q_B}{Q_{B0}} \qquad (2\text{-}81)$$

In the following analysis, all charge normalizations are with respect to Q_{B0} and are represented by q with an appropriate subscript.

The above definitions [Eqs. (2-79) to (2-81)] can be used to find the new definition of I_S. If Eq. (2-74) is multiplied and divided by Q_{B0} and I_n is replaced by $-I_{CT}$, then [5]

$$I_{CT} = \frac{qD_n n_i^2 A_J}{\displaystyle\int_{x'_E}^{x'_C} p(x)\, dx} \; \frac{qA_J \displaystyle\int_{x'_{E0}}^{x'_{C0}} N_A(x)\, dx}{qA_J \displaystyle\int_{x'_{E0}}^{x'_{C0}} N_A(x)\, dx} \left[(e^{qV_{BE}/kT} - 1) \right.$$
$$\left. - (e^{qV_{BC}/kT} - 1) \right] \qquad (2\text{-}82)$$

Combining the first term in the numerator with the second term in the denominator, and using Eqs. (2-77) to (2-81), gives

$$I_{CT} = \frac{I_{SS} Q_{B0}}{Q_B} \left[(e^{qV_{BE}/kT} - 1) - (e^{qV_{BC}/kT} - 1) \right]$$
$$= \frac{I_{SS}}{q_b} \left[(e^{qV_{BE}/kT} - 1) - (e^{qV_{BC}/kT} - 1) \right] \qquad (2\text{-}83)$$

Equation (2-83) is the new equation introduced by the Gummel-Poon model, and the new concept introduced is the fundamental importance of the (normalized) majority charge in the base, q_b. The old saturation current I_S (which was assumed constant in the Ebers-Moll model) has been replaced by the new term I_{SS}/q_b, where I_{SS} is the fundamental constant (defined at zero-bias condition) and q_b is a variable that still needs to be determined. The rest of the Gummel-Poon derivation involves the determination of q_b as a function of the bias conditions in terms of measurable parameters.

If the depletion approximation is removed, an updated expression for Q_B and Q_{B0}, including the space-charge layers (as seen for I_{SS}), follows:

$$Q_B = \int_{x_E}^{x_C} qA_J p(x)\, dx \qquad (2\text{-}84)$$

$$Q_{B0} = \int_{x_{E0}}^{x_{C0}} qA_J p_0(x) \, dx \simeq \int_{x'_{E0}}^{x'_{C0}} qA_J N_A(x) \, dx \qquad (2\text{-}85)$$

where $p_0(x)$ is the equilibrium hole concentration in the neutral base and space-charge regions. The second form of Eq. (2-85), which assumes that the depletion approximation is valid for the equilibrium case, naturally agrees with the previous definition given in Eq. (2-80).

2.5.4 Solution for q_b

When Q_B is normalized with respect to Q_{B0} and is split into its five components, it can be written in the following form:

$$q_b = \frac{Q_B}{Q_{B0}} = 1 + \frac{C_{JE}}{Q_{B0}} V_{BE} + \frac{C_{JC}}{Q_{B0}} V_{BC} + \frac{\tau_{BF}}{Q_{B0}} I_{CC} + \frac{\tau_{BR}}{Q_{B0}} I_{EC} \qquad (2\text{-}86)$$

(See Ref. [5] for a complete treatment of the components of Q_B; here only the SPICE user-oriented results are described.)

Equation (2-86) can be put into a more expanded form as

$$q_b = 1 + \underbrace{\frac{V_{BE}}{V_B}}_{q_e} + \underbrace{\frac{V_{BC}}{V_A}}_{q_c} + \underbrace{\frac{\tau_{BF}}{Q_{B0}} I_{SS} \frac{e^{qV_{BE}/kT} - 1}{q_b}}_{q_{bF}} + \underbrace{\frac{\tau_{BR}}{Q_{B0}} I_{SS} \frac{e^{qV_{BC}/kT} - 1}{q_b}}_{q_{bR}} \qquad (2\text{-}87)$$

where
$$\frac{Q_{B0}}{C_{JE}} = V_B \qquad \text{(reverse Early voltage)}$$

$$\frac{Q_{B0}}{C_{JC}} = V_A \qquad \text{(forward Early voltage)}$$

$$I_{CC} = \frac{I_{SS}}{q_b} (e^{qV_{BE}/kT} - 1) \qquad \text{[see Eqs. (2-6)]}$$

$$I_{EC} = \frac{I_{SS}}{q_b} (e^{qV_{BC}/kT} - 1) \qquad \text{[see Eqs. (2-7)]}$$

Note that q_e and q_c both model the effects of base-width modulation, while q_{bF} and q_{bR} both model the effects of high-level injection.

To simplify the algebra, the following definitions can be made:

$$q_b = q_1 + \frac{q_2}{q_b} \qquad (2\text{-}88)$$

where

$$q_1 = 1 + q_e + q_c = 1 + \frac{V_{BE}}{V_B} + \frac{V_{BC}}{V_A} \tag{2-89}$$

$$q_2 = \frac{\tau_{BF} I_{SS}(e^{qV_{BE}/kT} - 1) + \tau_{BR} I_{SS}(e^{qV_{BC}/kT} - 1)}{Q_{B0}}$$

By definition, q_1 is unity at zero bias; this reflects the fact that the physical base width and the electrical base width are equal to zero bias. If both junctions are forward-biased, the electrical base width is larger than the physical base width, and q_1 is larger than unity. Conversely, if both junctions are reverse-biased, then q_1 is less than unity, since the electrical base width is less than the physical base width. Thus q_1 models the effects of base-width modulation.

The q_2 component of base charge accounts for the excess majority-carrier base charge that results from injected minority carriers. Excess majority-carrier base charge is insignificant until the injected minority carrier is comparable to the zero-bias majority-carrier charge, that is, until high-level injection is reached [7]. Thus q_2 models the effects of high-level injection.

Equation (2-88) results in a quadratic expression for q_b,

$$q_b^2 - q_b q_1 - q_2 = 0 \tag{2-90}$$

which gives

$$q_b = \frac{q_1}{2} + \sqrt{\left(\frac{q_1}{2}\right)^2 + q_2} \tag{2-91}$$

where the negative solution has been ignored because q_b is greater than zero.

From Eq. (2-91) it follows that

$$\begin{aligned} q_b \simeq q_1 \qquad &\text{if } q_2 \ll \frac{q_1^2}{4} \qquad \text{(low-level injection)} \\ q_b \simeq \sqrt{q_2} \qquad &\text{if } q_2 \gg \frac{q_1^2}{4} \qquad \text{(high-level injection)} \end{aligned} \tag{2-92}$$

a. High-level injection solution. In the normal active region at high-injection levels, $q_b \simeq \sqrt{q_2}$. To simplify the situation, consider the case of $V_{BC} = 0$ (that is, $q_{bR} = 0$). The extension of the result to nonzero q_{bR} will be made by analogy.

For zero q_{bR}

$$q_b = \sqrt{q_2} \simeq \sqrt{\frac{\tau_{BF} I_{SS}}{Q_{B0}} e^{qV_{BE}/kT}} = \sqrt{\frac{\tau_{BF} I_{SS}}{Q_{B0}}} e^{qV_{BE}/2kT} \quad (2\text{-}93)$$

Therefore, solving for I_C ($I_C = I_{CC}$ in this situation),

$$I_C = \frac{I_{SS}(e^{qV_{BE}/kT} - 1)}{\sqrt{\tau_{BF} I_{SS}/Q_{B0}} \, e^{qV_{BE}/2kT}} \simeq \sqrt{\frac{Q_{B0} I_{SS}}{\tau_{BF}}} e^{qV_{BE}/2kT} \quad (2\text{-}94)$$

Therefore

$$I_C \propto e^{qV_{BE}/2kT} \quad (2\text{-}95)$$

b. Simplification of q_2. The simplification of the coefficient $\sqrt{Q_{B0} I_{SS}/\tau_{BF}}$ arises from a consideration of the $\ln I_C$ vs. V_{BE} characteristics at the two extremes: high- and low-level injection.

Figure 2-23 shows the variation of $\ln I_C$ as a function of qV_{BE}/kT.

The low-current asymptote is given approximately by the following (for $q_e \simeq q_c \simeq 0$):

$$I_C \simeq I_{SS} e^{qV_{BE}/kT} \quad (2\text{-}96)$$

The high-current asymptote is given by Eq. (2-94). The intersection of those two asymptotes defines the knee current I_{KF} and the *knee voltage* V_{KF}.

From Eq. (2-94), it follows that

$$I_{KF} = \sqrt{\frac{Q_{B0} I_{SS}}{\tau_{BF}}} e^{qV_{KF}/2kT} \quad \text{(high-level)} \quad (2\text{-}97)$$

while from Eq. (2-96), it follows that

$$I_{KF} = I_{SS} e^{qV_{KF}/kT} \quad \text{(low-level)} \quad (2\text{-}98)$$

The solution of Eqs. (2-97) and (2-98) yields

$$I_{KF} = \frac{Q_{B0}}{\tau_{BF}} \quad (2\text{-}99)$$

A similar analysis for the inverse region defines I_{KR}, the knee current for $\ln I_E$ vs. V_{BC} in the inverse region, as

$$I_{KR} = \frac{Q_{B0}}{\tau_{BR}} \tag{2-100}$$

Additional understanding of the Gummel-Poon parameters is obtained by inspection of the *asymptotic* behavior of the short-circuit current gain as shown in Fig. 2-20. The current gain is essentially constant at a value of β_{FM} for collector currents greater than I_L, falls off with a slope of $1 - 1/n_{EL}$ for collector currents less than I_L, and falls off with a slope of -1 for collector currents greater than I_{KF}. I_L, the low-current breakpoint, is given by the equation

$$I_L = I_S(C_2\beta_{FM})^{n_{EL}/(n_{EL}-1)} \tag{2-101}$$

This, of course, is only an asymptotic relation.

c. Final solution. As a result of Eqs. (2-99) and (2-100), q_b can be finally written as

$$q_b = \frac{q_1}{2} + \sqrt{\left(\frac{q_1}{2}\right)^2 + q_2} \tag{2-102}$$

where

$$q_1 = 1 + \frac{V_{BE}}{V_B} + \frac{V_{BC}}{V_A} \tag{2-103}$$

$$q_2 = \frac{I_{SS}}{I_{KF}}(e^{qV_{BE}/kT} - 1) + \frac{I_{SS}}{I_{KR}}(e^{qV_{BC}/kT} - 1)$$

It must be noticed that all parameters in the above expressions (V_B, V_A, I_{SS}, I_{KF}, and I_{KR}) are measurable from plots of $\ln I_C$ vs. V_{BE} in the normal active region, from plots of $\ln I_E$ vs. V_{BC} in the inverse region, and from I_C vs. V_{CE} and I_E vs. V_{CE} characteristics. Once these parameters are known, q_1, q_2, and, therefore, q_b can be easily determined by the above expressions.

The preceding derivation of the Gummel-Poon model differs from that of Ref. [2] in the integration limits used. Whereas the preceding derivation integrated from the outside of the space-charge layers, Gummel and Poon integrated over virtually the entire transistor.

The reason for this difference is widely explained in Ref. [5]. Two basic differences result. They are different definitions of I_{SS} and the knee currents I_{KF} and I_{KR}.

In the definition of I_{SS} [see Eq. (2-78)], the Gummel-Poon approach results in the denominator integration being performed over virtually the entire transistor. Since the base majority carriers are minority carriers in the neutral emitter and collector regions, this difference should have a negligible effect. This difference in definition is not important for the device characterization, since I_{SS} is determined experimentally [5].

In the equations for I_{KF} and I_{KR} [Eqs. (2-99) and (2-100)], the terms τ_{BF} and τ_{BR} occur. If the integration is performed over the entire transistor, the definitions of τ_{BF} and τ_{BR} would include the emitter delay τ_E and the collector delay τ_C, respectively (since Q_F would be added to Q_{BF}, and Q_C would be added to Q_{BR}). Again, since I_{KF} and I_{KR} are both experimentally determined, this difference in τ_{BF} and τ_{BR} is not important for the device characterization.

2.5.5 Current dependence of the base resistance

An improvement to the Gummel-Poon model [2] has been introduced in SPICE. It takes into account the dependence of the base resistance on the current *(current crowding)*.

In SPICE, the base resistance between the external and the internal base nodes comes from two separate resistances. The external constant resistance r_B *(extrinsic base resistance)* consists of the contact resistance and the sheet resistance of the external base region. The resistance r_{BM} of the internal base (active base) region *(intrinsic base resistance),* which is the part of the base lying directly under the emitter, is a function of the base current. The dependence of this resistance on the device current arises from nonzero base-region resistivity, which in turn precipitates nonuniform biasing of the base-emitter junction. (A reasonable approach entails a consideration of the average power dissipated in the intrinsic base.)

It is possible to analyze this effect exactly, but the mathematics become relatively tedious and might obscure the relevant physical mechanisms; therefore, we will consider only the results of this analysis.

Thus, it can be shown that the total base resistance can be expressed as [10]

$$r_{BB'} = r_{BM} + 3(r_B - r_{BM})\left(\frac{\tan z - z}{z \tan^2 z}\right) \tag{2-104}$$

where r_{BM} is the minimum base resistance that occurs at high currents; r_B is the base resistance at zero bias (small base currents); and z is a variable of base resistivity, thermal voltage, and intrinsic base length.

In order to reduce the computational complexity in calculating z, an approximation method is used to represent $\cos z$ by the first two terms of

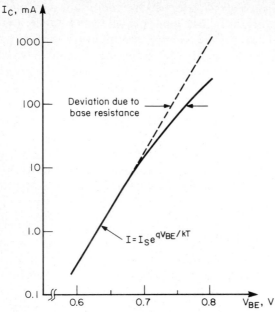

Figure 2-25 Collector current as a function of base-emitter voltage showing the deviation from ideal behavior at high currents due to the base resistance.

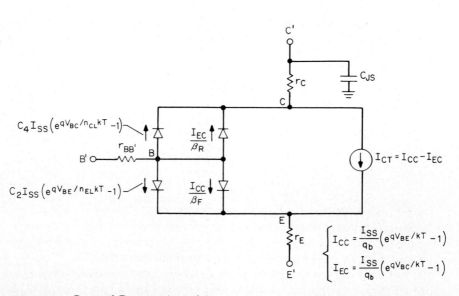

Figure 2-26 Gummel-Poon static model.

its McLaurin series. The value of z from this approximation is

$$z = \frac{-1 + \sqrt{1 + 144I_B/\pi^2 I_{rB}}}{(24/\pi^2)\ \sqrt{I_B/I_{rB}}} \qquad (2\text{-}105)$$

where I_{rB} is the current where the base resistance falls halfway to its minimum value.

In Fig. 2-25 the collector current is shown as a function of the base-emitter voltage, giving the deviation from ideal behavior at high currents due to the base resistance.

2.5.6 Static model summary

The improvements to the Ebers-Moll static model provided by the Gummel-Poon static model influence the following effects.

1. *Low-current drop in* β.   This drop is caused by extra components of I_B [see Eq. (2-64)], which can be described by four model parameters, C_2, n_{EL} (for β_F) and C_4, n_{CL} (for β_R). Two nonideal diodes were added to the Ebers-Moll static circuit model (see Fig. 2-22).

2. *Base-width modulation.*   Base-width modulation is due to a change in the collector-base and emitter-base junction voltages, which causes a variation of the collector-base and emitter-base space-charge layer. This effect is described by two parameters, V_A (forward Early voltage) and V_B (reverse Early voltage), which are identified in the q_c and q_e components of q_b [see Eq. (2-87)], respectively.

3. *High-level injection.*   At high-level injection, the injection of minority carriers into the base region is significant with respect to the majority-carrier concentration. Since space-charge neutrality is maintained in the base, the total majority-carrier concentration is increased by the same amount as the total minority-carrier concentration. The effect of the excess majority carriers is represented by a changing in the slope of the I_C vs. V_{BE} characteristic (see Fig. 2-23) above the so-called knee current I_{KF} (for the normal active region) [see Eq. (2-97)]. Another parameter, I_{KR}, is defined for the reverse active region.

4. *Base resistance.*   The base resistance is demonstrated to be current-dependent and is modeled by a combination of r_B, r_{BM}, and I_{rB} [see Eq. (2-104)].

It must be noted that involved in the derivation of the Gummel-Poon static model is a new definition of $I_S = I_{SS}/q_b$. Another feature provided by the Gummel-Poon model is the different approach to the device that is used, namely, how the analysis of the majority carriers in the base can be used to generate the preceding equations.

The complete Gummel-Poon model that accounts for the above-described effects is shown in Fig. 2-26.

2.5.7 Static model and its implementation in SPICE

This section shows how the Gummel-Poon static model is implemented in SPICE [7].

The dc characteristics of the intrinsic BJT are determined by the nonlinear current sources I_C and I_B. The values of I_C and I_B are defined by the following equations when the GMIN term is also considered (see Sec. 1.2.1a).

Normal active region

For $V_{BE} > -5 \dfrac{n_F kT}{q}$ and $V_{BC} \le -5 \dfrac{n_R kT}{q}$:

$$I_C = \frac{I_S}{q_b}\left(e^{qV_{BE}/n_F kT} + \frac{q_b}{\beta_R}\right) + C_4 I_S + \left[\frac{V_{BE}}{q_b} - \left(\frac{1}{q_b} + \frac{1}{\beta_R}\right)V_{BC}\right]\text{GMIN}$$

$$I_B = I_S\left[\frac{1}{\beta_F}\left(e^{qV_{BE}/n_F kT} - 1\right) - \frac{1}{\beta_R}\right] + C_2 I_S(e^{qV_{BE}/n_{EL}kT} - 1) \tag{2-106}$$

$$- C_4 I_S + \left(\frac{V_{BE}}{\beta_F} + \frac{V_{BC}}{\beta_R}\right)\text{GMIN}$$

Inverse region

For $V_{BE} \le -5 \dfrac{n_F kT}{q}$ and $V_{BC} > -5 \dfrac{n_R kT}{q}$:

$$I_C = -\frac{I_S}{q_b}\left[e^{qV_{BC}/n_R kT} + \frac{q_b}{\beta_R}\left(e^{qV_{BC}/n_R kT} - 1\right)\right] - C_4 I_S(e^{qV_{BC}/n_{CL}kT} - 1)$$

$$+ \left[\frac{V_{BE}}{q_b} - \left(\frac{1}{q_b} + \frac{1}{\beta_R}\right)V_{BC}\right]\text{GMIN}$$

$$I_B = -I_S\left[\frac{1}{\beta_F} - \frac{1}{\beta_R}\left(e^{qV_{BC}/n_R kT} - 1\right)\right] - C_2 I_S \tag{2-107}$$

$$+ C_4 I_S(e^{qV_{BC}/n_{CL}kT} - 1) + \left(\frac{V_{BE}}{\beta_F} + \frac{V_{BC}}{\beta_R}\right)\text{GMIN}$$

Saturated region

For $V_{BE} > -5 \dfrac{n_F kT}{q}$ and $V_{BC} > -5 \dfrac{n_R kT}{q}$:

$$I_C = \frac{I_S}{q_b} \left[\left(e^{qV_{BE}/n_F kT} - e^{qV_{BC}/n_R kT} \right) - \frac{q_b}{\beta_R} \left(e^{qV_{BC}/n_R kT} - 1 \right) \right]$$

$$- C_4 I_S (e^{qV_{BC}/n_{CL}kT} - 1) + \left[\frac{V_{BE}}{q_b} - \left(\frac{1}{q_b} + \frac{1}{\beta_R} \right) V_{BC} \right] \text{GMIN}$$

$$I_B = I_S \left[\frac{1}{\beta_F} (e^{qV_{BE}/n_F kT} - 1) + \frac{1}{\beta_R} (e^{qV_{BC}/n_R kT} - 1) \right]$$

(2-108)

$$+ C_2 I_S (e^{qV_{BE}/n_{EL}kT} - 1) + C_4 I_S (e^{qV_{BC}/n_{CL}kT} - 1) + \left(\frac{V_{BE}}{\beta_F} + \frac{V_{BC}}{\beta_R} \right) \text{GMIN}$$

Off region

For $V_{BE} \leq -5 \dfrac{n_F kT}{q}$ and $V_{BC} \leq -5 \dfrac{n_R kT}{q}$:

$$I_C = \frac{I_S}{\beta_R} + C_4 I_S + \left[\frac{V_{BE}}{q_b} - \left(\frac{1}{q_b} + \frac{1}{\beta_R} \right) V_{BC} \right] \text{GMIN}$$

(2-109)

$$I_B = -I_S \left(\frac{\beta_F + \beta_R}{\beta_F \beta_R} \right) - C_2 I_S - C_4 I_S + \left(\frac{V_{BE}}{\beta_F} + \frac{V_{BC}}{\beta_R} \right) \text{GMIN}$$

The SPICE BJT Gummel-Poon model is the same as in Fig. 2-26 when the new expressions for I_C and I_B [see Eqs. (2-106) to (2-109)] are used.

The extra parameters required for the Gummel-Poon model, as well as those for the Ebers-Moll model, can be specified in the .MODEL card as:

C2	Forward low-current nonideal base current coefficient (C_2)
C4	Reverse low-current nonideal base current coefficient (C_4)
NE	Nonideal low-current base-emitter emission coefficient (n_{EL})
NC	Nonideal low-current base-collector emission coefficient (n_{CL})
VAF	Forward Early voltage (V_A)
VAR	Reverse Early voltage (V_B)
IKF	Corner for forward β high-current roll-off (I_{KF})
IKR	Corner for reverse β high-current roll-off (I_{KR})
RB	Zero-bias resistance (r_B)
RBM	Minimum base resistance at high currents (r_{BM})
IRB	Current where base resistance falls halfway to its minimum value (I_{rB})
ISE	Nonideal base-emitter saturation current ($I_{SE} = C_2 I_S$)
ISC	Nonideal base-collector saturation current ($I_{SC} = C_4 I_S$)
NF	Forward current emission coefficient (n_F)
NR	Reverse current emission coefficient (n_R)

The ISE and ISC parameters are used in version 2G of SPICE instead of the C_2 and C_4 parameters. The parameter n_F models the forward collector current at low currents; it is the collector current exponential factor that specifies the slope of the I_C vs. V_{BF} on the logarithmic scale; n_R has exactly the same effect in the reverse mode.

The base charge q_b, as explained in the previous sections, is the total majority-carrier charge in the base region divided by the zero-bias majority-carrier charge and is defined by Eqs. (2-102) and (2-103).

SPICE makes the following modification in Eq. (2-102) to separate the effect of q_1 and q_2. The modification has a net effect of slightly more β roll-off at high collector current [10].

Thus

$$q_b = \frac{q_1}{2} [1 + \sqrt{1 + 4q_2}] \qquad (2\text{-}110)$$

In SPICE, q_1 is approximated by the equation

$$q_1 = \left(1 - \frac{V_{BC}}{V_A} - \frac{V_{BE}}{V_B}\right)^{-1} \qquad (2\text{-}111)$$

while q_2 has different expressions according to the regions of operation, as follows.

Normal active region

For $V_{BE} > -5\,\dfrac{n_F kT}{q}$ and $V_{BC} \le -5\,\dfrac{n_R kT}{q}$:

$$q_2 = \frac{I_S}{I_{KF}}\,(e^{qV_{BE}/n_F kT} - 1) - \frac{I_S}{I_{KR}} + \left(\frac{V_{BE}}{I_{KF}} + \frac{V_{BC}}{I_{KR}}\right)\text{GMIN} \qquad (2\text{-}112)$$

Inverse region

For $V_{BE} \le -5\,\dfrac{n_F kT}{q}$ and $V_{BC} > -5\,\dfrac{n_R kT}{q}$:

$$q_2 = \frac{I_S}{I_{KR}}\,(e^{qV_{BC}/n_R kT} - 1) - \frac{I_S}{I_{KF}} + \left(\frac{V_{BE}}{I_{KF}} + \frac{V_{BC}}{I_{KR}}\right)\text{GMIN} \qquad (2\text{-}113)$$

Saturated region

For $V_{BE} > -5 \dfrac{n_F kT}{q}$ and $V_{BC} > -5 \dfrac{n_R kT}{q}$:

$$q_2 = \frac{I_S}{I_{KF}} \left(e^{qV_{BE}/n_F kT} - 1\right) + \frac{I_S}{I_{KR}} \left(e^{qV_{BC}/n_R kT} - 1\right)$$
$$+ \left(\frac{V_{BE}}{I_{KF}} + \frac{V_{BC}}{I_{KR}}\right) \text{GMIN} \tag{2-114}$$

Off region

For $V_{BE} \leq -5 \dfrac{n_F kT}{q}$ and $V_{BC} \leq -5 \dfrac{n_R kT}{q}$:

$$q_2 = -I_s \left(\frac{I_{KF} + I_{KR}}{I_{KF} I_{KR}}\right) + \left(\frac{V_{BE}}{I_{KF}} + \frac{V_{BC}}{I_{KR}}\right) \text{GMIN} \tag{2-115}$$

It must be noted that since V_A, V_B, I_{KF}, and I_{KR} cannot be zero-valued, SPICE interprets a zero value for these parameters as an infinite value.

The current dependence of the base resistance is modeled in SPICE as follows:

$$r_{BB'} = \begin{cases} r_{BM} + \dfrac{r_B - r_{BM}}{q_b} & \text{for } I_{rB} = 0 \\[2ex] r_{BM} + 3(r_B - r_{BM}) \left(\dfrac{\tan z - z}{z \tan^2 z}\right) & \text{for } I_{rB} \neq 0 \end{cases} \tag{2-116}$$

where z is given by Eq. (2-105).

Figure 2-27 shows SPICE output plots of I_C vs. V_{CE} characteristics of a BJT with (a) $C_2 = 10^3$, (b) $I_{KF} = 5$ A, and (c) $n_F = 1.3$. The simulation has been made with $I_B = 0.5$ A, while the other parameter values (for each case) have been kept to their SPICE default values.

2.6   Gummel-Poon Large-Signal Model

2.6.1   Large-signal model and its implementation in SPICE

The SPICE Gummel-Poon large-signal circuit model is topologically the same as that of Fig. 2-18.

The nonlinear charge elements, or equivalently the voltage-dependent capacitances, are determined by Eqs. (2-36) to (2-43) (used also for the

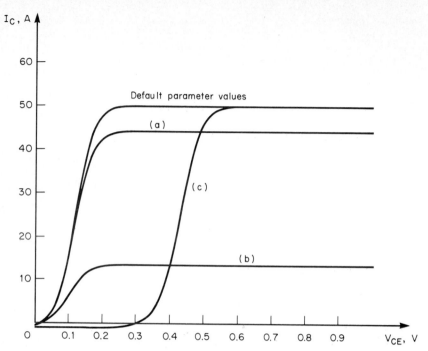

Figure 2-27 SPICE output plots of I_C vs. V_{CE} characteristics of a BJT with (a) $C_2 = 10^3$, (b) $I_{KF} = 5$ A, and (c) $n_F = 1.3$. The simulation has been made with $I_B = 0.5$ A, while the other parameter values (for each case) have been kept to their SPICE default values.

Ebers-Moll model) when the grading coefficients m_E, m_C, and m_S are taken into account (usually they vary between 0.33 and 0.5). Besides, in the Gummel-Poon large-signal model, three added effects have to be accounted for with respect to the Ebers-Moll model: distributed base-collector capacitance, τ_F modulation (transit charge), and distributed phenomena in the base region (excess phase).

a. Distributed base-collector capacitance. In order to find a better approximation of the distributed resistance and capacitance network at the base-collector junction, the junction capacitance is divided into two sections. Parameter X_{CJC}, which is always 0 and 1, specifies the ratio of this partitioning. The capacitance $X_{CJC}C_{JC}$ is placed between the internal base node and the collector. $(1 - X_{CJC})C_{JC}$ is the capacitance from external base to collector, while C_{JC} is the total base-collector capacitance. This parameter is usually important only at very high frequencies [10].

The implementation in SPICE is determined by the following relations:

$$
Q_{BX} = \begin{cases}
C_{JC}(0)(1 - X_{CJC}) \displaystyle\int_0^{V_{BX}} \left(1 - \frac{V}{\phi_C}\right)^{-m_C} dV & \text{for } V_{BX} < FC \times \phi_C \\[2em]
C_{JC}(0)(1 - X_{CJC})F_1 \\[0.5em]
\quad + \dfrac{C_{JC}(0)(1 - X_{CJC})}{F_2} \displaystyle\int_{FC \times \phi_C}^{V_{BX}} \left(F_3 + \frac{m_C V}{\phi_C}\right) dV \\[2em]
\hspace{8em} \text{for } V_{BX} \geq FC \times \phi_C
\end{cases}
$$

$$(2\text{-}117)$$

These charge-storage elements can be equivalently represented in SPICE by the following voltage-dependent capacitance equations:

$$
C_{JX} = \begin{cases}
C_{JC}(0)(1 - X_{CJC}) \left(1 - \dfrac{V_{BX}}{\phi_C}\right)^{-m_C} & \text{for } V_{BX} < FC \times \phi_C \\[2em]
\dfrac{C_{JC}(0)(1 - X_{CJC})}{F_2} \left(F_3 + \dfrac{m_C V_{BX}}{\phi_C}\right) & \text{for } V_{BX} \geq FC \times \phi_C
\end{cases}
$$

$$(2\text{-}118)$$

where F_1 and F_2 are given by Eq. (2-43).

Figure 2-28 shows the complete BJT large-signal model with the addition of the effect of the distributed base-collector capacitance.

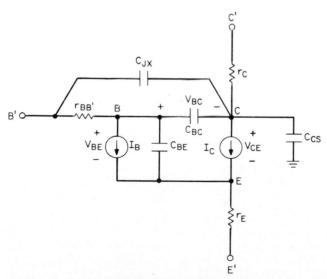

Figure 2-28 Gummel-Poon large signal model showing the distributed base-collector capacitance C_{JX}.

b. τ_F **modulation (transit charge).** The behavior of τ_F vs. I_C is shown in Fig. 2-29. The variation of τ_F at high collector currents is usually determined by an empirical equation derived from the gain–band-width product, f_T, vs. the collector current, I_C, at various collector-emitter voltages V_{CE}. Three distinct regions are observed in the f_T characteristics.

At low currents, f_T is dominated by the junction capacitances and the forward transconductance g_m; since g_m increases with current, f_T goes up with an increase in I_C. This variation has already been taken into account.

At the midrange, f_T is at its peak value and is almost constant; here the transit time is the time required by the carriers to transverse the base region and collector space-charge region. The base-emitter diffusion capacitance increases with current, canceling the increase in forward g_m, resulting in a definite limit for f_T. Thus, the ideal maximum τ_F is

$$\tau_F = \frac{1}{2\pi f_T} \tag{2-119}$$

At high currents, f_T and then τ_F become a function of I_C and V_{CE} and do not remain constant (see Fig. 2-29). Physically, anomalies such as base pushout, lateral spreading, space-charge-limited current flow, and quasi-saturation increase the transit time and decrease f_T. This effect is modeled by the following empirical function [10]:

$$ATF \equiv 1 + X_{\tau F}e^{V_{BC}/1.44V_{\tau F}}\left(\frac{I_{CC}}{I_{CC} + I_{\tau F}}\right)^2 \tag{2-120}$$

This function multiplies τ_F in the charge equations. The constant 1.44 simply gives the interpretation to $V_{\tau F}$ as the value of V_{BC} where the exponential equals ½. $X_{\tau F}$ (a SPICE parameter) controls the total fall-off of f_T; $V_{\tau F}$ (a SPICE parameter) dominates the change in f_T with respect to V_{CE}; $I_{\tau F}$ (a SPICE parameter) dominates the change in f_T with respect to current.

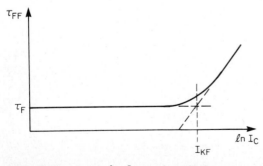

Figure 2-29 τ_{FF} vs. ln I_C to show high-current effects on charge storage.

It can be demonstrated [10] that

$$f_T = \cfrac{1}{2\pi\tau_F\left[ATF \;+\; \underbrace{\cfrac{2(ATF-1)I_{\tau F}}{I_{CC}+I_{\tau F}}}_{\dfrac{dATF}{dI_{CC}}} \;+\; \underbrace{\cfrac{kT}{q}\,n_F\,\cfrac{ATF-1}{1.44V_{\tau F}}}_{\dfrac{dATF}{dV_{BC}}}\right]}$$

(2-121)

At low currents or high V_{CE} (ATF = 1), Eq. (2-121) reduces to

$$f_T = \frac{1}{2\pi\tau_F}$$

(2-122)

which is the familiar Ebers-Moll expression given in Eq. (2-119).

At large I_C, such that $I_{CC}/(I_{CC}+I_{\tau F}) \simeq 1$ and V_{CE} is moderate, Eq. (2-121) reduces to

$$f_T\bigg|_{I_C\to\infty} = \frac{1}{2\pi\tau_F[1 + X_{\tau F}(e^{V_{BC}/1.44V_{\tau F}} + qe^{V_{BC}/1.44V_{\tau F}}/n_F kT \cdot 1.44V_{\tau F})]}$$

(2-123)

Thus, the high-current asymptote for a given V_{BC} is determined by parameters $X_{\tau F}$ and $V_{\tau F}$. Likewise, the asymptotic dependence of Eq. (2-121) in the extremes of V_{BC} are

$$f_T = \frac{1}{2\pi\tau_F\,[1 + X_{\tau F}(1 + q/n_F kT \cdot 1.44V_{\tau F}]} \qquad \text{for } I_C \to \infty,\ V_{BC}\to 0$$

(2-124)

where $\qquad ATF \simeq 1 + X_{\tau F} \qquad$ for $I_C\to\infty$, $V_{BC}\to 0 \qquad$ (2-125)

Furthermore, $\qquad f_T = \dfrac{1}{2\pi\tau_F} \qquad$ for $V_{BC}\to\infty \qquad$ (2-126)

Thus, the maximum possible fall-off in f_T is controlled by parameter $X_{\tau F}$.

In SPICE this effect is expressed by the following expression for charge or, equivalently, for its capacitance:

$$Q_{\tau FF} = \tau_{FF}\,\frac{I_{CC}}{q_b}$$

$$C_{\tau FF} = \frac{dQ_{\tau FF}}{dV_{BC}}$$

(2-127)

where τ_{FF} is the modulated transit time given by

$$\tau_{FF} = \tau_F \left[1 + X_{\tau F} \left(\frac{I_{CC}}{I_{CC} + I_{\tau F}} \right)^2 e^{V_{BC}/1.44V_{\tau F}} \right] \qquad (2\text{-}128)$$

and τ_F is the ideal forward transit time.

Thus the forward and reverse transit charge is modeled by τ_F, $X_{\tau F}$, $V_{\tau F}$, $I_{\tau F}$, and τ_R.

c. Distributed phenomena in the base region (excess phase). Measured phase shift in the forward transverse current from actual devices often exceeds the phase shift predicted by the finite lumped set of poles and zeros within the model. This is due to the distributed phenomena in the base region. Therefore, an extra phase delay is inserted in the model:

$$I_{FX} = I_{CC} \text{ (with excess phase)} = I_{CC}\phi(s) \qquad (2\text{-}129)$$

Unnecessary small time steps during transient analysis can be avoided by modeling the excess phase with a second-order *Bessel function*. This function is an all-pass filter in the frequency domain with no appreciable effect on magnitude. In the time domain, this polynomial resembles a time-domain delay for a Gaussian curve, which is similar to the physical phenomenon exhibited by the transistor action [10]:

$$\phi(s) = \frac{3\omega_0^2}{s^2 + 3\omega_0 s + 3\omega_0^2} \qquad (2\text{-}130)$$

It is important to realize that the equation for $\phi(s)$ is given in the frequency domain; s is the complex frequency variable in the transfer domain (s plane) and ω_0 is a parameter. The phase shift of this function is

$$\theta = \arctan \frac{3\omega_0\omega}{3\omega_0^2 - \omega^2} \qquad (2\text{-}131)$$

For $\omega < \omega_0$, the phase may be written from the binominal expansion of the arctangent as $\theta = \omega/\omega_0$. Assuming $P_{\tau F}$ is the phase delay at idealized maximum band width ($\omega = 1/\tau_F$), it can now be written

$$\omega_0 = \frac{1}{P_{\tau F}\tau_F} \qquad (2\text{-}132)$$

In SPICE ac analysis, this phase shift is implemented simply by adding the appropriate linear phase to the forward transverse term of the collector current I_{CC}. The transient analysis, however, is more complicated.

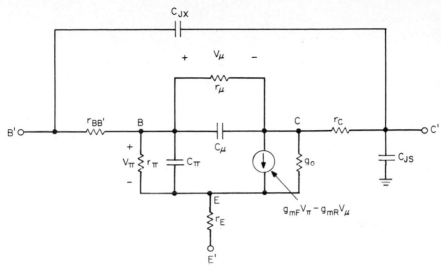

Figure 2-30 Gummel-Poon small-signal model.

Considering the inverse Laplace transform of $\phi(s)$, it can be written

$$\frac{d^2 I_{FX}}{dt^2} + 3\omega_0 \frac{dI_{FX}}{dt} + 3\omega_0^2 I_{FX} = 3\omega_0^2 I_{CC} \qquad (2\text{-}133)$$

This equation is then implemented by performing a numerical integration algorithm *(backward-Euler integration)*.

2.7 Gummel-Poon Small-Signal Model

2.7.1 Small-signal model and its implementation in SPICE

The Gummel-Poon small-signal model is drawn for an *npn* transistor in Fig. 2-30. It is identical in form to that used for the Ebers-Moll model (Fig. 2-19), but the expressions for Eqs. (2-53) to (2-60) are different here.

Taking into account Eqs. (2-44) to (2-51), the relations for the small-signal parameters defining the SPICE Gummel-Poon small-signal model can be written as follows.

For $V_{BE} > -5 \dfrac{n_F kT}{q}$ and $V_{BC} \leq -5 \dfrac{kT}{q}$:

$$g_m = \cfrac{\dfrac{qI_S}{n_F kT}e^{qV_{BE}/n_F kT} + \text{GMIN} - \dfrac{1}{q_b}[I_S e^{qV_{BE}/n_F kT} + (V_{BE} - V_{BC})\text{GMIN}]\dfrac{dq_b}{dV_{BE}}}{q_b}$$
$$- g_o \qquad (2\text{-}134a)$$

For $V_{BE} \le -5\,\dfrac{n_F kT}{q}$ and $V_{BC} > -5\,\dfrac{n_R kT}{q}$:

$$g_m = \frac{\dfrac{-I_S}{V_{BE}} + \text{GMIN} + \dfrac{1}{q_b}[I_S e^{qV_{BC}/n_R kT} - (V_{BE} - V_{BC})\text{GMIN}]\dfrac{dq_b}{dV_{BE}}}{q_b}$$
$$- g_o$$

$$(2\text{-}134b)$$

For $V_{BE} > -5\,\dfrac{n_F kT}{q}$ and $V_{BC} > -5\,\dfrac{n_R kT}{q}$:

$$g_m = \frac{\dfrac{qI_S}{n_F kT}e^{qV_{BE}/n_F kT} + \text{GMIN} - \dfrac{1}{q_b}[I_S(e^{qV_{BE}/n_F kT} - e^{qV_{BC}/n_R kT}) + (V_{BE} - V_{BC})\text{GMIN}]\dfrac{dq_b}{dV_{BE}}}{q_b}$$
$$- g_o$$

$$(2\text{-}134c)$$

For $V_{BE} \le -5\,\dfrac{n_F kT}{q}$ and $V_{BC} \le -5\,\dfrac{n_R kT}{q}$:

$$g_m = \frac{\dfrac{-I_S}{V_{BE}} + \text{GMIN} - \dfrac{1}{q_b}(V_{BE} - V_{BC})\text{GMIN}\dfrac{dq_b}{dV_{BE}}}{q_b} - g_o \qquad (2\text{-}134d)$$

For $V_{BE} > -5\,\dfrac{n_F kT}{q}$:

$$g_\pi = \frac{I_S}{\beta_F}\frac{q}{n_F kT}\,e^{qV_{BE}/n_F kT} + \frac{qC_2 I_S}{n_{EL} kT}\,e^{qV_{BE}/n_{EL} kT} + \frac{\text{GMIN}}{\beta_F} \qquad (2\text{-}135a)$$

For $V_{BE} \le -5\,\dfrac{n_F kT}{q}$:

$$g_\pi = -\frac{I_S}{\beta_F V_{BE}} - \frac{C_2 I_S}{V_{BE}} + \frac{\text{GMIN}}{\beta_F} \qquad (2\text{-}135b)$$

For $V_{BE} > -5\,\dfrac{n_F kT}{q}$ and $V_{BC} \le -5\,\dfrac{n_R kT}{q}$:

$$g_o = \frac{\dfrac{-I_S}{V_{BC}} + \text{GMIN} + \dfrac{1}{q_b}[I_S e^{qV_{BE}/n_F kT} + (V_{BE} - V_{BC})\text{GMIN}]\dfrac{dq_b}{dV_{BC}}}{q_b}$$

$$(2\text{-}136a)$$

For $V_{BE} \le -5 \dfrac{n_F kT}{q}$ and $V_{BC} > -5 \dfrac{n_R kT}{q}$:

$$g_{\scriptscriptstyle o} = \frac{\dfrac{qI_S}{n_R kT}e^{qV_{BC}/n_R kT} + \text{GMIN} - \dfrac{1}{q_b}[I_S e^{qV_{BC}/n_R kT} - (V_{BE} - V_{BC})\text{GMIN}]\dfrac{dq_b}{dV_{BC}}}{q_b}$$

$$(2\text{-}136b)$$

For $V_{BE} > -5 \dfrac{n_F kT}{q}$ and $V_{BC} > -5 \dfrac{n_R kT}{q}$:

$$g_{\scriptscriptstyle o} = \frac{\dfrac{qI_S}{n_R kT}e^{qV_{BC}/n_R kT} + \text{GMIN} + \dfrac{1}{q_b}[I_S(e^{qV_{BE}/n_F kT} - e^{qV_{BC}/n_R kT}) + (V_{BE} - V_{BC})\text{GMIN}]\dfrac{dq_b}{dV_{BC}}}{q_b}$$

$$(2\text{-}136c)$$

For $V_{BE} \le -5 \dfrac{n_F kT}{q}$ and $V_{BC} \le -5 \dfrac{n_R kT}{q}$:

$$g_{\scriptscriptstyle o} = \frac{\dfrac{-I_S}{V_{BC}} + \text{GMIN} + \dfrac{1}{q_b}(V_{BE} - V_{BC})\text{GMIN}\dfrac{dq_b}{dV_{BC}}}{q_b} \qquad (2\text{-}136d)$$

For $V_{BC} > -5 \dfrac{n_R kT}{q}$:

$$g_\mu = \frac{I_S}{\beta_R} \frac{q}{n_R kT} e^{qV_{BC}/n_R kT} + \frac{qC_4 I_S}{n_{CL} kT} e^{qV_{BC}/n_{CL} kT} + \frac{\text{GMIN}}{\beta_R} \qquad (2\text{-}137a)$$

For $V_{BC} \le -5 \dfrac{n_R kT}{q}$:

$$g_\mu = -\frac{I_S}{\beta_R V_{BC}} - \frac{C_4 I_S}{V_{BC}} + \frac{\text{GMIN}}{\beta_R} \qquad (2\text{-}137b)$$

For $I_{rB} = 0$:

$$g_x = \frac{1}{r_{BM} + (r_B - r_{BM})/q_b} \qquad (2\text{-}138a)$$

For $I_{rB} \ne 0$:

$$g_x = \frac{1}{r_{BM} + 3(r_B - r_{BM})[(\tan z - z)/(z \tan^2 z)]} \qquad (2\text{-}138b)$$

For $V_{BE} < FC \times \phi_E$:

$$C_\pi \equiv C_{BE} = \tau_F \frac{qI_S}{kT} e^{qV_{BE}/kT} + C_{JE}(0) \left(1 - \frac{V_{BE}}{\phi_E} \right)^{-m_E} \qquad (2\text{-}139a)$$

For $V_{BE} \geq FC \times \phi_E$:

$$C_\pi \equiv C_{BE} = \tau_F \frac{qI_S}{kT} e^{qV_{BE}/kT} + \frac{C_{JE}(0)}{F_2} \left(F_3 + \frac{m_E V_{BE}}{\phi_E} \right) \qquad (2\text{-}139b)$$

For $V_{BC} < FC \times \phi_C$:

$$C_\mu \equiv C_{BC} = \tau_R \frac{qI_S}{kT} e^{qV_{BC}/kT} + C_{JC}(0) \left(1 - \frac{V_{BC}}{\phi_C} \right)^{-m_C} \qquad (2\text{-}140a)$$

For $V_{BC} \geq FC \times \phi_C$:

$$C_\mu \equiv C_{BC} = \tau_R \frac{qI_S}{kT} e^{qV_{BC}/kT} + \frac{C_{JC}(0)}{F_2} \left(F_3 + \frac{m_C V_{BC}}{\phi_C} \right) \qquad (2\text{-}140b)$$

For $V_{CS} < 0$:

$$C_{CS} = C_{JS}(0) \left(1 - \frac{V_{CS}}{\phi_S} \right)^{-m_S} \qquad (2\text{-}141a)$$

For $V_{CS} \geq 0$:

$$C_{CS} = C_{JS}(0) \left(1 + \frac{m_s V_{CS}}{\phi_S} \right) \qquad (2\text{-}141b)$$

For $V_{BX} < FC \times \phi_C$:

$$C_{JX} = C_{JC}(0)(1 - X_{CJC}) \left(1 - \frac{V_{BX}}{\phi_C} \right)^{-m_C} \qquad (2\text{-}142a)$$

For $V_{BX} \geq FC \times \phi_C$:

$$C_{JX} = \frac{C_{JC}(0)(1 - X_{CJC})}{F_2} \left(F_3 + \frac{m_C V_{BX}}{\phi_C} \right) \qquad (2\text{-}142b)$$

where F_2 and F_3 are given by Eqs. (2-42) and (2-43) and q_b is given by Eqs. (2-110) to (2-115).

The preceding equations are not as simple, and therefore not as intuitive, as those for the Ebers-Moll model; however, they are basically equivalent.

2.8   Temperature and Area Effects on the BJT Model

This section analyzes the effects that temperature and area factors have on the BJT model implemented in SPICE. In order to more easily understand the relations introduced, refer to Sec. 1.6.1 (remember that a BJT can be considered an interacting pair of *pn* junctions).

2.8.1   Temperature dependence of the BJT model parameters

All input data for SPICE are assumed to have been measured at 27°C (300 K). The simulation also assumes a nominal temperature (TNOM) of 27°C that can be changed with the TNOM option of the .OPTIONS control card. The circuits can be simulated at temperatures other than TNOM by using a .TEMP control card.

Temperature appears explicitly in the exponential term of the BJT model equations.

In addition, SPICE modifies BJT parameters to reflect changes in the temperature. They are I_S, the saturation current; C_2 (or I_{SE}), the base-emitter leakage saturation current; C_4 (or I_{SC}), the base-collector leakage saturation current; ϕ_E, the base-emitter built-in potential; ψ_C, the base-collector built-in potential; β_F, the forward current gain; β_R, the reverse current gain; C_{JC}, the zero-bias base-collector depletion capacitance; C_{JE}, the zero-bias base-emitter depletion capacitance; and FC, the coefficient for forward-bias depletion capacitance formula.

The temperature dependence of I_S in the SPICE BJT model is determined by

$$I_S(T_2) = I_S(T_1) \left(\frac{T_2}{T_1}\right)^{X_{TI}} e^{[qE_g(300)/kT_2](1-T_2/T_1)} \tag{2-143}$$

where two new model parameters are introduced: X_{TI}, the saturation current temperature exponent, and E_g, the energy gap.

The effect of temperature on ϕ_J is modeled as follows:

$$\phi_E(T_2) = \frac{T_2}{T_1} \phi_E(T_1) - 2 \frac{kT_2}{q} \ln\left(\frac{T_2}{T_1}\right)^{1.5}$$

$$- \left[\frac{T_2}{T_1} E_g(T_1) - E_g(T_2)\right] \tag{2-144}$$

$$\phi_C(T_2) = \frac{T_2}{T_1} \phi_C(T_1) - 2 \frac{kT_2}{q} \ln\left(\frac{T_2}{T_1}\right)^{1.5}$$

$$- \left[\frac{T_2}{T_1} E_g(T_1) - E_g(T_2)\right]$$

where $E_g(T_1)$ and $E_g(T_2)$ are described by the following general expression [11]:

$$E_g(T) = E_g(0) - \frac{\alpha T^2}{\beta - T} \tag{2-145}$$

Experimental results give for Si

$$\alpha = 7.02 \times 10^{-4}$$

$$\beta = 1108$$

$$E_g(0) = 1.16 \text{ eV}$$

The temperature dependence of β_F and β_R is determined by

$$\beta_F(T_2) = \beta_F(T_1) \left(\frac{T_2}{T_1}\right)^{X_{T\beta}}$$

$$\beta_R(T_2) = \beta_R(T_1) \left(\frac{T_2}{T_1}\right)^{X_{T\beta}} \tag{2-146}$$

Through the parameters I_{SE} and I_{SC}, respectively, the temperature dependence of C_2 and C_4 is determined by

$$I_{SE}(T_2) = I_{SE}(T_1) \left(\frac{T_2}{T_1}\right)^{-X_{T\beta}} \left[\frac{I_S(T_2)}{I_S(T_1)}\right]^{1/n_{EL}}$$

$$I_{SC}(T_2) = I_{SC}(T_1) \left(\frac{T_2}{T_1}\right)^{-X_{T\beta}} \left[\frac{I_S(T_2)}{I_S(T_1)}\right]^{1/n_{CL}} \tag{2-147}$$

The temperature dependence of C_{JE} and C_{JC} is determined by

$$C_{JE}(T_2) = C_{JE}(T_1) \left\{1 + m_E \left[400 \times 10^{-6}(T_2 - T_1)\right.\right.$$

$$\left.\left. - \frac{\phi_E(T_2) - \phi_E(T_1)}{\phi_E(T_1)}\right]\right\}$$

$$C_{JC}(T_2) = C_{JC}(T_1) \left\{1 + m_C \left[400 \times 10^{-6}(T_2 - T_1)\right.\right.$$

$$\left.\left. - \frac{\phi_C(T_2) - \phi_C(T_1)}{\phi_C(T_1)}\right]\right\} \tag{2-148}$$

The temperature dependence of the coefficient for forward-bias depletion capacitance formula is determined by

$$FCPE(T_2) = FC \times \phi_E = FCPE(T_1) \frac{\phi_E(T_2)}{\phi_E(T_1)}$$

$$F_1(T_2) = F_1(T_1) \frac{\phi_E(T_2)}{\phi_E(T_1)}$$

(2-149)

$$FCPC(T_2) = FC \times \phi_C = FCPC(T_1) \frac{\phi_C(T_2)}{\phi_C(T_1)}$$

$$F_1(T_2) = F_1(T_1) \frac{\phi_C(T_2)}{\phi_C(T_1)}$$

(2-150)

In the previous equations, T_1 must be considered as follows:

- If TNOM in the .OPTIONS card is not specified, then $T_1 = $ TNOM = TREF $= 27°C$ (300 K); this is the default value for TNOM.

- If a new $TNOM_{new}$ value is specified, then $T_1 = $ TNOM $= 27°C$; $T_2 = TNOM_{new}$.

These two cases are valid if only one temperature request is made. If more than one temperature is requested (.TEMP card), then $T_1 = $ TNOM $= 27°C$ and T_2 is the first temperature specified in the .TEMP card. Afterward, T_1 is the last working temperature (T_2) and T_2 is the next temperature in the .TEMP card to be calculated.

It is very important to note that the temperature T is regarded as an operating condition and *not* as a model parameter, so that it is not possible to simulate the behavior of devices in the same circuit at different temperatures at the same time. In other words, the *same* temperature T is used in the *whole* circuit to be simulated.

2.8.2 Area dependence of the BJT model parameters

The AREA factor used in SPICE for the BJT model determines the number of equivalent parallel devices of a specified model. The BJT model parameters affected by the AREA factor and specified in the device card are I_S, the saturation current; I_{SE}, the base-emitter leakage saturation current; I_{SC}, the base-collector leakage saturation current; I_{KF}, the forward knee current; I_{KR}, the reverse knee current; I_{rB}, the current where base resistance falls halfway to its minimum value; I_{rF}, the high-current parameter for effect on τ_F; r_B, the zero-bias resistance; r_{BM}, the minimum base resistance at high currents; r_E, the emitter resistance; r_C, the collector resistance; C_{JC}, the zero-bias base-collector depletion capacitance; C_{JE}, the zero-bias base-emitter depletion capacitance; and C_{JS}, the zero-bias collector-substrate capacitance. In summary,

$$I_S = I_S \times \text{AREA}$$

$$I_{SE} = I_{SE} \times \text{AREA}$$

$$I_{SC} = I_{SC} \times \text{AREA}$$

$$I_{KF} = I_{KF} \times \text{AREA}$$

$$I_{KR} = I_{KR} \times \text{AREA}$$

$$I_{rB} = I_{rB} \times \text{AREA}$$

$$I_{\tau F} = I_{\tau F} \times \text{AREA} \tag{2-151}$$

$$C_{JC}(0) = C_{JC}(0) \times \text{AREA}$$

$$C_{JE}(0) = C_{JE}(0) \times \text{AREA}$$

$$C_{JS}(0) = C_{JS}(0) \times \text{AREA}$$

$$r_B = r_B/\text{AREA}$$

$$r_{BM} = r_{BM}/\text{AREA}$$

$$r_E = r_E/\text{AREA}$$

$$r_C = r_C/\text{AREA}$$

The default value for the AREA parameter is 1.

2.9   Diode Connections for the BJT†

As pointed out in Sec. 1.2.2a, junction diodes can be formed by various connections of the npn and pnp transistor structures, as illustrated in Fig. 2-31. Here the attention is focused on the diode connections for an npn transistor; similar considerations hold for a pnp transistor.

When the diode is forward-biased in the diode connections (Fig. 2-31), the collector-base junction becomes forward-biased as well. Because these connections represent integrated circuit diodes, when the forward biasing occurs, the collector-base junction injects holes into the epiregion that can be collected by the reverse-biased epi-isolation junction (or by other devices in the same isolation region). A similar phenomenon occurs when a transistor enters saturation. As a result, substrate currents can flow that

---

† The material in this section is taken from P. R. Gray and R. G. Meyer, *Analysis and Design of Analog Integrated Circuits,* copyright © 1977 by John Wiley & Sons, Inc. Reprinted by permission.

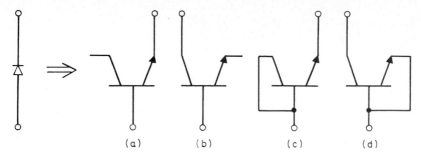

Figure 2-31 Diode connections for an *npn* transistor. *(From P. R. Gray and R. G. Meyer, Analysis and Design of Analog Integrated Circuits, copyright © 1977 by John Wiley & Sons. Used by permission.)*

can cause voltage drops in the high-resistivity substrate material, and other epi-isolation junctions within the circuit can become inadvertently forward-biased. Thus the diode connections of Fig. 2-31 are usually preferable since they keep the base-collector junction at zero bias. These connections have the additional advantage of resulting in the smallest amount of minority-charge storage within the diode under forward-bias conditions [12].

Figure 2-32 shows SPICE I_D vs. V_D characteristics for the diode connection of Fig. 2-31c for different BJT parameter values, while the others are kept to the default values.

2.10   Power Transistor Model

This section presents a model for the power transistor and shows how it can be implemented in SPICE.

2.10.1   Power transistor overview

Power transistors must be designed to withstand high voltage, current, and power ratings. Wide operation areas in forward or reverse base-driving condition and a good switching speed are also often required.

The need to support high collector-emitter voltages V_{CE} implies the presence of a high reverse collector-base voltage V_{BC}; a high breakdown voltage base-collector junction is therefore needed.

A significant improvement in the breakdown voltage can be obtained with an $n^+pn^-n^+$ structure, as shown in Fig. 2-33.† Here the additional

---

† The material in this paragraph is taken in part from S. K. Ghandhi, *Semiconductor Power Devices,* copyright © 1977 by John Wiley & Sons, Inc. Reprinted by permission.

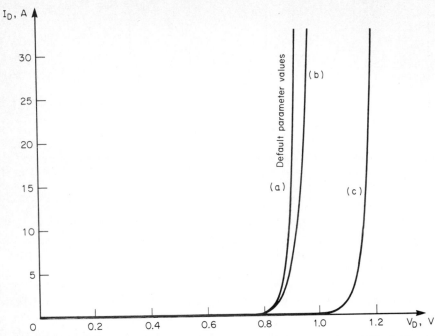

Figure 2-32 SPICE I_D vs. V_D characteristics for the diode connection of Fig. 2-31c: (a) default parameter values, (b) $I_{KF} = 5$ A, and (c) $n_F = 1.3$. For each case, the other parameter values have been kept to their SPICE default values, with $I_S = 10^{-14}$ A, to have the same condition as for the diode.

reverse voltage that can be supported because the n^- region is approximately equal to the breakdown voltage of a reverse-biased $p^+n^-n^+$ diode of comparable width. Furthermore, since the movement of the collector depletion layer into the diffused base is now inhibited by the grading, failure by punch-through seldom occurs in this device.

In normal operation, the n^- region is fully depleted, and the device behaves like a double diffused n^+pn^+ transistor, with a relatively narrow base width. The frequency response of the structure is thus considerably superior to that of the single diffused transistor. As a result of these many advantages, the $n^+pn^-n^+$ structure is the basic modern high-voltage power transistor [13].

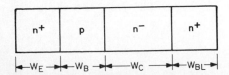

Figure 2-33 Power transistor structure.

Power transistor I_C vs. V_{CE} characteristics are divided into three regions, as shown in Fig. 2-34:

1. *Nonsaturation region* (a) to the right of the line marked $R_P + R_{SC}$. R_{SC} is the *collector series resistance* given by

$$R_{SC} = \frac{\rho_C W_C}{A_E} \qquad (2\text{-}152)$$

where ρ_C is the collector resistivity and A_E is the emitter area. R_P is the parasitic series resistance, which consists of internal device resistances, lead and connection resistances, and the voltage difference between V_{BE} and V_{BC}, which are both positive, since emitter-base and collector-base junctions are both forward-biased in saturation.

2. *Quasi-saturation region* (b), between R_P and $R_P + R_{SC}$.

3. *Saturation region* (c), formed by the points lying on R_P.

Transistor behavior in the three regions is discussed in the following sections for constant collector current I_C and decreasing V_{CE} [14].

a. Nonsaturation region. At point A of Fig. 2-34, V_{CE} is the sum of the voltage drop in the depletion region and of the ohmic drops in the unde-

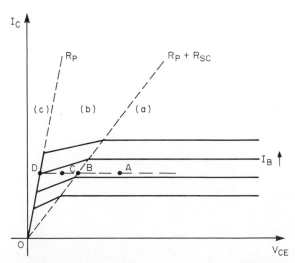

Figure 2-34 Power transistor I_C vs. V_{CE} characteristics, showing (a) nonsaturation region, (b) quasi-saturation region, and (c) hard-saturation region.

pleted collector and across R_P; that is,

$$V_{CE} = \frac{\epsilon_0(E_0 - E_C)^2}{2qN_D} + E_C W_C + R_P I_C \qquad (2\text{-}153)$$

where
$$E_C = \frac{\rho_C I_C}{A_E} \qquad (2\text{-}154)$$

Figure 2-35 is a representation of the electric field and the electron charge distribution. V_{CE} is given by the dashed area (plus $I_C R_P$ drop) [14].

b. Quasi-saturation region. At point B of Fig. 2-34,

$$V_{CE} = I_C(R_P + R_{SC}) \qquad (2\text{-}155)$$

The depletion layer disappears and V_{CE} is determined only by the ohmic drop on R_{SC}. The electron distribution is the same as in the nonsaturation region, while the electric field varies according to Fig. 2-36.

When V_{CE} is further decreased (point C in Fig. 2-34), the following phenomena occur [14]:

1. The collector-base junction becomes forward-biased.

2. Holes are injected from the base into some part of the collector region (from $x = W_B$ to $x = W_{CIB}$), which behaves as an extended base (see

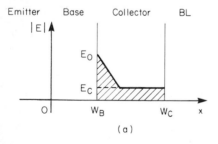

(a)

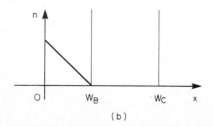

(b)

Figure 2-35 Nonsaturation region: (*a*) electric field and (*b*) electron stored charge. *(From Antognetti [14]. Used by permission.)*

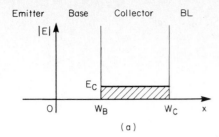

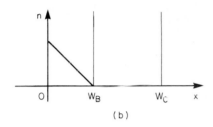

Figure 2-36 Boundary between nonsaturation and quasi-saturation regions: (a) electric field and (b) electron stored charge. *(From Antognetti [14]. Used by permission.)*

Fig. 2-37). The sum of this extended base and of the metallurgical base is usually called *current-induced base (CIB)*.

3. In the extended base, the quasi-neutrality condition must be applied.

4. The normally small N_D value is negligible with respect to $n(x)$, and the extended base can be considered as a real base region with $N_A = 0$.

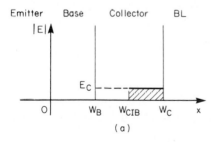

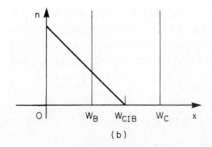

Figure 2-37 Quasi-saturation region: (a) electric field and (b) electron stored charge. *(From Antognetti [14]. Used by permission.)*

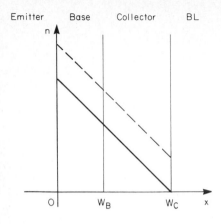

Figure 2-38 Electron stored charge in hard saturation (solid line, $I_B = I_{B,\text{sat}}$) and in oversaturation (dashed line, $I_B > I_{B,\text{sat}}$).

It can be shown [14] that V_{CE} is the sum of ohmic drops on the residual collector and on R_P; that is,

$$V_{CE} = R_P I_C + \frac{\rho_C(W_C - W_{CIB})}{A_E} I_C \qquad (2\text{-}156)$$

c. Saturation region. When the condition $V_{CE} = I_C R_P$ holds, the collector voltage drop is zero and $W_{CIB} = W_C$. The electron distribution, corresponding to point D in Fig. 2-34, is given by the solid line in Fig. 2-38 for $I_B = I_{B,\text{sat}}$, which is the minimum I_B required for the transistor to be in saturation.

Details of power transistor theory are described and developed in Refs. [13–15].

2.10.2   Power transistor quasi-saturation model†

Physical phenomena which determine power transistor performances and which affect the model implementation are briefly discussed before the model itself is described.

Two distinct possibilities are present when the current in an $n^+pn^-n^+$ transistor is increased. First, the increasing voltage drop across the n^- region reduces the available reverse bias at the collector-base junction. Consequently, its depletion layer shrinks and eventually collapses. Further increase of collector current drives the device into quasi-saturation,

---

† The material in this section is taken in part from S. K. Ghandhi, *Semiconductor Power Devices*, copyright © 1977 by John Wiley & Sons, Inc. Reprinted by permission.

accompanied by an expansion of the base width. This process is commonly encountered for low supply voltages and is known as the *low-field case (LFC)*.

Conversely, increasing collector current reduces the space-charge density in the pn^- depletion layer, causing it to expand until it fills the n^- region. Further increase in collector current results in space-charge limited current flow in this region. Still further increase leads to growth in the effective base width, by one or more mechanisms, described by the one- and two-dimensional models.

These processes are commonly encountered with higher supply voltages, which means that the device does not enter quasi-saturation: this process is known as the *high-field case (HFC)*.

Typically, $NW_C = 10^{11}$ cm^{-2} represents the boundary between the low-field case and the high-field case [13]. In both situations, however, it can be shown that the end result is an effective widening of the base width when the collector current density exceeds a specific critical value.

The model proposed here [16–19], which is an extension of the classic Gummel-Poon model used in SPICE, affects only the low-field case.

Consider an $n^+pn^-n^+$ transistor, fed from a constant collector-base supply voltage [13]. With increasing collector current, the ohmic drop across the undepleted part of the n^- region increases, until all the supply voltage is used up across this region and the depletion layer collapses. If I_C^* is the critical current at which this occurs, then from Ohm's law it can be written, if R_P is omitted,

$$I_C^* = \frac{A_E(V_{CB} + \phi_C)}{\rho_C W_C} = \frac{V_{CB} + \phi_C}{R_{SC}} \qquad (2\text{-}157)$$

where ϕ_C is the contact potential of the pn^- junction. This situation is commonly encountered with low supply voltages (LFC).

Transistor action at $I_C = I_C^*$ consists of diffusion through W_B, drift through a low-field n^- region of width W_C, and eventual collection. The pn^- junction becomes forward-biased when the current exceeds I_C^*. This is accompanied by hole injection into the n^- region, together with conductivity modulation of part of it, as in Fig. 2-37.

Thus for $I_C > I_C^*$, the drift region shrinks away from the pn^- junction, reducing its width from W_C to $W_C - W_{CIB}$. For this situation, I_C is obtained by modifying Eq. (2-157), so that

$$I_C = \frac{A_E(V_{CB} + \phi_C)}{\rho_C(W_C - W_{CIB})} \qquad (2\text{-}158)$$

where W_{CIB} is the current-induced base width or, in other words, the modulated part of the n^- region.

Combining Eqs. (2-157) and (2-158) yields:

$$W_{CIB} = W_C \left(1 - \frac{I_C^*}{I_C} \right) \tag{2-159}$$

Transistor action at current $I_C > I_C^*$ is thus one of diffusion through a neutral region of width $W_B + W_{CIB}$, drift through a region of width $W_C - W_{CIB}$, and eventual collection.

Gummel introduced the effective base width, $W_{B,\text{eff}}$, as follows:

$$W_{B,\text{eff}} = \begin{cases} W_B & \text{for } I_C < I_C^* \\ W_B + W_{CIB} = W_B + W_C \left(1 - \dfrac{I_C^*}{I_C} \right) & \text{for } I_C > I_C^* \end{cases} \tag{2-160}$$

Knowing that the base transit time varies as the square of the base width, Gummel then defines an effective base transit time given by the following:

$$\tau_{\text{eff}} = B\tau_F \tag{2-161}$$

where

$$B = \left(\frac{W_{B,\text{eff}}}{W_B} \right)^2 \tag{2-162}$$

B, called the *base pushout factor*, models the effective increase in the base width at high-current levels. Thus, from what was explained before, it characterizes the behavior of the power transistor in the quasi-saturation region.

From Eqs. (2-160) and (2-162), it follows that

$$B = 1 \qquad \text{for } I_C < I_C^*$$

$$B = \left[1 + \frac{W_C}{W_B} \left(1 - \frac{I_C^*}{I_C} \right) \right]^2 \tag{2-163}$$

$$= \left[1 + \frac{W_C}{W_B} \left(1 + \frac{V_{CB} + \phi_C}{R_{SC} I_C} \right) \right]^2 \qquad \text{for } I_C > I_C^*$$

Various expressions of B can be considered [5], but all of them are very complicated with respect to Eq. (2-163) and thus have never been implemented in any version of SPICE.

Now B has been defined as a base-widening factor; its implementation in SPICE requires only three new parameters:

R_{SC} n^--region resistance
W_C n^--region width
W_B p-base width

2.10.3 A possible implementation of the B factor in SPICE

The base-widening effect modifies both static behavior and dynamic behavior of the power transistor.

Furthermore, the analysis and modeling of this phenomenon on the dynamic behavior can become very complicated, since in SPICE there is no *explicit* expression for $\tau_{F,\text{eff}}$ (total forward transit time), which is *implicitly* obtained via the following relation:

$$\tau_{F,\text{eff}} = \frac{Q_{DE}(V_{BE})}{I_{CC}(V_{BE})} \qquad (2\text{-}164)$$

In Eq. (2-164), Q_{DE} and I_{CC} are separate functions of V_{BE}. By modeling each one independently, an implicit expression for $\tau_{F,\text{eff}}$ can be obtained.

Static and dynamic analyses follow separately, and deviations from the basic Gummel-Poon model in SPICE will be pointed out.

Dynamic analysis. Referring to Eq. (2-26), and assuming that Q_{JE} and Q_{JC} are negligible (which is often reasonable), then we have the following equation:

$$Q_{DE} \simeq Q_E + Q_{BF} \qquad (2\text{-}165)$$

Now Q_E, the mobile minority charge stored in the neutral emitter region, is unaffected by B, but Q_{BF}, the minority mobile charge stored in the neutral base region, is affected by B.

Equation (2-165) can now be rewritten as

$$Q_{DE} \simeq Q_E + Q_{BF}(B) \qquad (2\text{-}166)$$

To implement B in Q_{DE}, the preceding expression must be replaced by the following one:

$$Q_{DE} = \underbrace{\tau_E I_{SS}(e^{qVBE/kT} - 1)}_{Q_E} + \underbrace{B\tau_{BF}I_{SS}(e^{qVBE/kT} - 1)}_{Q_{BF}} \qquad (2\text{-}167)$$

or $$Q_{DE} = \underbrace{(\tau_E + B\tau_{BF})}_{\tau_{F,\text{eff}}} I_{SS}(e^{qVBE/kT} - 1) \qquad (2\text{-}168)$$

Naturally, if $B = 1$, $\tau_{F,\text{eff}} = \tau_E + \tau_{BF} = \tau_F$, and the original Gummel-Poon model used in SPICE can be found again according to

$$Q_{DE} = \tau_F I_{SS}(e^{qV_{BE}/kT} - 1) \qquad (2\text{-}169)$$

b. Static analysis.   To discuss the influence of B on the static distribution of carriers stored in the base region, the first term of I_{CT} in Eq. (2-83) must be considered. We make

$$I_{CC} = \frac{I_{SS}}{q_b}(e^{qV_{BE}/kT} - 1) \qquad (2\text{-}170)$$

Combining Eq. (2-169) with Eq. (2-170) yields

$$Q_{DE} = \tau_F I_{CC} q_b \qquad (2\text{-}171)$$

and it immediately follows that

$$\frac{Q_{DE}}{I_{CC}} = \tau_F q_b \qquad (2\text{-}172)$$

Finally, combining Eq. (2-164) with Eq. (2-172) yields

$$\tau_F q_b(B) = \tau_{F,\text{eff}} \qquad (2\text{-}173)$$

Now, q_b represents an important parameter in modeling the static behavior of the power transistor. In particular, q_{bF} can be regarded as the component of q_b[see Eq. (2-87)], which is affected by the base pushout effect.

By the definition of q_{bF},

$$q_{bF} = \frac{\tau_{BF}I_{CC}}{Q_{B0}} \qquad (2\text{-}174)$$

and taking into account Eq. (2-170),

$$q_{bF} = \frac{\tau_{BF} I_{SS}(e^{qV_{BE}/kT} - 1)}{Q_{B0} q_b} \qquad (2\text{-}175)$$

Therefore, from Eq. (2-167),

$$q_{bF} = \frac{B\tau_{BF} I_{SS}(e^{qV_{BE}/kT} - 1)}{Q_{B0} q_b} \qquad (2\text{-}176)$$

The complete solution for q_b vs. B comes from a combination of Eqs. (2-176) and (2-87).

$$q_b = 1 + \frac{V_{BE}}{V_B} + \frac{V_{BC}}{V_A} + \frac{B\tau_{BF}}{Q_{B0} q_b} I_{SS}(e^{qV_{BE}/kT} - 1)$$

$$+ \frac{\tau_{BR}}{Q_{B0} q_b} I_{SS}(e^{qV_{BC}/kT} - 1) \qquad (2\text{-}177)$$

If Eqs. (2-99) and (2-100) are taken into account, then

$$q_b = 1 + \frac{V_{BE}}{V_B} + \frac{V_{BC}}{V_A} + \frac{B}{I_{KF} q_b} I_{SS}(e^{qV_{BE}/kT} - 1)$$

$$+ \frac{1}{I_{KR} q_b} I_{SS}(e^{qV_{BC}/kT} - 1) \qquad (2\text{-}178)$$

Equations (2-163), (2-168), and (2-178) cover the implementation of the base pushout factor B in SPICE, even if there is no simple method (other than calculation from process and geometric data) for determining τ_F and τ_{BF} parameters. The implementation of Eq. (2-178) in SPICE is also subjected to the simplifications pointed out in Eqs. (2-110) and (2-111).

REFERENCES

1. R. S. Muller and T. Kamins, *Device Electronics for Integrated Circuits,* Wiley, New York, 1977.
2. H. K. Gummel and H. C. Poon, An Integral Charge Control Model of Bipolar Transistors, *Bell Syst. Tech. J.,* **49,** 1970.
3. J. J. Ebers and J. L. Moll, Large-Signal Behavior of Junction Transistors, *Proc. IRE,* **42,** 1954.
4. W. G. Oldham and S. E. Schwarz, *An Introduction to Electronics,* Holt, Rinehart and Winston, New York, 1972.
5. I. E. Getreu, *Modeling the Bipolar Transistor,* Tektronix, Inc., Beaverton, Ore., 1976.
6. B. R. Chawla and H. K. Gummel, Transition Region Capacitance of Diffused pn Junctions, *IEEE Trans. Electron Devices,* **ED-18,** 1971.
7. L. W. Nagel, SPICE2: A Computer Program to Simulate Semiconductor Circuits, Electronics Research Laboratory Rep. No. ERL-M520, University of California, Berkeley, 1975.

8. *SPICE Version 2G User's Guide.*
9. H. K. Gummel and H. C. Poon, Modeling of Emitter Capacitance, *IEEE Proc. (Lett.)* **57,** 1969.
10. E. Khalily, Hewlett-Packard Co., private communication.
11. S. M. Sze, *Physics of Semiconductor Devices,* Wiley, New York, 1969.
12. P. R. Gray and R. G. Meyer, *Analysis and Design of Analog Integrated Circuits,* Wiley, New York, 1977.
13. S. K. Ghandhi, *Semiconductor Power Devices,* Wiley, New York, 1977.
14. P. Antognetti (ed.), *Power Integrated Circuits,* McGraw-Hill, New York 1986.
15. Sescosem, *Le Transistor de Puissance dans son Environment,* Thomson-CSF.
16. P. Antognetti, G. Massobrio, G. Sciutto, B. Gabriele, and B. Cotta, Modeling and Simulation of Power Transistors in Electronic Power Converter, *Eurocon '80,* Stuttgart, 1980.
17. P. Antognetti, G. Massobrio, M. Mazzucchelli, and G. Sciutto, Optimum Design of a Power Transistor Inverter Controlled by PWM Technique, *Motorcontrol,* 1981.
18. G. Massobrio, Modello del Transistore di Potenza per Applicazioni CAD, *Pixel,* **2,** 1983.
19. G. Massobrio, Vincoli all'Implementazione in SPICE del Modello del Transistore di Potenza, *Pixel,* **4,** 1984.
20. P. E. Gray and C. L. Searle, *Electronic Principles: Physics, Models, and Circuits,* Wiley, New York, 1969.
21. J. Choma, Jr., A Process-Oriented Model for the Simulation of Base Pushout in Integrated Bipolar Devices, *IEEE Trans. Electron Devices,* **ED-22,** 1975.
22. P. L. Hower, Application of a Charge-Control Model to High-Voltage Power Transistors, *IEEE Trans. Electron Devices,* **ED-23,** 1976.
23. G. Rey, F. Dupuy, and J. P. Bailbe, A Unified Approach to the Base Widening Mechanisms in Bipolar Transistors, *Solid-State Electron.,* **18,** 1975.
24. H. K. Gummel and H. C. Poon, A Compact Bipolar Transistor Model, *IEEE Int. Solid-State Circuit Cont.,* 1970.
25. F. A. Lindholm and D. J. Hamilton, Incorporation of the Early Effect in the Ebers-Moll, *Proc. IEEE,* 1971.
26. W. E. Drobish, Higher Order Models for Computer-Aided Circuit Design, *IEEE J. Solid-State Circuits,* **SC-7,** 1972.
27. J. G. Ruch, Modeling Bipolar Transistors, *Journées d'Electronique,* Lausanne, 1977.
28. P. Antognetti and G. Massobrio, Thyristor Model for CAD Applications, *Journées d'Electronique,* Lausanne, 1977.
29. P. Antognetti, G. Massobrio, M. La Regina, and S. Parodi, Computer Aided Analysis of Power Electronic Circuits Containing Thyristors, *Computer-Aided Design and Manufacture of Electronic Components, Circuits and Systems,* London, 1979.
30. G. Rey and J. P. Bailbe, Some Aspects of Current Gain Variations in Bipolar Transistors, *Solid-State Electron.,* **17,** 1974.
31. R. B. Schilling, A Bipolar Transistor Model for Device and Circuit Design, *RCA Rev.,* **32,** 1971.
32. R. A. Aubrey and H. L. Kraus, Nonlinear Large-Signal Modeling of Transistor Collector Characteristics for Computer-Aided Circuit Design, *IEEE J. Solid-State Circuits,* **SC-5,** 1970.
33. B. R. Chawla, Circuit Representation of the Integral Charge-Control Model of Bipolar Transistors, *IEEE J. Solid-State Circuits,* **SC-6,** 1971.
34. F. A. Perner, Quasi-Saturation Region Model of an npn^-n Transistor, *International Electron Devices Meeting,* Washington, D.C., 1974.

Junction Field-Effect Transistor (JFET)

Giuseppe Massobrio

*Department of Electronics (DIBE), University of
Genoa, Genoa, Italy*

The *junction field-effect transistor (JFET)* is a semiconductor device
that depends for its operation on the control of current by an electric field.
The structure of an *n-channel* field-effect transistor is shown in Fig. 3-
1a. It consists of a conductive channel that has two ohmic contacts, one
acting as the cathode *(source)* and the other as the anode *(drain)*, with
an appropriate voltage applied between drain and source. The third electrode *(gate)* forms a rectifying junction with the channel. Thus the JFET
is basically a voltage-controlled resistor, and its resistance can be varied
by changing the width of the depletion region extending into the channel.

Since the conduction process involves predominantly one kind of carrier, the JFET is also called a *unipolar* transistor in order to distinguish
it from the bipolar transistor (BJT), in which both types of carriers are
involved.

The symbols and sign convention for a *p*-channel and an *n*-channel
JFET are indicated in Fig. 3-1b. The direction of the arrow at the gate of
the JFET indicates the direction in which gate current would flow if the
gate junction were forward-biased.

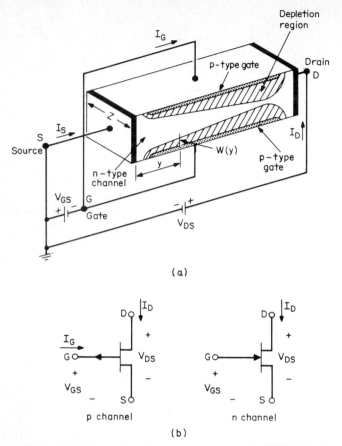

(a)

(b)

Figure 3-1 (a) Basic structure of an n-channel JFET and (b) JFET symbols and sign convention. *(Adapted from Millman [2]. Used by permission.)*

Note that the n-channel JFET requires zero or negative gate bias and positive drain voltage. The p-channel JFET requires opposite voltage polarities [1, 2].

SPICE JFET behavior is directly regulated by the subroutines *JFET* (it computes all the required parameters and loads them into the coefficient matrix), *FETLIM* (it limits, for each iteration, the voltage variation of the MOSFET devices), and *TMPUPD* (it updates the temperature-dependent parameters), besides the subroutines concerning each type of analysis.

3.1   Static Model

This section emphasizes the basic aspects of the behavior of a junction field-effect transistor in order to obtain its static model. The following sections deal with the derivation of the large- and small-signal models for the JFET and indicate how they can be modeled by equivalent circuits in SPICE.

3.1.1 DC characteristics†

To analyze the JFET, consider first a very small bias V_{DS} applied to the *drain* electrode (while the *source* is grounded) of the n-channel field-effect transistor of Fig. 3-1a. Under this condition the gate-channel bias (and therefore the width of the gate depletion region) is uniform along the entire channel. The voltage at the gate is V_{GS}. An expanded view of the channel region is shown in Fig. 3-2.

A one-dimensional structure with a gate length L between the source and drain regions and a width Z perpendicular to the plane of the paper is assumed (usually $Z \gg L$). Drain current flows along the dimension L.

A *one-sided step junction* at the gate with N_A in the p region much greater than N_D in the channel is also assumed. Therefore, the depletion layer extends primarily into the n channel. The distance between the p-type gate and the substrate is d, the thickness of the gate depletion region in the n-type channel is W, and the thickness of the neutral portion of the channel is x_W. In order to focus on the role of the gate, assume that the depletion region at the substrate junction extends primarily into the substrate so that $x_W \simeq d - W$. This is generally true in practice.

The resistance of the channel region can be written as

$$R = \frac{\rho L}{x_W Z} \tag{3-1}$$

where ρ is the *resistivity* of the channel. Hence, the drain current is

$$I_D = \frac{V_{DS}}{R} = \frac{Z}{L}(q\mu_n N_D x_W V_{DS}) \tag{3-2}$$

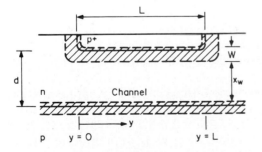

Figure 3-2 Channel region of a JFET showing depletion regions. *(From Muller and Kamins [3]. Copyright © 1977 by John Wiley & Sons, Inc. Used by permission.)*

---

† The material in this section is based in part on R. S. Muller and T. I. Kamins, *Device Electronics for Integrated Circuits*, copyright © 1977 by John Wiley & Sons, Inc. Reprinted by permission.

The dependence on gate voltage is incorporated in Eq. (3-2) by expressing $x_W = d - W$, where W from Eq. (A-36), when $N_A \gg N_D$, is

$$W = \sqrt{\frac{2\epsilon_s}{qN_D}(\phi_0 - V_{GS})} \qquad (3\text{-}3)$$

and ϕ_0 is the built-in potential. The current can now be written as a function of the gate and drain voltages.

$$I_D = \frac{Z}{L}q\mu_n N_D d \left[1 - \sqrt{\frac{2\epsilon_s}{qN_D d^2}(\phi_0 - V_{GS})} \right] V_{DS} \qquad (3\text{-}4)$$

The factors in front of the bracketed terms represent the *conductance* G_0 of the n region if it were completely undepleted (the so-called metallurgical channel) so that Eq. (3-4) may be rewritten as

$$I_D = G_0 \left[1 - \sqrt{\frac{2\epsilon_s}{qN_D d^2}(\phi_0 - V_{GS})} \right] V_{DS} \qquad (3\text{-}5)$$

Hence, at a given gate voltage, a linear relationship between I_D and V_{DS} can be found. This is a consequence of having assumed small applied drain voltages. The square-root dependence on gate voltage in Eq. (3-5) arises from the assumption of an *abrupt gate-channel junction.*

From Eq. (3-5) it can be seen that the current is maximum at zero applied gate voltage and decreases as $|V_{GS}|$ increases. The equation predicts zero current when the gate voltage is large enough to deplete the entire channel region [3].

Now to see the physics behind the device, the restriction of small drain voltages is removed and the problem for arbitrary V_{DS} and V_{GS} values (with the restriction that the gate must always remain reverse-biased) is considered.

With V_{DS} arbitrary, the voltage between the channel and the gate is a function of position y. Consequently, the depletion-region width and therefore the channel cross section also vary with position. The voltage across the depletion region is higher near the drain than near the source in this n-channel device. Therefore, the depletion region is wider near the drain, as shown in Fig. 3-3.

The *gradual-channel approximation* will be used. This approximation assumes that the channel and depletion-region widths vary slowly from source to drain so that the depletion region is influenced only by fields in the vertical dimension and not by fields extending from drain to source. In other words, the field in the y direction is much less than that in the x direction in the depletion region, and the depletion-region width from a one-dimensional analysis may be found [3].

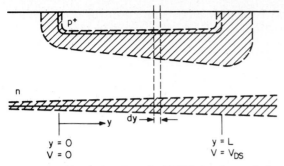

Figure 3-3   Channel region of a JFET showing variation of the width of depletion regions along the channel when the drain voltage is significantly higher than the source voltage. *(From Muller and Kamins [3]. Copyright © 1977 by John Wiley & Sons, Inc. Used by permission.)*

Within this approximation, an expression for the increment of voltage across a small section of the channel of length dy at y may be written as

$$dV = I_D \, dR = \frac{I_D \, dy}{Zq\mu_n N_D(d - W)} \tag{3-6}$$

The width W of the depletion region is now controlled by the voltage $[\phi_0 - V_{GS} + V(y)]$, where $V(y)$ is the potential in the channel at point y, so that, from Eq. (A-36),

$$W = \sqrt{\frac{2\epsilon_s}{qN_D} [\phi_0 - V_{GS} + V(y)]} \tag{3-7}$$

This expression may be used in Eq. (3-6), which is then integrated from source to drain to obtain the current-voltage relationship for the JFET:

$$\frac{I_D \int_0^L dy}{Zq\mu_n N_D} = \int_0^{V_D} \left[d - \sqrt{\frac{2\epsilon_s}{qN_D} (\phi_0 - V_{GS} + V)} \right] dV \tag{3-8}$$

After integrating and rearranging,

$$I_D = G_0 \left\{ V_{DS} - \frac{2}{3} \sqrt{\frac{2\epsilon_s}{qN_D d^2}} [\sqrt{(\phi_0 - V_{GS} + V_{DS})^3} - \sqrt{(\phi_0 - V_{GS})^3}] \right\} \tag{3-9}$$

At low drain voltages, Eq. (3-9) reduces to the simpler expression of Eq. (3-5); at large drain voltages, Eq. (3-9) indicates that the current reaches a maximum and begins decreasing with increasing drain voltage, but this maximum corresponds to the limit of validity of the analysis. Indeed, from Fig. 3-4 it can be seen that, as the drain voltage increases, the width of the conducting channel near the drain decreases, until finally the channel is completely depleted in this region (Fig. 3-4b).

When this occurs, Eq. (3-6) becomes indeterminate ($W \to d$). The equations are, therefore, valid only for V_{DS} below the drain voltage that *pinches off* the channel. Current continues to flow when the channel has been pinched off because there is no barrier to the transfer of electrons traveling down the channel toward the drain. As they arrive at the edge of the pinched-off zone, they are pulled across it by the field directed from the drain toward the source. If the drain bias is increased further, any additional voltage is dropped across a depleted, high-field region near the drain electrode, and the point at which the channel is entirely depleted moves slightly toward the source (Fig. 3-4c) [3].

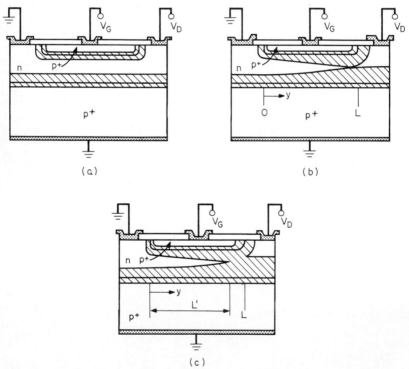

Figure 3-4 Behavior of the depletion regions in a JFET. (*a*) For very small drain voltage, the channel is nearly equipotential and the dimensions of the depletion regions are uniform. (*b*) When V_{DS} is increased to $V_{D,\text{sat}}$, the depletion regions on both sides of the channel meet at the pinch-off point at $y = L$. (*c*) When $V_{DS} > V_{D,\text{sat}}$, the pinch-off point at $y = L'$ moves slightly closer to the source. *(From Muller and Kamins [3]. Copyright © 1977 by John Wiley & Sons, Inc. Used by permission.)*

If this slight movement is neglected, the drain current remains constant (saturates) as the drain voltage is increased further, and the bias condition is referred to as *saturation*. The drain voltage at which the channel is entirely depleted near the drain electrode is found from Eq. (3-7) to be

$$V_{D,\text{sat}} = \frac{qN_Dd^2}{2\epsilon_s} - (\phi_0 - V_{GS})$$
$$= V_P - \phi_0 + V_{GS} = V_{T0} + V_{GS} \tag{3-10}$$

where $V_P = qN_Dd^2/2\epsilon_s$ is referred to as the *pinch-off voltage* and $V_{T0} = V_P - \phi_0$ is referred to as the *threshold voltage* (V_{T0} is a SPICE JFET model input parameter).

From Eqs. (3-9) and (3-10) the drain saturation current is found to be

$$I_{D,\text{sat}} = G_0 \left\{ \frac{qN_Dd^2}{6\epsilon_s} - (\phi_0 - V_{GS}) \left[1 - \frac{2}{3} \sqrt{\frac{2\epsilon_s(\phi_0 - V_{GS})}{qN_Dd^2}} \right] \right\}$$

$$= \frac{G_0 V_P}{3} \left[1 - 3\frac{\phi_0 - V_{GS}}{V_P} + 2 \sqrt{\left(\frac{\phi_0 - V_{GS}}{V_P}\right)^3} \right] \tag{3-11}$$

The maximum value of $I_{D,\text{sat}}$ (designated I_{DSS}) occurs for $V_{GS} = 0$: if $I_{D,\text{sat}}$ is normalized to I_{DSS} and plotted as a function of V_{GS}/V_P, the curve shown in Fig. 3-5 results. Also plotted in Fig. 3-5 is a square-law transfer

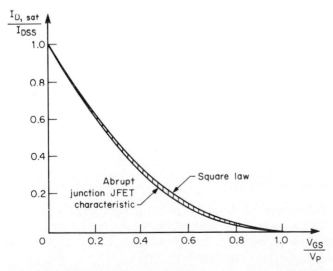

Figure 3-5 Normalized transfer characteristics of an abrupt-junction JFET compared with a square-law characteristic. *(Adapted from Sze [1]. Copyright © 1969 by John Wiley & Sons, Inc. Used by permission.)*

characteristic given by

$$I_{D,\text{sat}} = I_{DSS}\left(1 - \frac{V_{GS}}{V_P}\right)^2 \qquad (3\text{-}12)$$

The two curves agree quite closely, and Eq. (3-12) is commonly used as an approximation of the JFET characteristic in the saturation region.

The I_D vs. V_{DS} characteristic may be divided into three regions (see Fig. 3-6): (1) the linear region at low drain voltages, (2) a region with less than linear increase of current with drain voltage, and (3) a saturation region where the current remains relatively constant as the drain voltage is increased further.

Because of the physics of the device, Eq. (3-11) predicts the current to be maximum for zero gate bias and to decrease as negative gate voltage is applied [3].

As the gate voltage becomes more negative, the drain saturation voltage and the corresponding current decrease. At a sufficiently negative value of gate voltage, the saturation drain current becomes zero. This *turn-off* voltage V_T is found from Eq. (3-11) to be

$$V_T = \phi_0 - \frac{qN_D d^2}{2\epsilon_s} \qquad (3\text{-}13)$$

Field-effect transistors are often operated in the saturation region where the output current is not appreciably affected by the output (drain) voltage but only by the input (gate) voltage. For this bias condition, the JFET is almost an ideal current source controlled by an input voltage. The *transconductance* g_m of the transistor expresses the effectiveness of the control of the drain current by the gate voltage; it is defined as

$$g_m = \left.\frac{dI_D}{dV_{GS}}\right|_{op} \qquad (3\text{-}14)$$

The subscript *op* denotes that the independent variables assume the values at operating-point bias.

3.1.2   Limitations of the ideal theory

Previous analysis has included several simplifying assumptions. In practical devices, however, some of these assumptions may not be sufficiently valid to obtain a good match between theory and experiment. This section considers some deviations from the simple theory.

a. Control of depletion-layer width.   One assumption affecting the theory in the previous section is that the depletion-layer width is controlled by the gate-channel junction and not by the channel-substrate junction.

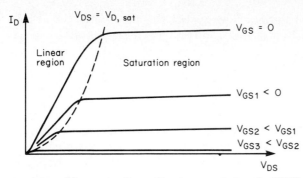

Figure 3-6 The output I_D vs. V_{DS} characteristics of a JFET as a function of the gate voltage.

There will be a variation of the potential across the channel-substrate junction along the channel, with the maximum potential and depletion-layer thickness near the drain. Consequently, the channel becomes completely depleted at lower drain voltage than is indicated by Eq. (3-10) [3].

b. Channel-length modulation. It has already been seen in the previous sections that in the saturation region the potential at the end of the channel, at point L of Fig. 3-4b, will be fixed at precisely the value of $V_{D,\text{sat}}$ corresponding to the applied gate voltage. This is so because point L itself is where the two depletion regions just touch. Hence, the reverse bias across the gate junctions at this point is fixed by the condition that $W = d$ there.

As the drain voltage is increased further, the reverse bias between the gate and the drain regions is also increased; hence, the width of the depletion region near the drain will also increase. As a result, the point will move toward the source as indicated in Fig. 3-4c. The voltage at point L' remains at the same value, but the length from the source to point L' shortens. It is evident that the drain current will increase at a given gate voltage as the drain voltage is increased; the end point for the integration in Eq. (3-8) now becomes L' rather than L, where L' is the point at which the channel becomes completely depleted [$V(L') = V_{D,\text{sat}}$].

Then for $V_{DS} > V_{D,\text{sat}}$, the expression for $I_{D,\text{sat}}$ in Eq. (3-11) should be multiplied by the factor L/L'.

This phenomenon, which is particularly important in devices with short channel lengths, is quite analogous to the Early effect discussed in connection with junction transistors. In both cases, the increase in current takes place because the current path is shortened by the widening of a reverse-biased depletion region [4].

Thus λ (a SPICE JFET model input parameter) can be defined as a *channel-length modulation parameter* which is equivalent to the inverse of the Early voltage for the BJT and which is a measure of the JFET output conductance in saturation. If we specify this parameter, the JFET will have a finite but constant output conductance in saturation. λ can be defined as

$$\lambda = \frac{L'}{L V_{DS}} \tag{3-15}$$

c. Effect of series resistance.† In the preceding calculation, we considered only the resistance of that portion of the channel which can be modulated by the application of a reverse bias to the gate. In reality, there are series resistances present, both near the source and near the drain, which interpose an IR drop between the source and drain contacts and the channel.

The effect of these series resistances on the channel conductance in the linear region can be readily calculated by noting that

$$\frac{1}{g(\text{obs})} = \frac{1}{g} + r_S + r_D \tag{3-16}$$

where g is the true channel conductance while $g(\text{obs})$ is the conductance observed experimentally; r_S and r_D are the series resistances near the source and the drain, respectively. Thus,

$$g(\text{obs}) = \frac{g}{1 + (r_S + r_D)g} \tag{3-17}$$

which shows that the observed conductance will be reduced due to the two series resistances.

The effect of the series resistance near the source region, r_S, on the transconductance in the saturation region is now considered. Because of this resistance, the potential at the beginning of the channel will not be zero but will have some finite value V_s. Thus the *effective* gate voltage will be

$$V_{GS} = V_{GS,\text{appl}} - V_s \tag{3-18}$$

As a result, the observed transconductance is given by

$$g_m(\text{obs}) = \frac{dI_D}{dV_{GS,\text{appl}}} = \frac{dI_D}{d(V_{GS} + V_s)} \tag{3-19}$$

---

† The material in this section is taken from A. S. Grove, *Physics and Technology of Semiconductor Devices,* copyright © 1967 by John Wiley & Sons, Inc. Reprinted by permission.

which, in turn, yields

$$g_m(\text{obs}) = \frac{1}{dV_{GS}/dI_D + dV_s/dI_D} \qquad (3\text{-}20)$$

and hence

$$g_m(\text{obs}) = \frac{g_m}{1 + r_S g_m} \qquad (3\text{-}21)$$

This last equation shows that the observed transconductance in the saturation region will be reduced due to the presence of a series resistance near the source from that attainable in the absence of such a series resistance.

The series resistance near the drain will act in a different manner. Because of the IR drop across this resistance, the drain voltage required to bring about saturation of the drain current will be larger than without it. However, since beyond that voltage, i.e., for $V_{DS} > V_{D,\text{sat}}$, the magnitude of V_{DS} has no significant effect on the drain current, the drain series resistance will have no further effect either [5].

Both r_S and r_D are taken into account in SPICE as JFET model parameters.

d. Breakdown. The pn junctions between gate and channel of a JFET are subject to *avalanche breakdown*. Since the JFET is a majority-carrier device whose operation does not depend on minority-carrier concentrations, the breakdown process is quite straightforward. Breakdown occurs when the voltage between gate and channel exceeds a critical value at which an abrupt increase of the drain current occurs.

The breakdown characteristic of the JFET can be described by specifying the breakdown voltage from drain to gate with the source open. This process, however, is not taken into account in SPICE; thus it cannot be simulated.

3.2 Static Model and Its Implementation in SPICE

The SPICE JFET static model follows from the quadratic FET model of Shichman and Hodges [6].

The dc characteristics are defined by the parameters V_{T0} and β. These parameters determine the variation of drain current with gate voltage, λ, which determines the output conductance, and I_S, the saturation current of the two gate junctions.

The results derived in the previous sections can be used to form the static model of the JFET.

The SPICE model for an n-channel JFET is shown in Fig. 3-7.

For a p-channel device, the polarities of the terminal voltages V_{GD}, V_{GS}, and V_{DS} must be reversed, the direction of the two gate junctions must be reversed, and the direction of the nonlinear current source I_D must be reversed.

The ohmic resistances of the drain and source regions of the JFET are modeled by the two linear resistors r_D and r_S, respectively. The dc characteristics of the JFET are represented by the nonlinear current source I_D. The value of I_D is determined by the equations shown in the following sections.

Forward dc characteristic. The forward region, or the normal mode operation, is characterized in SPICE by the following relations (for $V_{DS} \geq 0$):

$$
I_D = \begin{cases}
0 & \text{for } V_{GS} - V_{T0} \leq 0 \\[2ex]
\beta(V_{GS} - V_{T0})^2(1 + \lambda V_{DS}) & \text{for } 0 < V_{GS} - V_{T0} \leq V_{DS} \\[2ex]
\beta V_{DS}[2(V_{GS} - V_{T0}) - V_{DS}](1 + \lambda V_{DS}) & \\
& \text{for } 0 < V_{DS} < V_{GS} - V_{T0}
\end{cases}
$$

$$(3\text{-}22)$$

Reverse dc characteristic. The reverse region, or the inverted mode operation, is characterized in SPICE by the following relations (for $V_{DS} < 0$):

$$
I_D = \begin{cases}
0 & \text{for } V_{GD} - V_{T0} \leq 0 \\[2ex]
-\beta(V_{GD} - V_{T0})^2(1 - \lambda V_{DS}) & \text{for } 0 < V_{GD} - V_{T0} \leq -V_{DS} \\[2ex]
\beta V_{DS}[2(V_{GD} - V_{T0}) + V_{DS}](1 - \lambda V_{DS}) & \\
& \text{for } 0 < -V_{DS} < V_{GD} - V_{T0}
\end{cases}
$$

$$(3\text{-}23)$$

In SPICE the JFET drain current is then modeled by a simple square-law characteristic that is determined by the parameters β and V_{T0} [7]. The convention in SPICE is that V_{T0} is negative for all JFETs regardless of polarity. The parameters V_{T0} and β are usually determined from a

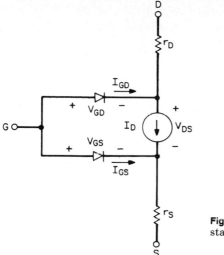

Figure 3-7 n-Channel JFET
static model.

graph of $\sqrt{I_D}$ vs. V_{GS}. An example for an n-channel device is shown in
Fig. 3-8. The parameter V_{T0} is the x-axis intercept of this graph, whereas
the parameter β, or its square root, is the slope of the $\sqrt{I_D}$ vs. V_{GS}
characteristic.

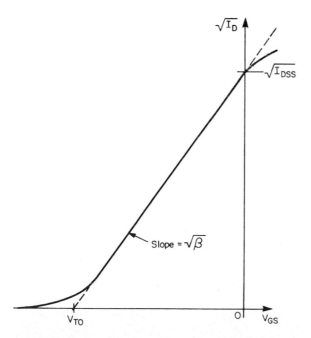

Figure 3-8 How to determine V_{T0} and β parameters.

The parameter λ (which has typical values in the range 0.1 to 0.01 V^{-1}) determines the effect of *channel-length modulation* (see Sec. 3.1.2b) on the JFET characteristic. As the output conductance of the JFET, in the forward saturation region, is given by

$$g_{D,\text{sat}} \equiv \frac{dI_D}{dV_{DS}} = \beta\lambda(V_{GS} - V_{T0})^2 \simeq \lambda I_D \qquad (3\text{-}24)$$

the saturation conductance is directly proportional to the drain current.

The two diodes shown in Fig. 3-7 are modeled by the following equations:

$$I_{GD} = \begin{cases} -I_S + V_{GD}\text{GMIN} & \text{for } V_{GD} \leq -5\dfrac{kT}{q} \\[3ex] I_S(e^{qV_{GD}/kT} - 1) + V_{GD}\text{GMIN} & \text{for } V_{GD} > -5\dfrac{kT}{q} \end{cases}$$

$$(3\text{-}25)$$

$$I_{GS} = \begin{cases} I_S(e^{qV_{GS}/kT} - 1) + V_{GS}\text{GMIN} & \text{for } V_{GS} > -5\dfrac{kT}{q} \\[3ex] -I_S + V_{GS}\text{GMIN} & \text{for } V_{GS} \leq -5\dfrac{kT}{q} \end{cases}$$

$$(3\text{-}26)$$

The saturation current I_S is a JFET model parameter.

To aid convergence, a small conductance, GMIN, is inserted in SPICE in parallel with every junction. The value of this conductance is a program parameter that can be set by the user by selection of the .OPTIONS card. The default value for GMIN is 10^{-12} mho; moreover, the user cannot set GMIN to zero.

Figure 3-9 shows the drain current vs. drain voltage of an n-channel JFET using SPICE parameter default values for a particular value of gate voltage. This figure also shows curves obtained by varying the λ, β, and V_{T0} parameters one at a time (keeping the default values for the other model parameters).

3.2.1 Relation between β and V_P

Typical values of V_P are in the range 1 to 3 V; this leads us to consider $V_{T0} = V_P$ (see Sec. 3.1.1). That being granted, a relation between β and V_P can be obtained.

Taking into account Eq. (A-43) and the fact that V_P is negative for an

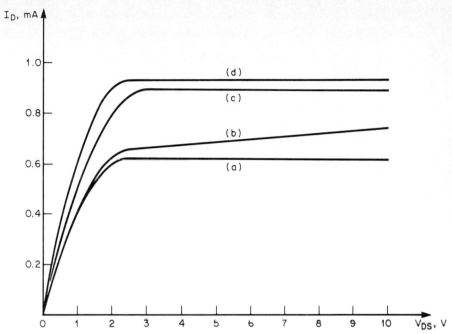

Figure 3-9 SPICE I_D vs. V_{DS} characteristics: (a) parameter default values, (b) $\lambda = 0.02$ V^{-1}, (c) $V_{T0} = -2.5$ V, and (d) $\beta = 0.15 \times 10^{-3}$ A/V^2. The characteristics have been derived for $V_{GS} = -0.5$ V.

n-channel JFET, under the condition that $V_{D,\text{sat}} \gg V_P$ (saturation region), we can derive a more suitable expression of Eq. (3-11), expanding it into a Taylor series in the neighborhood of V_P and truncated after the second-order term. Thus, Eq. (3-11) reduces to the form [8]

$$I_{D,\text{sat}} = \beta(V_{GS} - V_P)^2 \tag{3-27}$$

[See Eq. (3-22) for the saturation region when the output conductance is not included.]

An alternative drain-source current expression is given by Eq. (3-12), which is rewritten here for convenience as

$$I_{D,\text{sat}} = I_{DSS}\left(1 - \frac{V_{GS}}{V_P}\right)^2 \tag{3-28}$$

where I_{DSS} is the drain-source current at $V_{GS} = 0$. From Eqs. (3-27) and (3-28) the following expression for β is obtained [8]:

$$\beta = \frac{I_{DSS}}{V_P^2} \tag{3-29}$$

From Fig. 3-1a, the channel conductance can be written as shown in the following equation [8]:

$$g_{DS} = (d - W)q\mu_n N_D \tag{3-30}$$

It is also well known [5] that the transconductance in the saturation region exactly equals the conductance on the linear region. From Eq. (3-27) we can find the transconductance from

$$g_m = \frac{dI_{D,\text{sat}}}{dV_{GS}} = 2\beta(V_{GS} - V_P) \tag{3-31}$$

Combining Eqs. (3-30) and (3-31) for $V_{GS} = 0$ yields [8]

$$(d - W)\mu_n q N_D = 2\beta V_P \tag{3-32}$$

Before calculating β from Eq. (3-32), a suitable expression may be derived for $d - W$ to be used in Eq. (3-32). From Eq. (3-3),

$$W = \sqrt{\frac{2\epsilon_s}{qN_D}\phi_0} \tag{3-33}$$

Besides, $d = W$ when $V_{GS} = V_P$, so that

$$d = \sqrt{\frac{2\epsilon_s}{qN_D}(\phi_0 - V_P)} \tag{3-34}$$

A first-order calculation shows that for low values of V_P [8],

$$d - W = \sqrt{\frac{2\epsilon_s}{qN_D}\phi_0}\frac{V_P}{2\phi_0} \tag{3-35}$$

Combining Eqs. (3-32) and (3-35) yields

$$\sqrt{\frac{2\epsilon_s}{qN_D}\phi_0}\frac{1}{4\phi_0}q\mu_n N_D = \beta \tag{3-36}$$

Since the zero-voltage gate-channel capacitance for unit area C'_{GC} is given by

$$C'_{GC} = \sqrt{\frac{\epsilon_s q N_D}{2\phi_0}} \tag{3-37}$$

then

$$\beta = \frac{\mu_n C'_{GC}}{2} \tag{3-38}$$

Equation (3-38) shows that β is a constant for any value of V_P if the same channel concentration is used.

3.3 Large-Signal Model and Its Implementation in SPICE

Charge storage in a JFET occurs in the two gate junctions. This charge storage is modeled by the nonlinear elements Q_{GS} and Q_{GD}. Since neither gate junction normally is forward-biased, the diffusion charge component of these junctions can be ignored.

The elements Q_{GS} and Q_{GD} are defined by the equations

$$Q_{GS} = C_{GS}(0) \int_0^{V_{GS}} \frac{dV}{\sqrt{1 - V/\phi_0}} \tag{3-39}$$

$$Q_{GD} = C_{GD}(0) \int_0^{V_{GD}} \frac{dV}{\sqrt{1 - V/\phi_0}} \tag{3-40}$$

which are implemented in SPICE as follows:

$$Q_{GS} = \begin{cases} 2\phi_0 C_{GS}(0) \left(1 - \sqrt{1 - \dfrac{V_{GS}}{\phi_0}} \right) & \text{for } V_{GS} < FC \times \phi_0 \\[2em] C_{GS}(0)F_1 + \dfrac{C_{GS}(0)}{F_2} \left[F_3(V_{GS} - FC \times \phi_0) + \dfrac{V_{GS}^2 - FC^2 \times \phi_0^2}{4\phi_0} \right] & \\[1em] & \text{for } V_{GS} \geq FC \times \phi_0 \end{cases}$$

$$\tag{3-41}$$

$$Q_{GD} = \begin{cases} 2\phi_0 C_{GD}(0) \left(1 - \sqrt{1 - \dfrac{V_{GD}}{\phi_0}} \right) & \text{for } V_{GD} < FC \times \phi_0 \\[2em] C_{GD}(0)F_1 + \dfrac{C_{GD}(0)}{F_2} \left[F_3(V_{GD} - FC \times \phi_0) + \dfrac{V_{GD}^2 - FC^2 \times \phi_0^2}{4\phi_0} \right] & \\[1em] & \text{for } V_{GD} \geq FC \times \phi_0 \end{cases}$$

$$\tag{3-42}$$

Alternatively, these two charges can be expressed as voltage-dependent capacitors with values determined by the following expressions:

$$
C_{GS} = \begin{cases} \dfrac{C_{GS}(0)}{\sqrt{1 - V_{GS}/\phi_0}} & \text{for } V_{GS} < FC \times \phi_0 \\[4mm] \dfrac{C_{GS}(0)}{F_2}\left(F_3 + \dfrac{V_{GS}}{2\phi_0}\right) & \text{for } V_{GS} \geq FC \times \phi_0 \end{cases} \tag{3-43}
$$

$$
C_{GD} = \begin{cases} \dfrac{C_{GD}(0)}{\sqrt{1 - V_{GD}/\phi_0}} & \text{for } V_{GD} < FC \times \phi_0 \\[4mm] \dfrac{C_{GD}(0)}{F_2}\left(F_3 + \dfrac{V_{GD}}{2\phi_0}\right) & \text{for } V_{GD} \geq FC \times \phi_0 \end{cases} \tag{3-44}
$$

where F_1, F_2, and F_3 are program constants defined as follows:

$$
F_1 = \frac{\phi_0}{1 - m}\left[1 - (1 - FC)^{1-m}\right]
$$

$$
F_2 = (1 - FC)^{1+m} \tag{3-45}
$$

$$
F_3 = 1 - FC(1 + m)
$$

From the above, it follows that the JFET large-signal model can be modeled by the following parameters in the .MODEL card:

CGS, CGD
Zero-bias gate-source and gate-drain junction capacitances, respectively [$C_{GS}(0)$, $C_{GD}(0)$]

PB
Gate junction potential (ϕ_0)

FC
Forward-bias depletion capacitance coefficient (FC); along with ϕ_0, it is used in matching the transition point and voltage characteristics of the junction capacitance when bias changes from reverse to forward

m
Junction grading coefficient (m); in SPICE it is set to 0.5 and it cannot be varied

The JFET equivalent large-signal model is shown in Fig. 3-10.

3.4   Small-Signal Model and Its Implementation in SPICE

The small-signal linearized model for the JFET is given in Fig. 3-11. The conductances g_{GS} and g_{GD} are the conductances of the two gate junctions.

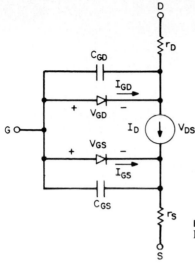

Figure 3-10 n-Channel JFET large-signal model.

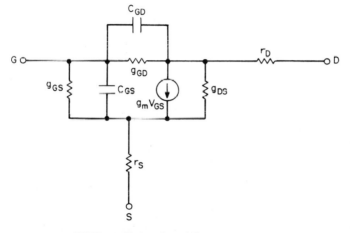

Figure 3-11 JFET small-signal model.

Since neither gate junction is forward-biased in normal operation, both g_{GS} and g_{GD} usually are very small. The transconductance g_m and the output conductance g_{DS} are defined by the equations

$$g_m = \left. \frac{dI_D}{dV_{GS}} \right|_{op} \tag{3-46}$$

$$g_{DS} = \left. \frac{dI_D}{dV_{DS}} \right|_{op} \tag{3-47}$$

$$g_{GS} = \left. \frac{dI_{GS}}{dV_{GS}} \right|_{op} \tag{3-48}$$

$$g_{GD} = \left. \frac{dI_{GD}}{dV_{DS}} \right|_{op} \tag{3-49}$$

The values of the preceding equations are determined in SPICE by the following equations for the two mode operations [taking into account Eqs. (3-22) to (3-26)].

Normal mode operation ($V_{DS} \geqslant 0$)

$$g_m = \begin{cases} 0 & \text{for } V_{GS} - V_{T0} \leq 0 \\ 2\beta(1 + \lambda V_{DS})(V_{GS} - V_{T0}) & \text{for } 0 < V_{GS} - V_{T0} \leq V_{DS} \\ 2\beta(1 + \lambda V_{DS})V_{DS} & \text{for } 0 < V_{DS} < V_{GS} - V_{T0} \end{cases} \tag{3-50}$$

$$g_{DS} = \begin{cases} 0 & \text{for } V_{GS} - V_{T0} \leq 0 \\ \lambda\beta(V_{GS} - V_{T0})^2 & \text{for } 0 < V_{GS} - V_{T0} \leq V_{DS} \\ 2\beta(1 + \lambda V_{DS})(V_{GS} - V_{T0} - V_{DS}) & \\ \quad + \lambda\beta V_{DS}[2(V_{GS} - V_{T0}) - V_{DS}] & \text{for } 0 < V_{DS} < V_{GS} - V_{T0} \end{cases} \tag{3-51}$$

Inverted mode operation ($V_{DS} < 0$)

$$g_m = \begin{cases} 0 & \text{for } V_{GD} - V_{T0} \leq 0 \\ -2\beta(1 - \lambda V_{DS})(V_{GD} - V_{T0}) & \text{for } 0 < V_{GD} - V_{T0} \leq -V_{DS} \\ 2\beta(1 - \lambda V_{DS})V_{DS} & \text{for } 0 < -V_{DS} \leq V_{GD} - V_{T0} \end{cases} \tag{3-52}$$

$$g_{DS} = \begin{cases} 0 & \text{for } V_{GD} - V_{T0} \leq 0 \\ \lambda\beta(V_{GD} - V_{T0})^2 - g_m & \text{for } 0 < V_{GD} - V_{T0} \leq -V_{DS} \\ 2\beta(1 - \lambda V_{DS})(V_{GD} - V_{T0}) & \\ \quad -\lambda\beta V_{DS}[2(V_{GD} - V_{T0}) + V_{DS}] & \text{for } 0 < -V_{DS} < V_{GD} - V_{T0} \end{cases} \tag{3-53}$$

When I_{GS} and I_{GD} are considered, we can write

$$
g_{GS} = \begin{cases} \dfrac{q}{kT} I_S e^{qV_{GS}/kT} + \text{GMIN} & \text{for } V_{GS} > -5\dfrac{kT}{q} & (3\text{-}54) \\[4mm] -\dfrac{I_S}{V_{GS}} + \text{GMIN} & \text{for } V_{GS} \le -5\dfrac{kT}{q} & (3\text{-}55) \end{cases}
$$

$$
g_{GD} = \begin{cases} \dfrac{q}{kT} I_S e^{qV_{GD}/kT} + \text{GMIN} & \text{for } V_{GD} > -5\dfrac{kT}{q} & (3\text{-}56) \\[4mm] -\dfrac{I_S}{V_{GD}} + \text{GMIN} & \text{for } V_{GD} \le -5\dfrac{kT}{q} & (3\text{-}57) \end{cases}
$$

The capacitances C_{GS} and C_{GD} are defined in Eqs. (3-43) and (3-44), respectively.

The JFET small-signal equivalent circuit of Fig. 3-11 is quite similar to that of the bipolar transistor, and much of the theory developed for bipolar transistor circuits applies equally to JFET circuits if the appropriate parameter values are used. One major difference is the absence of any input shunt resistance in the JFET because of the reverse biased pn junction in series with the gate electrode. This makes the JFET very attractive for circuits requiring high input impedance. An additional difference between the JFET and the bipolar transistor is the absence of any resistance in the JFET, which is equivalent to base resistance in the bipolar transistor.

See Table 3-1 for a list of SPICE JFET model parameters.

3.5   Temperature and Area Effects on the JFET Model

3.5.1   Temperature dependence of the JFET model parameters

All input data for SPICE is assumed to have been measured at 27°C (300 K). The simulation also assumes a nominal temperature (TNOM) of 27°C, which can be changed with the TNOM option of the .OPTIONS control card. The circuits can be simulated at temperatures different from TNOM by using a .TEMP control card.

Temperature appears explicitly in the exponential terms of the JFET model equations. In addition, saturation current I_S, gate-junction potential ϕ_0, zero-bias gate-source and gate-drain junction capacitances C_{GS} and C_{GD}, and the coefficient for forward-bias depletion capacitance formula

TABLE 3-1 SPICE JFET Model Parameters

Symbol	SPICE 2G keyword	Parameter name	Default value	Unit
V_{TO}	VTO	Threshold voltage	-2	V
β	BETA	Transconductance parameter	10^{-4}	A/V^2
λ	LAMBDA	Channel-length modulation	0	V^{-1}
r_D	RD	Drain ohmic resistance	0	Ω
r_S	RS	Source ohmic resistance	0	Ω
C_{GS}	CGS	Zero-bias gate-source junction capacitance	0	F
C_{GD}	CGD	Zero-bias gate-drain junction capacitance	0	F
ϕ_0	PB	Gate-junction potential	1	V
I_S	IS	Gate-junction saturation current	10^{-14}	A
FC	FC	Coefficient for forward-bias depletion capacitance formula	0.5	
k_f	KF	Flicker-noise coefficient	0	
a_f	AF	Flicker-noise exponent	1	
T	T†	Nominal temperature for simulation and at which all input data is assumed to have been measured	27	°C

† Remember, the temperature T is regarded as an operating condition and not as a model parameter.

FC have built-in temperature dependence. The temperature dependence of I_S in the JFET model is determined by

$$I_S(T_2) = I_S(T_1)e^{[-qE_g(300)/kT_2](1-T_2/T_1)}$$

(3-58)

where $E_g(300) = 1.11$ eV for Si.

The effect of temperature on ϕ_0 is modeled by the formula

$$\phi_0(T_2) = \frac{T_2}{T_1}\phi_0(T_1) - 2\frac{kT_2}{q}\ln\left(\frac{T_2}{T_1}\right)^{1.5}$$
$$- \left[\frac{T_2}{T_1}E_g(T_1) - E_g(T_2)\right]$$

(3-59)

where $E_g(T_1)$ and $E_g(T_2)$ are described by the relations [1]

$$E_g(T_1) = E_g(0) - \frac{\alpha T_1^2}{\beta + T_1}$$

$$E_g(T_2) = E_g(0) - \frac{\alpha T_2^2}{\beta + T_2}$$

(3-60)

where experimental results give, for Si,

$$\alpha = 7.02 \times 10^{-4}$$

$$\beta = 1108$$

$$E_g(0) = 1.16 \text{ eV}$$

The temperature dependence of C_{GS} and C_{GD} is determined by

$$C_{GS}(T_2) = C_{GS}(T_1) \left\{ 1 + m \left[400 \times 10^{-6}(T_2 - T_1) \right. \right.$$
$$\left. \left. - \frac{\phi_0(T_2) - \phi_0(T_1)}{\phi_0(T_1)} \right] \right\}$$

(3-61)

$$C_{GD}(T_2) = C_{GD}(T_1) \left\{ 1 + m \left[400 \times 10^{-6}(T_2 - T_1) \right. \right.$$
$$\left. \left. - \frac{\phi_0(T_2) - \phi_0(T_1)}{\phi_0(T_1)} \right] \right\}$$

where m is set to 0.5.

The temperature dependence of FC is determined by

$$FCPB(T_2) = FC \times \phi_0 = FCPB(T_1) \frac{\phi_0(T_2)}{\phi_0(T_1)}$$

(3-62)

$$F_1(T_2) = F_1(T_1) \frac{\phi_0(T_2)}{\phi_0(T_1)}$$

In the preceding equations, T_1 must be considered as follows:

- If TNOM in the .OPTIONS card is not specified, then T_1 = TNOM = TREF = 27°C (300 K); this is the default value for TNOM.
- If a TNOM_{new} value is specified, then T_1 = TNOM = 27°C; T_2 = TNOM_{new}.

These two cases are valid only if one request for temperature is made.

If more than one temperature is requested (using the .TEMP card), then T_1 = TNOM = 27°C and T_2 is the first temperature in the .TEMP card. Afterward, T_1 is the last working temperature (T_2) and T_2 is the next temperature in the .TEMP card to be calculated.

3.5.2 Area dependence of the JFET model parameters

The AREA factor used in SPICE for the JFET model determines the number of equivalent parallel devices of a specified model. The *JFET* model parameters affected by the *AREA* factor and specified in the device card are β, r_D, r_S, C_{GS}, C_{GD}, and I_S; that is,

$$\beta = \beta \times \text{AREA}$$

$$r_D = r_D/\text{AREA}$$

$$r_S = r_S/\text{AREA}$$

$$C_{GS}(0) = C_{GS}(0) \times \text{AREA} \qquad (3\text{-}63)$$

$$C_{GD}(0) = C_{GD}(0) \times \text{AREA}$$

$$I_S = I_S \times \text{AREA}$$

REFERENCES

1. S. M. Sze, *Physics of Semiconductor Devices,* Wiley, New York, 1969.
2. J. Millman, *Microelectronics,* McGraw-Hill, New York, 1979.
3. R. S. Muller and T. I. Kamins, *Device Electronics for Integrated Circuits,* Wiley, New York, 1977.
4. P. R. Gray and R. G. Meyer, *Analysis and Design of Analog Integrated Circuits,* Wiley, New York, 1977.
5. A. S. Grove, *Physics and Technology of Semiconductor Devices,* Wiley, New York, 1967.
6. H. Shichman and D. A. Hodges, Modeling and Simulation of Insulated-Gate Field-Effect Transistor Switching Circuits, *IEEE J. Solid-State Circuits,* **SC-3**, 1968.
7. L. W. Nagel, SPICE2: A Computer Program to Simulate Semiconductor Circuits, Electronics Research Laboratory Rep. No. ERL-M520, University of California, Berkeley, 1975.
8. M. C. Sansen and J. M. Das, A Simple Model of Ion-Implanted JFETs Valid in both the Quadratic and the Subthreshold Regions, *IEEE J. Solid-State Circuits,* **SC-17** (4), 1982.
9. E. Cohen, Program Reference for SPICE2, Electronics Research Laboratory Rep. No. ERL-M592, University of California, Berkeley, 1976.
10. R. J. Brewer, The Barrier Mode Behavior of a Junction FET at Low Drain Currents, *Solid-State Electron.,* **18**, 1975.
11. M. Arai, Charge-Storage Junction Field-Effect Transistor, *IEEE Trans. Electron Devices,* **ED-22**, 1975.
12. J. T. Wallmark and H. Johnson, *Field-Effect Transistors,* Prentice-Hall, Englewood Cliffs, N.J., 1966.
13. T. A. DeMassa and S. R. Iyer, Closed Form Solution for the Linear Graded JFET, *Solid-State Electron.,* **18**, 1975.
14. V. Y. Stenin, The Large-Signal Equivalent Circuit of a Field-Effect Transistor, *Telecommun. Radio Eng. Part 2,* 1972.
15. B. D. Wedlock, Static Large-Signal Field Effect Transistor Models, *Proc. IEEE,* **58**, 1970.

16. K. Lehovec and W. G. Seely, On the Validity of the Gradual Channel Approximation for Junction Field Effect Transistors with Drift Velocity Saturation, *Solid-State Electron.*, **16,** 1973.
17. T. L. Chiu and H. N. Ghosh, Characteristics of the Junction-Gate Field Effect Transistor with Short Channel Length, *Solid-State Electron.*, **14,** 1971.
18. T. J. Drummond, H. Morkoc, K. Lee, and M. Shur, Model for Modulation Doped Field Effect Transistor, *IEEE Electron. Device Lett.*, **3,** 1982.

The MOS Transistor

Enrico Profumo

SGS Microelettronica, Milan, Italy

4.1 Structure and Operating Regions of the MOS Transistor

Before presenting the equations and the parameters of the three models implemented in SPICE, it is useful to explain the meaning of the terms used in this chapter as indicated in Figs. 4-1 and 4-2. See Table 4-1 for a list of the model parameters for the MOS transistor.

4.1.1 Introduction to MOS transistor fabrication

Figures 4-1 and 4-2 represent the structure of the MOS transistor; a brief description of its construction follows. In the case shown in Fig. 4-2a, the substrate is a p-type silicon wafer on which a layer of thermal oxide (t_{oxf}) about 1 μm thick is built up; this oxide is removed (see mask 1 in Fig. 4-2a) from the regions where the transistor will be made. Another layer of thermal oxide t_{ox}, in this case an extremely thin oxide (100 nm or less), is then built up. Afterward, a layer of conducting material, polysilicon in the example, is deposited; this acts as a control electrode and is called the *gate*.

In the old processes, this layer was made of aluminum, and the structure formed in this way was called *MOS (metal oxide semiconductor)*.

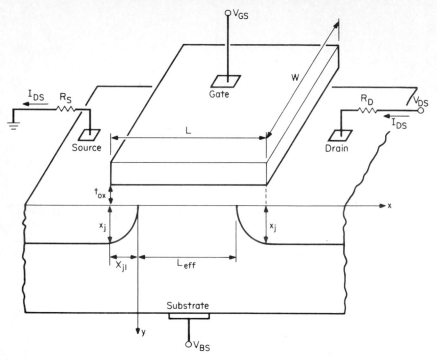

Figure 4-1 Structure of the MOST.

After the second mask has been used, the remaining polysilicon covers part of the thin layer of oxide, which prevents it from being removed by a later etching (Fig. 4-2b). Such etching, however, removes the thin oxide from the areas not covered by the polysilicon; in these regions a high concentration of n-type dopant is thermally diffused inside the silicon to obtain the so-called *source* and *drain* regions.

The region covered by the thin layer of oxide and by the gate represents the *channel,* and its length is (L_{nom}). This length is indicated in the SPICE input description of the MOS transistor and in this chapter as L. L_{eff} is the effective channel length defined as the distance measured on the silicon surface between the two diffused regions (Fig. 4-1). The width of the channel, indicated in SPICE as W, is the width of the area covered by the thin oxide between the thick oxide.

During diffusion, a new layer of thermal oxide is grown on both the silicon and the polysilicon (Fig. 4-2c); this layer is removed (mask 3) from the regions where the contacts are made. Lastly, the layer of aluminum interconnections is deposited and defined with mask 4 (Fig. 4-2d).

This structure is usually called *MOST (MOS transistor);* an often-used equivalent name is *MOSFET (MOS field-effect transistor).*

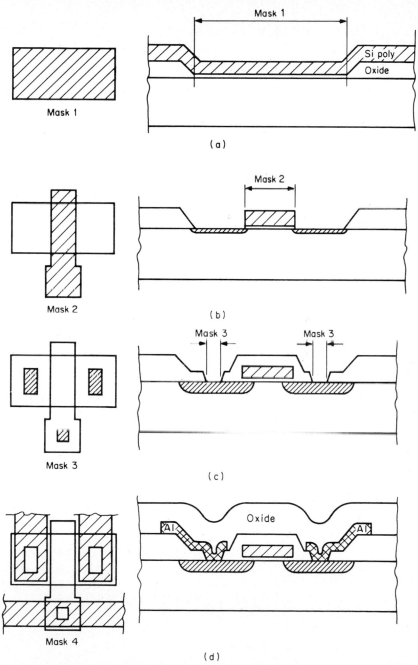

Figure 4-2 Plan of construction of a silicon-gate MOST.

TABLE 4-1 Model Parameters for the MOST

Symbol	SPICE keyword	LEVEL	Parameter description	Default value	Typical value	Units
			Parameters of the MOST			
V_{T0}	VTO	1-3	Zero-bias threshold voltage	1.0	1.0	V
KP	KP	1-3	Transconductance parameter	2×10^{-5}	3×10^{-5}	A/V^2
γ	GAMMA	1-3	Body-effect parameter	0.0	0.35	$V^{1/2}$
$2\phi_F$	PHI	1-3	Surface inversion potential	0.6	0.65	V
λ	LAMBDA	1,2	Channel-length modulation	0.0	0.02	V^{-1}
t_{ox}	TOX	1-3	Thin oxide thickness	1×10^{-7}	1×10^{-7}	m
N_b	NSUB	1-3	Substrate doping	0.0	1×10^{15}	cm^{-3}
N_{SS}	NSS	2,3	Surface state density	0.0	1×10^{10}	cm^{-2}
N_{FS}	NFS	2,3	Surface-fast state density	0.0	1×10^{10}	cm^{-2}
N_{eff}	NEFF	2	Total channel charge coefficient	1	5	
X_j	XJ	2,3	Metallurgical junction depth	0.0	1×10^{-6}	m
X_{jl}	LD	1-3	Lateral diffusion	0.0	0.8×10^{-6}	m
T_{PG}	TPG	2,3	Type of gate material	1	1	
μ_0	UO	1-3	Surface mobility	600	700	$cm^2/(V \cdot s)$
U_c	UCRIT	2	Critical electric field for mobility	1×10^4	1×10^4	V/cm
U_e	UEXP	2	Exponential coefficient for mobility	0.0	0.1	
U_t	UTRA	2	Transverse field coefficient	0.0	0.5	
v_{max}	VMAX	2,3	Maximum drift velocity of carriers	0.0	5×10^4	m/sec
	XQC	2,3	Coefficient of channel charge share	0.0	0.4	
δ	DELTA	2,3	Width effect on threshold voltage	0.0	1.0	

Symbol	Keyword	Parameter	Value	Example	Units	
η	ETA	Static feedback on threshold voltage	0.0	1.0	V^{-1}	3
θ	THETA	Mobility modulation	0.0	0.05		3
A_F	AF	Flicker-noise exponent	1.0	1.2		1-3
K_F	KF	Flicker-noise coefficient	0.0	1×10^{-26}		1-3

Parameters of parasitic effects

Symbol	Keyword	Parameter	Value	Example	Units	
I_s	IS	Bulk-junction saturation current	1×10^{-14}	1×10^{-15}	A	1-3
J_s	JS	Bulk-junction saturation current per square meter	0.0	1×10^{-8}	A	1-3
ϕ_j	PB	Bulk-junction potential	0.80	0.75	V	1-3
C_j	CJ	Zero-bias bulk capacitance per square meter	0.0	2×10^{-4}	F/m^2	1-3
M_j	MJ	Bulk-junction grading coefficient	0.5	0.5		1-3
C_{jsw}	CJSW	Zero-bias perimeter capacitance per meter	0.0	1×10^{-9}	F/m	1-3
M_{jsw}	MJSW	Perimeter capacitance grading coefficient	0.33	0.33		1-3
FC	FC	Bulk-junction forward-bias coefficient	0.5	0.5		1-3
C_{GBO}	CGBO	Gate-bulk overlap capacitance per meter	0.0	2×10^{-10}	F/m	1-3
C_{GDO}	CGDO	Gate-drain overlap capacitance per meter	0.0	4×10^{-11}	F/m	1-3
C_{GSO}	CGSO	Gate-source overlap capacitance per meter	0.0	4×10^{-11}	F/m	1-3
R_D	RD	Drain ohmic resistance	0.0	10.	Ω	1-3
R_S	RS	Source ohmic resistance	0.0	10.	Ω	1-3
R_{sh}	RSH	Source and drain sheet resistance	0.0	30.	Ω	1-3

It has been already seen from the example in Fig. 4-2 that technological developments have brought the use of materials other than aluminum for the gate electrode—e.g., polysilicon. The same has happened with the insulating layer between the gate and the semiconductor, where materials other than silicon oxide, e.g., silicon nitride, have been used. In these processes, the term *IGFET (insulated-gate field-effect transistor)* has therefore become more appropriate. However, in this text, the transistor will be referred to by the most commonly used name, MOST.

4.1.2 Operating regions

In n-channel devices, the drain potential is higher than the source one, and the voltages are referred to the source (Fig. 4-1). The voltage V_{GS} between gate and source determines the concentration of the carriers in the channel; the gate voltage for which the channel current becomes significant is called the *threshold voltage* V_{TH}. If V_{GS} is greater than V_{TH}, the current I_{DS} flows at the surface region from the drain to the source; this region is called the *channel*. The on-off behavior of the MOST is based on the fact that for gate voltages greater than the threshold voltage a current flows, whereas for smaller voltages the current is negligible.

The current vs. voltage characteristics for $V_{GS} > V_{TH}$ can be divided into a *linear region* and a *saturation region;* this is shown in Fig. 4-3. In the linear region the transfer characteristics are linear (Fig. 4-4), while in the saturation region I_{DS} does not depend on V_{DS}.

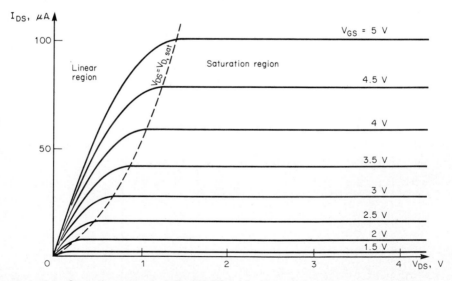

Figure 4-3 Operating regions of the MOST in the plan of the output characteristics.

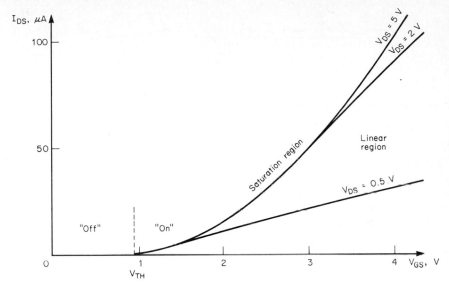

Figure 4-4   Operating regions of the MOST in the plan of the transfer characteristics.

There also exists a region near the threshold voltage where the current depends exponentially on V_{GS}; this is the region of *weak inversion*.

4.1.3   Types of MOSTs

The device described here is the n-channel type; i.e., conduction is maintained by the electrons in the channel. In the p-channel device the substrate is n-type and conduction is due to the holes. In both these MOSTs the carriers are attracted to the surface by the gate voltage; they are called *enhancement MOSTs*.

In the type of transistor called a *buried channel* or *depletion MOST* (Fig. 4-5), the carriers are available in the channel region also without the gate effect, because the surface is weakly doped by an n-type layer, obtained with ion implantation; in the MOST obtained in this way the threshold voltage is negative, and its electrical characteristics make it a useful substitute for load resistances in integrated circuits.

4.2   Equations for the LEVEL 1 Model

The first step in studying the theory of the MOST is the behavior of the surface of a semiconductor under the influence of an electric field. This electric field is perpendicular to the oxide-semiconductor interface, and it is produced by the voltage applied between the gate and the substrate. Later, the effect of the source and drain voltages will be introduced.

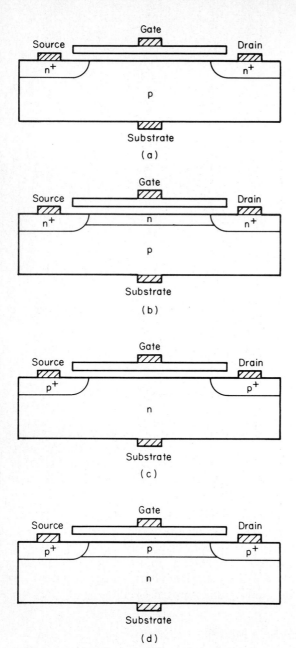

Figure 4-5 Types of MOSTs: (*a*) *n*-channel enhancement; (*b*) *n*-channel depletion; (*c*) *p*-channel enhancement; (*d*) *p*-channel depletion.

4.2.1 Surface behavior of the MOS structure

The theory outlined here is approximate, the purpose being to explain the equations of SPICE models; a more complete theory is described in Appendix B.

The reference condition of the surface is when the semiconductor has the same carrier concentrations at the surface and at the substrate. This state is called the *flat-band* condition; the voltage V_{GB} which is needed to obtain this condition is the *flat-band voltage V_{FB}*.

When $V_{GB} = V_{FB}$, all the voltage V_{GB} falls on the oxide; from Eq. (B-10)

$$V_{FB} = \phi_{MS} - \frac{Q_0'}{C_{ox}'} \qquad (4\text{-}1)$$

Q_0', which is always positive, is the charge at the oxide-silicon interface, and therefore it requires a negative charge on the gate to neutralize its effect (see Appendix B); C_{ox}' is the capacitance per unit area of the thin oxide layer (the terms for capacitances and charges per unit area in this text are shown with an apex, or prime).

Here MOSTs made with a p-type substrate and an n-type channel are considered; the behavior of $MOST$s made with an n-type substrate is dual when the sign of the voltages and the current is changed [1].

When $V_{GB} = V_{FB}$, the carrier concentration is constant in the semiconductor, and it is equal to N_A.

The condition $V_{GB} < V_{FB}$ increases the negative charge on the gate and increases the concentration of the holes near the surface in order to balance the gate charge. Thus the concentration of the p-type carriers is greater at the surface than in the substrate; in this case the surface is said to be *accumulated*.

The charge Q_{sc}' in the semiconductor is

$$Q_{sc}' = Q_G' - Q_0' = (V_{FB} - V_{GB})C_{ox}' \qquad (4\text{-}2)$$

where Q_G' is the charge on the gate.

When $V_{GB} > V_{FB}$, the holes (majority carriers) are pushed away from the surface so that the negative charge of the fixed ions restores the balance with the gate charge. The carrier concentration near the surface is less than that in the substrate; in this case the surface is said to be *depleted*. Using the same formulas as in the theory of the step junction (Chap. 1), the thickness X_B of the depleted region (Fig. 4-1b) is, from Eq. (B-20),

$$X_B = \sqrt{\frac{2\epsilon_S}{qN_A}\phi_s} \qquad (4\text{-}3)$$

where ϕ_s is the potential across the depleted region. If ϕ_s is small enough to neglect the minority carriers in the channel, the fixed charge Q_B' is equal to $Q_G' - Q_0'$. Therefore

$$Q_B' = N_A q X_B = \sqrt{2\epsilon_s q \phi_s} \tag{4-4}$$

and

$$Q_G' - Q_0' = (V_{GB} - V_{FB} - \phi_s) C_{ox}' \tag{4-5}$$

By substituting Eq. (4-3) in Eq. (4-4) and making it equal to Eq. (4-5), the surface potential ϕ_s can be obtained. When ϕ_s is known, the capacitance between the gate and the substrate follows. Therefore

$$\phi_s = \tfrac{1}{4}[\sqrt{\gamma^2 + 4(V_{GB} - V_{FB})} - \gamma]^2 \tag{4-6}$$

$$C_{GB}' = \frac{dQ_G'}{dV_{GB}} = \frac{C_{ox}'}{\sqrt{1 + (4/\gamma^2)(V_{GB} - V_{FB})}} \tag{4-7}$$

where

$$\gamma = \frac{\sqrt{2\epsilon_s q N_A}}{C_{ox}'} \tag{4-8}$$

These equations are valid as long as the depletion approximation is valid, i.e., as long as the carrier concentration remains negligible with respect to N_A in the channel. However, when the potential ϕ_s is sufficiently high, the concentration of the electrons at the surface can exceed that of the holes in the substrate. From Boltzmann's distribution [1], this condition is reached when $\phi_s = 2\phi_p$, that is, when

$$n = N_A = n_i e^{\phi_p q / kT} \tag{4-9}$$

or

$$\phi_p = \frac{kT}{q} \ln \frac{N_A}{n_i}$$

where ϕ_p has been defined in Eq. (A-20b). This approximated theory assumes arbitrarily that $\phi_s = 2\phi_p$ as the surface potential needed to change from a condition of depletion to one of inversion. In inversion, the carrier concentration at the surface is greater and of opposite type than the concentration in the substrate; the condition $V_{GB} = V_{TH}$ is therefore reached when $\phi_s = 2\phi_p$.

When $V_{GB} > V_{TH}$, it can be supposed that the surface potential ϕ_s and the fixed charge Q_B' are not altered, because the new charges are supplied by electrons near the surface. In fact, an increase in V_{GB} implies an increase in the charge Q_I' in the inversion layer; a slight increase in ϕ_s implies an increase in Q_I' according to the exponential law of Boltzmann's equation, while Q_B' remains almost constant according to Eq. (4-4) (see Appendix B). Therefore the mobile charge in the channel is

$$Q_I' = Q_G' - Q_0' - Q_B' \tag{4-10}$$

The charge Q_I' determines the conduction of the MOST. Substituting Eqs. (4-4) and (4-5) in Eq. (4-10) and making them equal yields

$$V_{TH} = V_{FB} + 2\phi_p + \gamma\sqrt{2\phi_p} \qquad (4\text{-}11)$$

4.2.2 Equations of the simple model

The first model of the MOST used in SPICE is basically the model proposed by Shichman and Hodges [2], and it is called the *LEVEL 1 model*. In order to extend the theory of the MOS capacitor seen in the preceding section to the MOST, the effect of the diffused regions of source and drain must be considered.

In this chapter the source potential will be used as a reference potential, and the voltage in the channel will be indicated as $V_C(x)$. At equilibrium—that is, when there is a zero voltage between drain and source—in order to make the concentration of minority carriers at the surface equal to that of the majority carriers in the substrate, it is necessary that the surface potential ϕ_s exceed the inverse bias V_{BS} between source and substrate; that is,

$$\phi_s = 2\phi_p - V_{BS} \qquad (4\text{-}12)$$

Substituting Eq. (4-12) in Eq. (4-5), the threshold voltage follows:

$$V_{TH} = V_{FB} + 2\phi_p + \gamma\sqrt{2\phi_p - V_{BS}} \qquad (4\text{-}13)$$

The threshold voltage is then a function of the square root of the substrate voltage (or *body bias*); γ is called a *body-effect* parameter, and it is defined in Eq. (4-8).

The condition $V_{GS} > V_{TH}$ allows channel formation, and, therefore, by applying a positive voltage to the drain, the electrons in the channel flow by drift from the source to the drain.

The current I_x in a section dx of the channel will be

$$I_x = \frac{dQ_I}{dt} \qquad (4\text{-}14)$$

where dQ_I is the mobile charge present in the dx element, and dt is the time necessary for this charge to cross dx. The value of dQ_I is

$$dQ_I = dQ_G - dQ_0 - dQ_B \qquad (4\text{-}15)$$

where

$$dQ_G - dQ_0 = C_{ox}' W\,dx[V_{GS} - V_{FB} - 2\phi_p - V_c(x)] \qquad (4\text{-}16)$$

$$dQ_B = C_{ox}' W\gamma\,dx\sqrt{2\phi_p - V_{BS}} \qquad (4\text{-}17)$$

The gate capacitance is

$$C_G' = \frac{dQ_G'}{dV_{GS}} = C_{ox}' \tag{4-18}$$

In the calculation of dQ_B, no account has been made of the voltage between the channel and the substrate $[V_{BS} + V_c(x)]$, and only V_{BS} has been considered. This approximation is valid only when the value of V_{DS} is small; otherwise the thickness X_B of the depleted region is noticeably larger near the drain than near the source. The value of dQ_B calculated in this way will be less than the real one, and the value of dQ_I will be overestimated.

By substituting Eqs. (4-16) and (4-17) in Eq. (4-15), it follows that

$$dQ_I = C_{ox}' W[V_{GS} - V_c(x) - V_{TH}] \tag{4-19}$$

The speed of the carriers $v(x)$ is linked to the electric field $E_x(x)$ by the transport equation

$$v(x) = \frac{dx}{dt} = -\mu E_x(x) = \mu \frac{dV_c(x)}{dx} \tag{4-20}$$

From Eq. (4-20),

$$\frac{1}{dt} = \mu \frac{dV_c(x)}{dx^2} \tag{4-21}$$

and by substituting Eqs. (4-19) and (4.21) in Eq. (4-14), it follows that

$$I_x = \mu W C_{ox}'[V_{GS} - V_{FB} - V_c(x)] \frac{dV_c(x)}{dx} \tag{4-22}$$

The integral between source and drain can be evaluated by using as a variable $V_C(x)$, which varies from zero for $x = 0$ to V_{DS} for $x = L_{eff}$. Therefore

$$\int_0^{L_{eff}} I_x \, dx = \mu W C_{ox}' \int_0^{V_{DS}} [V_{GS} - V_{FB} - V_c(x)] \, dV_c(x) \tag{4-23}$$

As I_x is constant in every section of the channel, it can be written

$$I_{DS} = \mu C_{ox}' \left(\frac{W}{L_{eff}}\right) \left[(V_{GS} - V_{TH})V_{DS} - \frac{V_{DS}^2}{2}\right] \tag{4-24}$$

The model is valid as long as a continuous channel exists between source and drain. However, when the voltage V_{DS} is such that there is a point in the channel where the voltage between the channel and the gate is equal to the threshold voltage, from this point to the drain, the conditions suitable for channel formation no longer exist; i.e.,

$$V_c(x) < V_{GS} - V_{TH} \qquad (4\text{-}25)$$

In this case the channel is formed only from $x = 0$ to $x = L'$, where L' is the point where the channel voltage reaches the *saturation voltage* $V_{D,\text{sat}}$

$$V_{D,\text{sat}} = V_c(L') = V_{GS} - V_{TH} \qquad (4\text{-}26)$$

For $V_{DS} > V_{D,\text{sat}}$, the current I_{DS} is not a function of V_{DS} because the voltage at the end of the channel still remains equal to $V_{D,\text{sat}}$. By substituting Eq. (4-26) in Eq. (4-24), the current for $V_{DS} > V_{D,\text{sat}}$ can be obtained. Thus

$$I_{DS} = \frac{\beta}{2} (V_{GS} - V_{TH})^2 \qquad (4\text{-}27)$$

where
$$\beta = \mu C'_{\text{ox}} \left(\frac{W}{L_{\text{off}}}\right) = KP \left(\frac{W}{L_{\text{eff}}}\right) \qquad (4\text{-}28)$$

KP is a SPICE parameter called the *transconductance parameter* [6]. The voltage $V_{DS} - V_{D,\text{sat}}$ falls in the region between $x = L'$ and $x = L_{\text{eff}}$, and the electric field, determined by such voltage difference, transports the carriers from the channel to the drain.

The simplifying assumptions used to formulate these equations are rather hard, and the precision obtained from this model is limited; however, experience has shown that the simplicity of this equation is useful in many situations.

4.2.3   Implementation of the LEVEL 1 model in SPICE

The equations used for the LEVEL 1 MOST model in SPICE are as follows.

Linear region

For $V_{GS} > V_{TH}$ and $V_{DS} < V_{GS} - V_{TH}$:

$$I_{DS} = KP \frac{W}{L - 2X_{jl}} \left(V_{GS} - V_{TH} - \frac{V_{DS}}{2}\right) V_{DS}(1 + \lambda V_{DS}) \qquad (4\text{-}29)$$

where X_{jl} is the lateral diffusion (see Fig. 4-1), and

$$V_{TH} = V_{T0} + \gamma(\sqrt{2\phi_p - V_{BS}} - \sqrt{2\phi_p}) \qquad (4\text{-}30)$$

being V_{T0}, the threshold voltage for $V_{BS} = 0$, a parameter that must be specified in the list of model parameters in input to SPICE with the .MODEL card [6].

Saturation region

For $V_{GS} > V_{TH}$ and $V_{DS} > V_{GS} - V_{TH}$:

$$I_{DS} = \frac{KP}{2} \frac{W}{L - 2X_{jl}} (V_{GS} - V_{TH})^2(1 + \lambda V_{DS}) \qquad (4\text{-}31)$$

where W and L are the values of the length and width of the channel specified in the SPICE input card of the individual MOST. The amount by which the gate electrode overlaps the source and drain regions must be subtracted from the nominal channel length. Thus

$$L_{\text{eff}} = L - 2X_{jl} \qquad (4\text{-}32)$$

The term $1 + \lambda V_{DS}$ introduced in the model is an empirical correction of the conductance in the saturation region. There are therefore five parameters which characterize this model: KP, V_{T0}, γ, $2\phi_p$, and λ. These parameters are electrical; i.e., they refer to the electrical behavior of the MOST. They can be specified directly in the .MODEL card, or they can be calculated from physical parameters, using the following equations:

$$KP = \mu C'_{\text{ox}} \qquad (4\text{-}33)$$

$$\gamma = \frac{\sqrt{2\epsilon_s q N_A}}{C'_{\text{ox}}} \qquad (4\text{-}34)$$

$$2\phi_p = 2\frac{kT}{q}\ln\frac{N_A}{n_i} \qquad (4\text{-}35)$$

where
$$C'_{\text{ox}} = \frac{\epsilon_{\text{ox}}}{t_{\text{ox}}} \qquad (4\text{-}36)$$

Thus it is possible to use the parameters μ, t_{ox}, and N_A instead of KP, γ, or $2\phi_p$, or a combination of the two types of parameters. In case of conflict (for example, if both μ and KP are present in the .MODEL card), the value of the electrical parameter is read from the input and not computed from Eqs. (4-33) and (4-35).

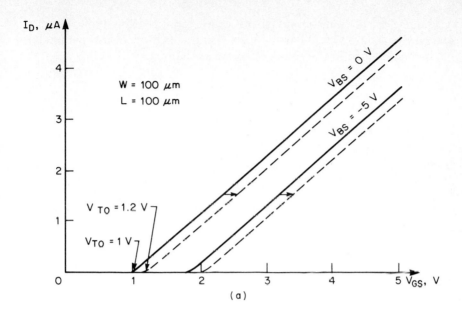

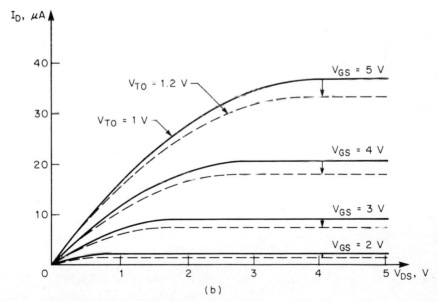

Figure 4-6 Variations of I_{DS} with V_{T0}, for the LEVEL 1 model: (*a*) transfer characteristics and (*b*) output characteristics.

Simulations of transfer and output characteristics using typical parameters are shown in Figs. 4-6 to 4-10. The values of the electrical parameters of the model used in the simulations are as follows:

$KP = 27.6 \ \mu A/V^2$

$V_{T0} = 1 \ V$

$\gamma = 0.526 \ V^{1/2}$

$2\phi_p = 0.58 \ V$

$\lambda = 0$

The values of the physical parameters are as follows:

$\mu = 800 \ cm^2/(V \cdot s)$

$t_{ox} = 100 \ nm$

$N_A = 10^{15} \ cm^{-3}$

$X_{jl} = 0.8 \ \mu m$

4.3   Equations for the LEVEL 2 Model

4.3.1   Equations for the basic model

To obtain a better model for I_{DS} it is necessary to eliminate the simplification made in Eq. (4-17) to find the value of Q_B by taking into account the effect of the voltage in the channel on Q_B. This problem was solved by Meyer [3], who started from the charge equation (Eq. 4-15), where the new value of Q_B is

$$dQ_B = W \ dx \gamma C'_{ox} \sqrt{2\phi_p - V_{BS} + V_c(x)} \qquad (4\text{-}37)$$

As in the previous theory, a new equation for the current can be obtained. Thus

$$I_{DS} = \beta \left\{ \left(V_{GS} - V_{FB} - 2\phi_p - \frac{V_{DS}}{2} \right) V_{DS} \right.$$

$$\left. - \tfrac{2}{3}\gamma[(V_{DS} - V_{BS} + 2\phi_p)^{1.5} - (-V_{BS} + 2\phi_p)^{1.5}] \right\} \qquad (4\text{-}38)$$

The saturation condition is reached when the charge in the channel for $x = L'$ is zero. By substituting Eqs. (4-16) and (4-37) for $V_{DS} = V_{D,\text{sat}}$ into Eq. (4.15), it follows that

$$Q_I(L') = (V_{GS} - V_{D,\text{sat}} - V_{FB})C'_{ox} - \gamma C'_{ox}\sqrt{V_{D,\text{sat}} - V_{BS} + 2\phi_p}$$

$$= 0 \qquad (4\text{-}39)$$

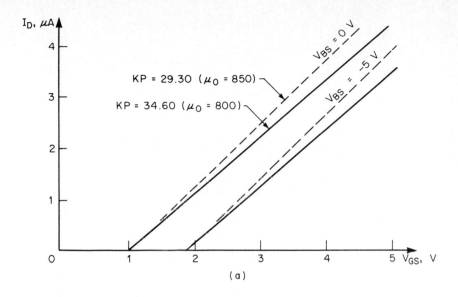

(a)

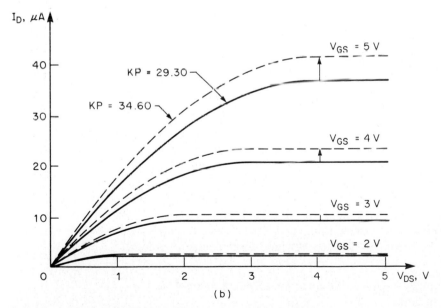

(b)

Figure 4-7 Variations of I_{DS} with KP or with μ, for the LEVEL 1 model: (a) transfer characteristics and (b) output characteristics.

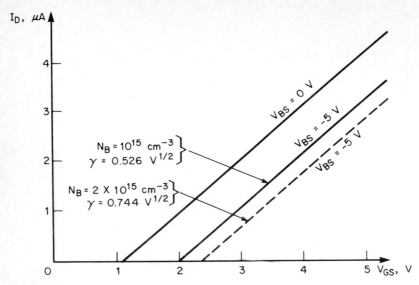

Figure 4-8 Variations of I_{DS} with γ or with N_A, for the LEVEL 1 model: (a) transfer characteristics and (b) output characteristics.

from which the value of $V_{D,\text{sat}}$ is

$$V_{D,\text{sat}} = V_{GS} - V_{FB} - 2\phi_p + \gamma^2 \left[1 - \sqrt{1 + \frac{2}{\gamma^2}(V_{GS} - V_{FB})} \right] \quad (4\text{-}40)$$

From Fig. 4-11 and Eq. (4-39), it can be noted that this model also includes a variation of the current with the parameter γ even if $V_{BS} = 0$; this was not included in the LEVEL 1 model.

Also, this model has validity limits that can be easily reached with the present technology. The method chosen to overcome these limitations is to adapt the model by applying semiempirical corrections to the basic equations rather than start again from the basis of the theory each time one needs to develop new and more accurate models.

4.3.2 Implementation of the LEVEL 2 model in SPICE

This section summarizes the basic equations of Meyer's model as they have been introduced in SPICE. The corrections made to the model to simulate effects not provided for in the theory of the basic model are described in the following sections.

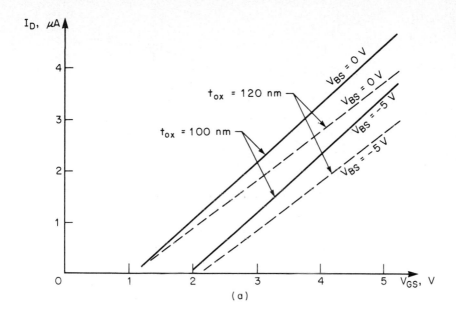

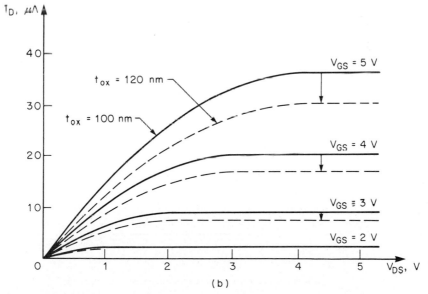

Figure 4-9 Variations of I_{DS} with t_{ox}, for the LEVEL 1 model: (*a*) transfer characteristics and (*b*) output characteristics.

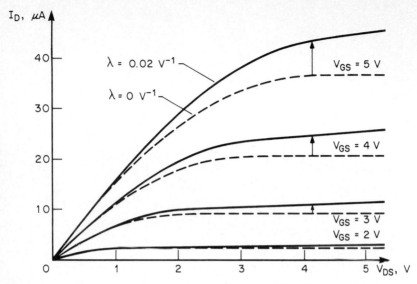

Figure 4-10 Variations of I_{DS} with λ, for the LEVEL 1 model: output characteristics.

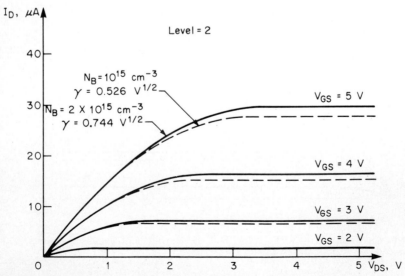

Figure 4-11 Variations of I_{DS} with γ, for the LEVEL 2 model: output characteristics.

The threshold voltage can be calculated from physical parameters through the equation

$$V_{T0} = \phi_{MS} - \frac{qN'_{SS}}{C'_{ox}} + 2\phi_p + \gamma\sqrt{2\phi_p} \qquad (4\text{-}41)$$

where

$$\phi_{MS} = -T_{PG}\frac{E_g}{2} - \frac{kT}{q}\ln\frac{N_A}{n_i} \qquad (4\text{-}42)$$

The parameter T_{PG} represents the type of gate and takes a value of 0 for the metal-gate MOST and a value of -1 or $+1$ for a MOST with the gate electrode made of polysilicon. The value -1 is used if the polysilicon is doped of the same type as the substrate, and $+1$ if it is of the opposite type. The current in the linear region is calculated through Eq. (4-38), where the corrective term of the channel-length modulation is also present. Therefore

$$I_{DS} = \frac{KP}{1 - \lambda V_{DS}}\frac{W}{L - 2X_{jl}}\left\{\left(V_{GS} - V_{FB} - 2\phi_p - \frac{V_{DS}}{2}\right)V_{DS}\right.$$

$$\left. - \tfrac{2}{3}\gamma[(V_{DS} - V_{BS} + 2\phi_p)^{1.5} - (-V_{BS} + 2\phi_p)^{1.5}]\right\} \qquad (4\text{-}43)$$

It should be noted that when V_{DS} is very small the values supplied by this model are very close to those of the LEVEL 1 model.

In the saturation region the current is

$$I_{DS} = I_{D,\text{sat}}\frac{1}{1 - \lambda V_{DS}} \qquad (4\text{-}44)$$

where $I_{D,\text{sat}}$ is calculated from Eq. (4-43) at $V_{DS} = V_{D,\text{sat}}$, and $V_{D,\text{sat}}$ is from Eq. (4-40). These equations give better results than the simple model, but they are still not sufficient for a good agreement with experimental data, even when short- and narrow-channel effects are absent.

The modifications that will improve precision for long and wide channels will be presented first; then the variations introduced in the model for short- and narrow-channel MOSTs will be presented.

4.3.3 How mobility varies according to the gate electric field

In the calculation of the current in the channel, which has led to the equations of the LEVEL 1 and 2 models, the mobility has been assumed constant with the applied voltage. This approximation is convenient in the calculation of the integral in Eq. (4-23), but the results do not agree with the experimental data; a reduction in mobility with an increase in the gate voltage is observed [4]. In order to simulate this effect, a variation of the parameter KP has been introduced in SPICE. This new expression is used in Eq. (4-43). Therefore

$$KP' = KP \left(\frac{\epsilon_s}{\epsilon_{ox}} \frac{U_c t_{ox}}{V_{GS} - V_{TH} - U_t V_{DS}} \right)^{U_e} \tag{4-45}$$

The value of the term in parentheses is limited to 1.

The parameter U_c is the gate-to-channel critical field; above this level, the mobility begins to decrease, while the term $(V_{GS} - V_{TH} - U_t V_{DS})/t_{ox}$ represents the average electric field perpendicular to the channel. The U_t parameter value is chosen between 0 and 0.5, and represents the contribution to the gate-to-channel electric field due to the drain voltage. The use of this formula allows a good agreement between SPICE and experimental data in absence of short- and narrow-channel effects, and in the strong inversion region only.

The effect of U_c and U_e parameters on the characteristics and on KP' is shown in Figs. 4-12 to 4-14.

4.3.4 Conduction in the weak inversion region

The basic model implemented in SPICE calculates the drift current when the surface potential is equal to or greater than $2\phi_p$. In reality, as explained in Appendix B, a concentration of electrons near the surface exists also for $V_{GS} < V_{TH}$, and therefore there is a current even when the surface is not in *strong* inversion. This current is due mainly to diffusion between the source and the channel.

The model [5] implemented in SPICE introduces an exponential dependence of the current I_{DS} and V_{GS} for the *weak* inversion region. This model has defined a voltage V_{on} which acts as a boundary between the regions of weak and strong inversion. This voltage is defined as

$$V_{on} = V_{TH} + \frac{nkT}{q} \tag{4-46}$$

where

$$n = 1 + \frac{qN_{FS}}{C'_{ox}} + \frac{C_d}{C'_{ox}} \tag{4-47}$$

C_d is the capacitance associated with the depleted region and is obtained from Eq. (4-37). Therefore

$$C_d = \frac{dQ_B}{dV_{BS}} = \frac{\gamma}{2\sqrt{2\phi_p - V_{BS}}} \tag{4-48}$$

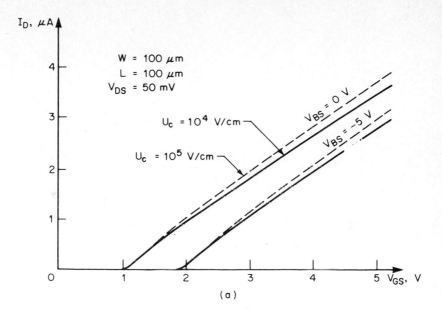

(a)

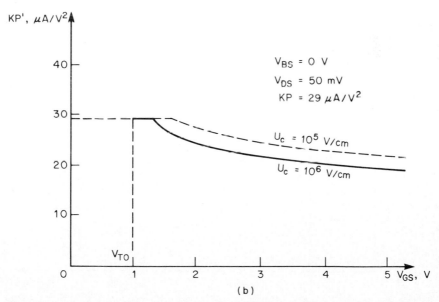

(b)

Figure 4-12 Variations of I_{DS} with U_c, for the LEVEL 2 model: (a) transfer characteristics and (b) the value of KP effective.

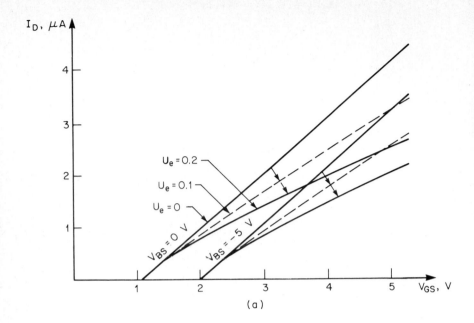

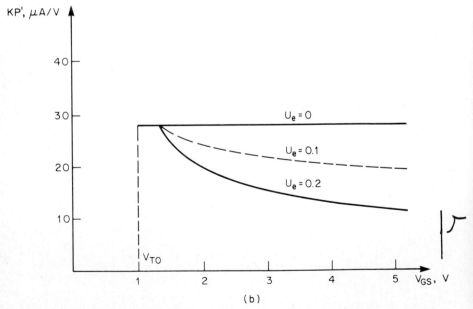

Figure 4-13 Variations of I_{DS} with U_c, for the LEVEL 2 model: (*a*) transfer characteristics and (*b*) the value of KP effective.

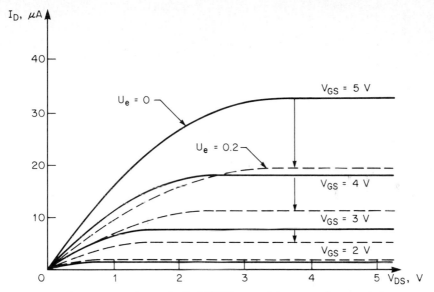

Figure 4-14 Variations of I_{DS} with U_e, LEVEL 2 model: output characteristics.

N_{FS} is a model parameter and is defined as the number of fast superficial states [6]. This parameter determines the slope of log I_{DS} vs. V_{GS} characteristics. The current in weak inversion is

$$I_{DS} = I_{on}e^{(V_{GS}-V_{on})(q/nkT)} \tag{4-49}$$

where I_{on} is the current in strong inversion for $V_{GS} = V_{on}$.

In Fig. 4-15, a simulation of the transfer characteristics of this model is shown; it is clear that the model introduces a discontinuity in the derivative for $V_{GS} = V_{on}$ and therefore the simulation of the transition region between strong and weak inversion is not very precise.

In fact, this model separates two operating regions in which only one mechanism of conduction, diffusion or drift, is considered; therefore, it is not possible, under this assumption, to correctly simulate the intermediate region (or *moderate inversion* region) in which the two conduction mechanisms make comparable contributions.

Another model for weak inversion has been proposed and implemented in the SGS version of SPICE [7]. This model is based on the separate calculation of the contributions of the drift and diffusion currents, which are then simply added together, without distinguishing between the regions of weak and strong inversion. Figure 4-16 shows the currents I_{drift} and I_{diff} described by this model.

The equation proposed for the diffusion current is

$$I_{diff} = \frac{qD_nX_{ch}W_{eff}}{L_n \tanh \dfrac{L_{eff}}{L_n}} \frac{n_sn_x}{n_s + n_x} (1 - e^{-qV_{DS}/kT}) \tag{4-50}$$

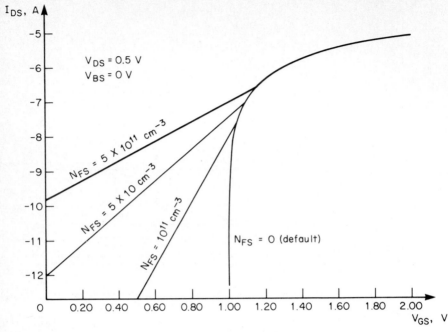

Figure 4-15 Variations of I_{DS} in the weak inversion region with N_{FS}, for the LEVEL 2 model: transfer characteristics.

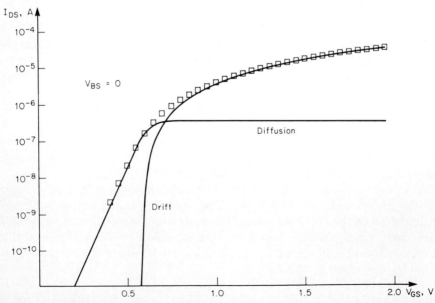

Figure 4-16 Drift and diffusion current components, computed according to the model [7].

where D_n and L_n are the diffusion constant and the diffusion length, X_{ch} is the average thickness of the channel, and n_x is an asymptotic value of the carrier concentration in the channel in the weak inversion condition. The expression for n_s is

$$n_s = N_{surf}\, \exp\left[(1 - \alpha V_{BS}) \frac{V_{GS} - V_{TH}}{(1 + qN_{FS}/C'_{ox})(kT/q)} \right] \qquad (4\text{-}51)$$

where α is an empirical parameter included to increase the slope of log I_{DS} vs. V_{GS}, when the body bias is greater than zero.

4.3.5   Variation of channel length in the saturation region

In Secs. 4.2.2 and 4.3.2, an empirical equation using the parameter λ has been derived for the LEVEL 1 and LEVEL 2 models to calculate the conductance in the saturation region (see Fig. 4-10).

The LEVEL 2 model also offers the possibility of using a physical model to calculate the channel length in saturation on the basis of the same corrective term in Eq. (4-44):

$$L' = L_{eff}(1 \quad \lambda V_{DS}) \qquad (4\text{-}52)$$

If the parameter λ is not specified in the .MODEL card, it is calculated using the equation [8]

$$\lambda = \frac{L_{eff} - L'}{L_{eff} V_{DS}} \qquad (4\text{-}53)$$

where

$$L_{eff} - L' = X_D \left[\frac{V_{DS} - V_{D,sat}}{4} + \sqrt{1 + \left(\frac{V_{DS} - V_{D,sat}}{4} \right)^2} \right] \qquad (4\text{-}54)$$

and

$$X_D = \sqrt{\frac{2\epsilon_s}{qN_A}} \qquad (4\text{-}55)$$

The value of the slope of $I_{D,sat}$ vs. V_{DS} calculated with this model, using in Eq. (4-55) the correct substrate doping N_A [the same used in Eq. (4-34) to calculate γ], is usually greater than the measured values (see Fig. 4-17). In this case it is possible to reduce the conductance G_{DS} in saturation by increasing N_A. However, if N_A is used to fit G_{DS}, it is not possible to use the same value to calculate $2\phi_p$ and γ through Eqs. (4-34) and (4-35); it is then necessary to specify these parameters in the .MODEL card.

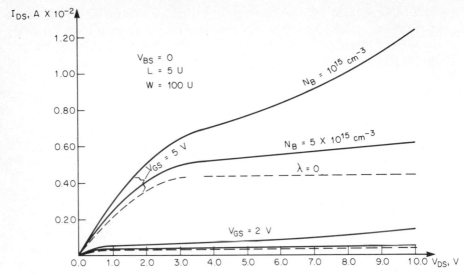

Figure 4-17 Variations of I_{DS} with $\lambda = 0$, for the LEVEL 2 model, or without λ in the .MODEL card, and with several values of N_A: output characteristics.

This model shows a correct dependence of the conductance in the saturation region on L_{eff}, while if the model with constant λ is used, G_{DS} does not vary with L_{eff} when $V_{DS} > V_{D,\text{sat}}$.

4.3.6 Effect of channel length on threshold voltage

Equation (4-30) has been obtained from a theory that does not take two-dimensional effects into account. Therefore, these equations do not include a link between the threshold voltage and the channel dimensions W and L. Experimental data, on the other hand, show that when the channel length is small enough to be comparable with the thickness of the depleted region, this relationship exists; it is negligible when L is large.

The models presented in the literature explain this phenomenon in two ways: a reduction in the charge Q_B due to the source and drain depleted regions [9–11] or an increase in the channel surface potential ϕ_s due to the effect of the voltage V_{DS} [12–15].

The model used in SPICE and described in this section is based on the first of the two hypotheses considered above [9].

The charge Q_B that contributes to the threshold voltage, according to this model, is shown in Fig. 4-18. This effect is introduced in Eq. (4-30) by modifying the value of γ as

$$\gamma' = \gamma \left[1 - \frac{X_j}{2L_{\text{eff}}} \left(\sqrt{1 + \frac{2W_S}{X_j}} + \sqrt{1 + \frac{2W_D}{X_j}} - 2 \right) \right] \quad (4\text{-}56)$$

where W_S and W_D are the widths of the depleted regions of source and drain, respectively, and are defined as

$$W_S = X_D \sqrt{2\phi_p - V_{BS}} \tag{4-57}$$

$$W_D = X_D \sqrt{2\phi_p - V_{BS} + V_{DS}} \tag{4-58}$$

If the real physical values of N_A and X_j are used in the .MODEL card, the model overestimates the reduction of V_{TH} for short L_{eff} with respect to long L_{eff}. The agreement with the experimental data can be improved by changing the values of X_j or N_A, but it is difficult to obtain satisfactory results over a large range of channel lengths (see Fig. 4-19). Moreover, this model does not adequately explain the dependence of V_{TH} on the drain voltage (see Fig. 4-20).

The effects of the parameter X_j on the transfer and output characteristics are shown in Figs. 4-21 and 4-22.

In conclusion, a correct analytical model of the two-dimensional effects on V_{TH} is in practice too cumbersome. On the other hand, it is possible to obtain a good model for the threshold voltage of a short-channel MOST with a two-dimensional numerical model, which takes into account the spatial configuration of the fields and potentials. This type of analysis is too time-consuming to be implemented in a circuit simulation program; it is used in other programs [16, 17], which can be used in the study of physical behavior of short-channel MOSTs.

4.3.7   Effect of speed limit of the carriers

The calculation of the saturation voltage using Eq. (4-40) is based on the hypothesis that the charge in the channel is zero when $x = L'$, i.e., near

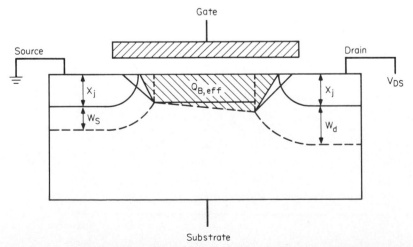

Figure 4-18 Geometrical calculation of the effective substrate charge Q_B with Yau's model [9].

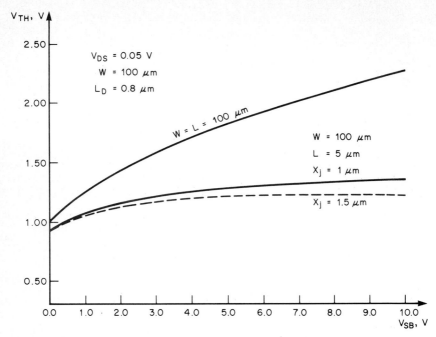

Figure 4-19 Body-effect simulation of short-channel MOSTs for the LEVEL 2 model.

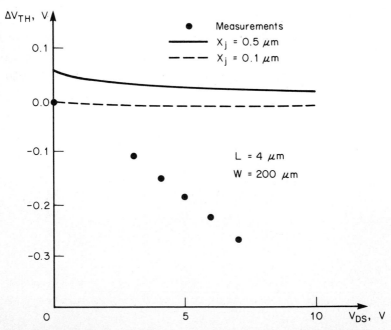

Figure 4-20 Simulation of V_{TH} vs. V_{DS} in a short-channel MOST device, LEVEL 2 model.

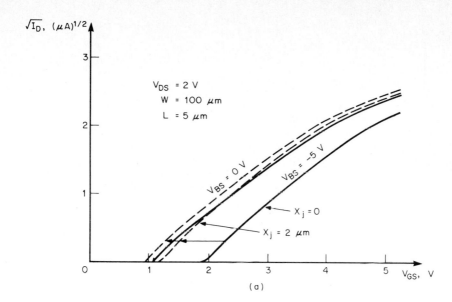

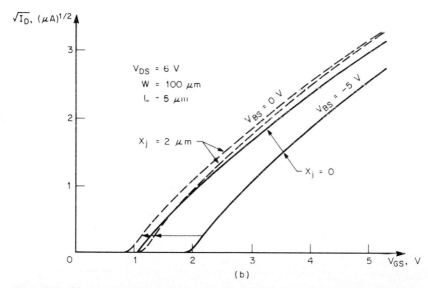

Figure 4-21 Variations of I_{DS} with X_j, for the LEVEL 2 model: transfer characteristics for (a) $V_{DS} = 2$ V and (b) $V_{DS} = 6$ V.

the drain. This hypothesis is false, because a minimum concentration greater than zero must exist in the channel because of the carriers that sustain the saturation current; this concentration depends on the speed at which the carriers are moving.

Moreover, the electric field between the drain and the channel end ($x = L'$) can be sufficiently high to drift the carriers at the speed limit, which

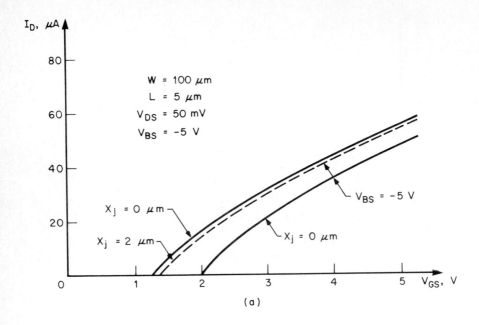

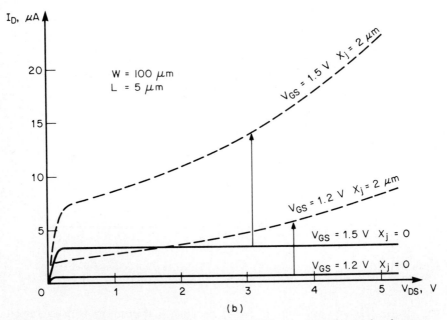

Figure 4-22 Variations of I_{DS} with X_j, for the LEVEL 2 model: (a) transfer characteristics and (b) output characteristics.

is the maximum speed allowed for the scattering effect within the crystal lattice. This value is indicated as v_{max}, and it is used to calculate the charge Q'_I for $V_{DS} = V_{D,sat}$. Therefore

$$Q'_I = \frac{I_{D,sat}}{W v_{max}} \tag{4-59}$$

The value of $V_{D,sat}$ is calculated from Eq. (4-59), but the solution of this equation is rather time-consuming. The effect of v_{max} on $V_{D,sat}$ is shown in Fig. 4-23.

If the parameter v_{max} is specified in the .MODEL card, SPICE does not use the model of Sec. 4.3.5 for the modulation of the channel length in the saturation region. In this case the model described by Baum and Beneking [19] is used:

$$L_{eff} - L' = X_D \sqrt{\left(\frac{X_D v_{max}}{2\mu}\right)^2 + V_{DS} - V_{D,sat}} - \frac{X_D^2 v_{max}}{2\mu} \tag{4-60}$$

In the calculation of X_D, the value of N_A can be modified through a coefficient N_{eff} that is used as a fitting parameter. Therefore

$$X_D = \sqrt{\frac{2\epsilon_s}{q N_A N_{eff}}} \tag{4-61}$$

Figure 4-24 shows the output characteristics simulated with this model.

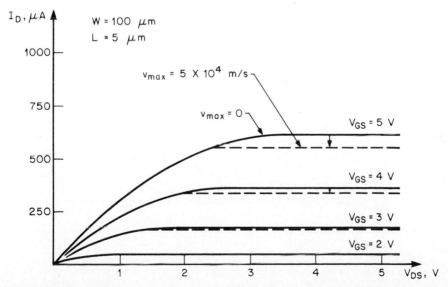

Figure 4-23 Variations of I_{DS} with v_{max}, for the LEVEL 2 model: output characteristics.

The model seen so far provides good results in the simulations of MOSTs with a minimum channel length of about 4 to 5 μm. This model, however, introduces a discontinuity in the derivative at the boundary between the saturation and the linear regions; this leads to a less precise calculation of the conductance and is sometimes the cause of difficulties in the convergence of the Newton-Raphson algorithm.

4.3.8   Effect of channel width on threshold voltage

In MOSTs with a small channel width W (for example, less than 5 or 6 μm), the value of the threshold voltage V_{TH} is greater than the one indicated by the previous theory. This effect is caused by the two-dimensional distribution of the charge Q_B at the edges of the channel (see Fig. 4-25) [10–20].

In the model used in SPICE, the thickness of the depleted region varies gradually from X_B under the channel to zero under the thick oxide. The empirical parameter δ allows the fitting of the experimental data.

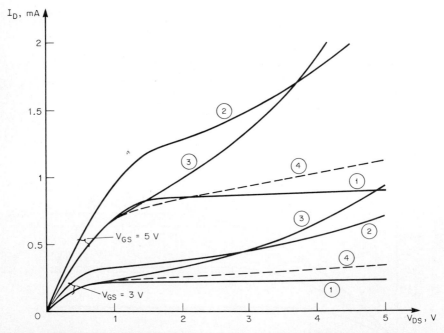

Figure 4-24 Comparison between the models for channel-length modulation in saturation: (1) empirical model (Sec. 4.2.2), $\lambda = 0.05$ V^{-1}; (2) flat-junction model (Sec. 4.3.5), $N_A = 10^{15}$ cm^{-3}; (3) Baum's model [19], $v_{max} = 5 \times 10^4$ m/s; (4) Baum's model, $v_{max} = 5 \times 10^4$ m/s, $N_A = 10^{15}$ cm^{-3}, $N_{eff} = 5$.

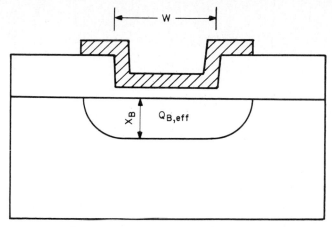

Figure 4-25 Geometrical calculation of the substrate charge Q_B reduced by the depleted regions at the channel edges.

Adding to the value of Q_B the contribution of the boundary regions (see Fig. 4-25), a modified equation for the threshold voltage can be obtained:

$$V_{TH} = V_{FB} + 2\phi_p + \gamma'\sqrt{2\phi_p - V_{BS}} + \frac{\epsilon_s \delta \pi}{4 C'_{ox} W}(2\phi_p - V_{BS}) \quad (4\text{-}62)$$

The calculation of I_{DS} is obviously affected by the increase in the threshold voltage (see Fig. 4-26).

This model is used in SPICE even if the parameter δ is not specified in the .MODEL card; to prevent the program from using this model, $\delta = 0$ must be specified.

4.3.9 Gate capacitance

SPICE uses a gate capacitance model similar to that proposed by Meyer [3]. In this simple model, the charge-storage effect is represented by three nonlinear two-terminal capacitors: C_{GB}, C_{GS}, and C_{GD}. The equations of this model are as follows (see Fig. 4-27).

Accumulation region

For $V_{GS} < V_{on} - 2\phi_p$:

$$C_{GB} = C_{ox} + C_{GBO}L_{eff} \quad (4\text{-}63)$$

$$C_{GS} = C_{GSO}W \quad (4\text{-}64)$$

$$C_{GD} = C_{GDO}W \quad (4\text{-}65)$$

Depletion region

For $V_{\text{on}} - 2\phi_p < V_{GS} < V_{\text{on}}$:

$$C_{GB} = C_{\text{ox}} \frac{V_{\text{on}} - V_{GS}}{2\phi_p} + C_{GBO}L_{\text{eff}} \tag{4-66}$$

$$C_{GS} = \tfrac{2}{3}C_{\text{ox}}\left(\frac{V_{\text{on}} - V_{GS}}{2\phi_p} + 1\right) + C_{GSO}W \tag{4-67}$$

$$C_{GD} = C_{GDO}W \tag{4-68}$$

Saturation region

For $V_{\text{on}} < V_{GS} < V_{\text{on}} + V_{DS}$:

$$C_{GB} = C_{GBO}L_{\text{eff}} \tag{4-69}$$

$$C_{GS} = \tfrac{2}{3}C_{\text{ox}} + C_{GSO}W \tag{4-70}$$

$$C_{GD} = C_{GDO}W \tag{4-71}$$

Linear region

For $V_{GS} > V_{\text{on}} + V_{DS}$:

$$C_{GB} = C_{GBO}L_{\text{eff}} \tag{4-72}$$

$$C_{GS} = C_{\text{ox}}\left\{1 - \left[\frac{V_{GS} - V_{DS} - V_{\text{on}}}{2(V_{GS} - V_{\text{on}}) - V_{DS}}\right]^2\right\} + C_{GSO}W \tag{4-73}$$

$$C_{GD} = C_{\text{ox}}\left\{1 - \left[\frac{V_{GS} - V_{\text{on}}}{2(V_{GS} - V_{\text{on}}) - V_{DS}}\right]^2\right\} + C_{GDO}W \tag{4-74}$$

$$C_{\text{ox}} = C'_{\text{ox}}WL_{\text{eff}}$$

The voltage V_{on} is calculated from Eq. (4-46) if the parameter N_{FS} is specified in the .MODEL card; otherwise V_{on} is equal to V_{TH}.

C_{GBO}, C_{GSO}, and C_{GDO} are the overlap capacitances among the gate electrode and the other terminals outside the channel region. The model introduced in SPICE differs from that developed by Meyer because discontinuity between the different operating regions causes nonconvergence in the Newton-Raphson algorithm. In particular, Eqs. (4-66) and (4-67) have been introduced as a link between the capacitance with the accumulated surface and with the depleted surface.

The transient current in the MOST terminals is the sum of the current in the dc model and the current of the capacitances.

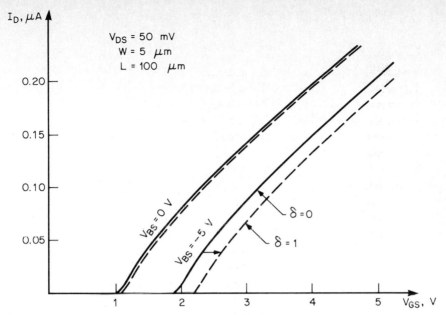

Figure 4-26 Variations of I_{DS} with σ, for the LEVEL 2 model: transfer characteristics.

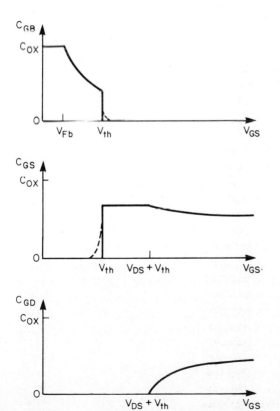

Figure 4-27 Meyer's model of the capacitance.

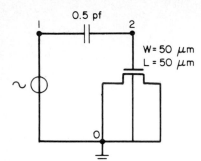

Figure 4-28 An example of a circuit critical for simulation of the capacitance with Meyer's model.

If Meyer's model is used, the average transient current in each time step $h = t_k - t_{k-1}$ and using trapezoidal integration is

$$i_S(t_k) = -i_S(t_{k-1}) + \frac{2}{h}[V_{GS}(t_k) - V_{GS}(t_{k-1})]\frac{C_{GS}(t_k) + C_{GS}(t_{k-1})}{2}$$

$$(4\text{-}75)$$

$$i_D(t_k) = -i_S(t_{k-1}) + \frac{2}{h}[V_{GD}(t_k) - V_{GD}(t_{k-1})]\frac{C_{GD}(t_k) + C_{GD}(t_{k-1})}{2}$$

$$(4\text{-}76)$$

$$i_B(t_k) = -i_S(t_{k-1}) + \frac{2}{h}[V_{GB}(t_k) - V_{GB}(t_{k-1})]\frac{C_{GB}(t_k) + C_{GB}(t_{k-1})}{2}$$

$$(4\text{-}77)$$

$$i_G = -i_S - i_D - i_B \qquad (4\text{-}78)$$

where $V(t_k)$ are the tentative voltages in the time step under analysis, and $V(t_{k-1})$ are the voltages in the last converged time step.

The transient currents calculated in Eqs. (4-75) to (4-77) are also functions of the time step, while in the quasi-static operation hypothesis, the charge is only a function of the voltages. This problem causes errors in the simulation of circuits where some nodes in the network cannot change their charge, as node 2 in Fig. 4-28 (see also Fig. 4-29).

A charge-control model for the transient behavior of the MOST has been introduced in SPICE to avoid the mistakes due to this model.

If the parameter X_{QC} is specified in the .MODEL card, SPICE selects a simplified version of the model proposed by Ward [21, 22] instead of that proposed by Meyer. Ward's model calculates analytically the charge in the gate and in the substrate. The charge in the channel is found by difference, and it is subdivided between the source and the drain with a formula determined by the parameter X_{QC}.

$$Q_{ch} = Q_D + Q_S = -(Q_G + Q_B) \qquad (4\text{-}79)$$

$$Q_D = X_{QC}Q_{ch} \qquad (4\text{-}80)$$

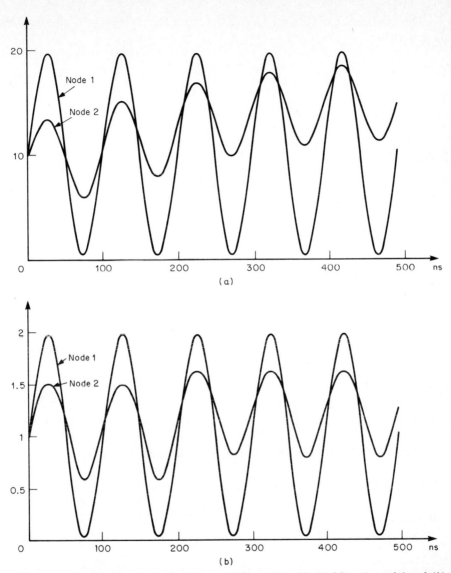

Figure 4-29 Simulation of the circuit seen in Fig. 4-28, with (*a*) Meyer's model and (*b*) Ward's model.

$$Q_S = (1 - X_{QC})Q_{ch} \qquad (4\text{-}81)$$

This approach, used by several authors [45], is not justified from the physical point of view, and sometimes causes errors and convergence problems (for example, when V_{DS} is changing sign).

The transient currents are calculated from the charges in the following way:

$$i_G = \frac{dQ_G}{dt} \tag{4-82}$$

$$i_B = \frac{dQ_B}{dt} \tag{4-83}$$

$$i_S + i_D = \frac{dQ_{ch}}{dt} \tag{4-84}$$

The capacitances between the MOST terminals are also requested for the definition of the companion model circuit; the model defines a matrix of capacitive terms, and each term is the derivative of the charge at the node x with respect to the voltage at the node y:

$$C_{xy} = \frac{dQ_x}{dV_y} \tag{4-85}$$

This matrix is not reciprocal; i.e.,

$$C_{xy} = \frac{dQ_x}{dV_y} \neq C_{yx} = \frac{dQ_y}{dV_x} \tag{4-86}$$

where x and y are two generic terminals of the MOST. In Fig. 4-30 the capacitance terms of the gate and of the substrate are shown. The same example of simulation that led to errors with the previous model is shown in Fig. 4-27b. With the new model, the effect of changing the average value of the voltage of the node has been eliminated.

It must be noted that some SPICE simulations cannot reach convergence when this model is used; moreover, solving the problem of finding an exact analytical model of the capacitance of the MOST requires several improvements [45, 47].

4.3.10   Junction capacitance

The capacitance of the diffused regions of the source and drain is simulated with the pn-junction model. A separate capacitance model is defined for the periphery of the junction; this is because the capacitance per unit area and its dependence on the reverse-bias voltage in the boundary regions of the diffusion are different from those associated with the flat junction. Moreover, below the thick oxide region, the doping is usually increased by ion implantation [31].

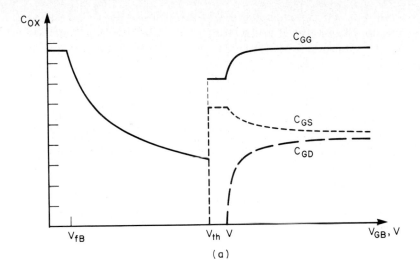

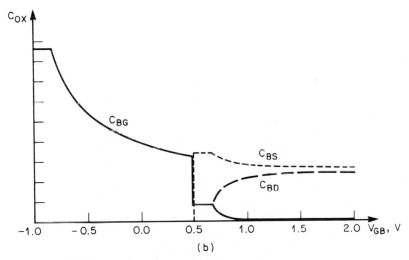

Figure 4-30 Ward's model of the capacitance.

The total capacitance of a diffused region is therefore calculated from the sum of an area and a perimeter capacitance:

$$C_{BS} = \frac{C_j A_S}{(1 - V_{BS}/\phi_j)^{M_j}} + \frac{C_{jsw} P_S}{(1 - V_{BS}/\phi_j)^{M_{jsw}}} \qquad (4\text{-}87)$$

$$C_{BD} = \frac{C_j A_D}{(1 - V_{BD}/\phi_j)^{M_j}} + \frac{C_{jsw} P_D}{(1 - V_{BD}/\phi_j)^{M_{jsw}}} \qquad (4\text{-}88)$$

where C_j and C_{jsw} are the capacitances at zero-bias voltage, for square meter of area and for meter of perimeter; and ϕ_j is the junction potential that can be specified in the .MODEL card or calculated from the physical parameters as

$$\phi_j = \frac{E_g}{2} + \frac{kT}{q} \ln \frac{N_A}{n_i} \qquad (4\text{-}89)$$

The exponential factors have the following default values: $M_j = 0.5$, as for the step-junction approximation, and $M_{jsw} = 0.33$, as for the linear gradient approximation. The variations of the junction capacitance with the parameters C_j and M_j are shown in Fig. 4-31.

The junction capacitances are very important not only for the correct description of the circuit to be simulated but also because the convergence of the algorithm is easier if all the nodes are connected to ground with a capacitance. For this reason, it is useful to define the areas of the source and drain diffusions even if the layout is not done when the designer is writing down the input for SPICE; the use of the default values DEFAD and DEFAS in the .OPTIONS card to avoid dangerous floating nodes is recommended.

4.4   Equations for the LEVEL 3 Model

The LEVEL 3 model has been developed to simulate short-channel MOSTs; it simulates quite precisely the characteristics of MOSTs which

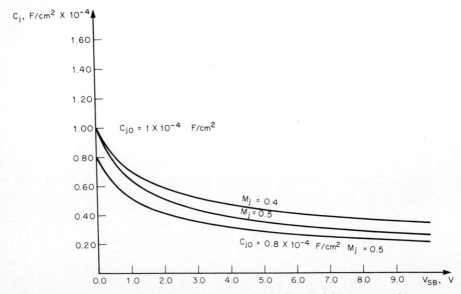

Figure 4-31   Variations of the junction capacitance with ϕ_j and M_j in reverse bias.

have a channel length up to 2 μm. The basic equations have been proposed by Dang [23]. The model for the weak inversion region is the same as that for the LEVEL 2 model.

4.4.1 The basic LEVEL 3 model

The equations for the LEVEL 3 model are formulated in the same way as for the LEVEL 2 model; however, a simplification of the current equation in the linear region has been obtained with a Taylor series expansion of Eq. (4-43). This approximation allows the development of more manageable basic equations than the LEVEL 2 model; the short-channel effects are introduced in the calculation of threshold voltage and mobility.

Many of the equations used in the model are empirical. The purpose is both to improve the precision of the model and to limit the complexity of the calculations and the resulting time needed to carry out the program. The basic equations are very simple. The current in the linear region is

$$I_{DS} = \beta \left(V_{GS} - V_{TH} - \frac{1 + F_B}{2} V_{DS} \right) V_{DS} \qquad (4\text{-}90)$$

where
$$F_B = \frac{\gamma F_s}{2\sqrt{2\phi_p - V_{BS}}} + F_n \qquad (4\text{-}91)$$

The effects of a short channel influence the parameters V_{TH}, F_s, and β, while the narrow-channel effects influence the term F_n.

The dependence of mobility on the gate electric field is simulated with a simpler equation than that used for the LEVEL 2 model, without any appreciable loss in precision [24, 25]. Thus

$$\mu_s = \frac{\mu}{1 + \theta(V_{GS} - V_{TH})} \qquad (4.92)$$

In fact, as the oxide thickness has been greatly reduced with modern processes, it is practically impossible to find a region where the surface electric field E_x is less than the critical field E_c, as hypothesized in Eq. (4-45).

The effects of the parameters KP (or μ) and γ (or N_A) on the characteristics simulated with this model are analogous to those seen for the other models. In Fig. 4-32 the effect of parameter θ is shown.

One suggestion to improve the model is to use V_{T0} instead of V_{TH} in the denominator of Eq. (4-92); in this way it is possible to introduce a mobility variation with V_{BS} very close to its actual measurements without adding any new parameter in the .MODEL card (see Fig. 4-33).

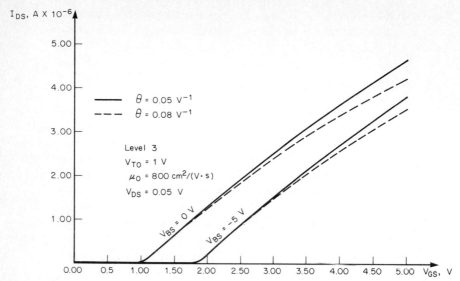

Figure 4-32 Variations of I_{DS} in the linear region with θ, LEVEL 3 model: transfer characteristics.

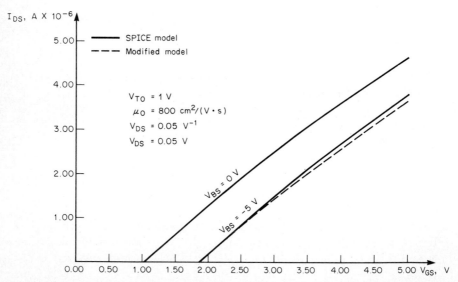

Figure 4-33 Variations of I_{DS} in the linear region with θ, modified model: transfer characteristics.

4.4.2 Effect of channel length on threshold voltage

The more complex parts of the equations of this model are employed in calculating the threshold voltage V_{TH}. The equations have been formulated with hypotheses similar to those used for the LEVEL 2 model (see

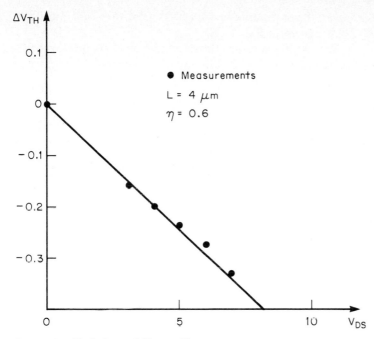

Figure 4-34   Variations of V_{TH} vs. V_{DS}.

Sec. 4.3.6); the effect of the different thickness of the depleted regions in the cylindrical junction region and under the channel has also been introduced in the calculation of the effective charge Q_B. The equation proposed by Dang [23] for the threshold voltage is

$$V_{TH} = V_{FB} + 2\phi_p - \sigma V_{DS} + \gamma F_s \sqrt{2\phi_p - V_{BS}}$$
$$+ F_n(2\phi_p - V_{BS}) \quad (4\text{-}93)$$

The parameter σ expresses empirically the dependence of the threshold voltage on V_{DS}, which otherwise would not be included in the model. In fact it is not possible to explain this effect with the reduction of Q_B caused by the source and drain depleted regions. Simulations on short-channel MOSTs carried out with two-dimensional programs have shown that the inversion potential $2\phi_p$ is reached at a smaller gate voltage than expected from Eq. (4-6) because of the presence of a strong drain potential, which causes a lower threshold voltage [13, 14]. This effect causes a linear variation of V_{TH} vs. V_{DS} (see Fig. 4-34) [27]. These experimental results allow the introduction of the term σV_{DS} seen in Eq. (4-93). The relationship between σ and L_{eff} is also empirical [28]:

$$\sigma = \eta \frac{8.15 \times 10^{-22}}{C'_{\text{ox}} L_{\text{eff}}^3} \quad (4\text{-}94)$$

The parameter η is an input parameter, and its typical value is 1; its effect on the current I_{DS} is shown in Fig. 4-35.

The term F_B in Eq. (4-90) expresses the dependence of Q_B on the three-dimensional geometry of the MOST. The term F_s expresses the effect of the short channel and is calculated as

$$F_s = 1 - \frac{X_j}{L_{eff}} \left(\frac{X_{jl} + W_c}{X_j} \sqrt{1 - \frac{W_p}{X_j + W_p}} - \frac{X_{jl}}{X_j} \right) \qquad (4\text{-}95)$$

W_p is the thickness of the depleted region on the flat source junction (see Fig. 4-36); i.e.,

$$W_p = X_D \sqrt{\phi_j - V_{BS}} \qquad (4\text{-}96)$$

where
$$X_D = \sqrt{\frac{2\epsilon_s}{qN_A}} \qquad (4\text{-}97)$$

W_c is the thickness of the depleted cylindrical region of the source-substrate junction; an empirical formula has been used again in this case, i.e.,

$$\frac{W_c}{X_j} = 0.0831353 + 0.8013929 \frac{W_p}{X_j} - 0.0111077 \left(\frac{W_p}{X_j} \right)^2 \qquad (4\text{-}98)$$

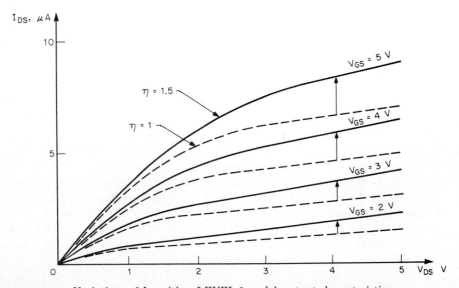

Figure 4-35   Variations of I_{DS} with η, LEVEL 3 model: output characteristics.

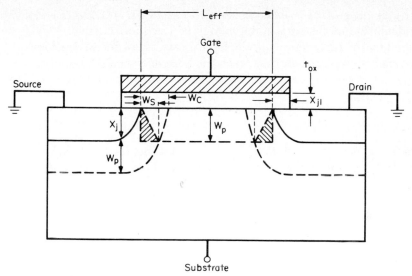

Figure 4-36 Geometrical calculation of the substrate charge Q_B reduced by the depleted regions at the channel edges.

The effects of the parameter X_j on the value of the threshold voltage and on the current are shown in Figs. 4-37 and 4-38.

4.4.3 Effect of channel width on threshold voltage

The term F_n in Eq. (4-93) introduces the narrow-channel effect in the model and is calculated with the same equation used in the LEVEL 2 model, i.e.,

$$F_n = \frac{\epsilon_s \delta \pi}{4 C'_{ox} W} \qquad (4\text{-}99)$$

4.4.4 Effect of channel length on mobility

The solution chosen for the LEVEL 3 model is simpler and more accurate than that of the LEVEL 2 model. This model includes a decrease in the effective mobility with the average electrical field between the source and the drain [24]. Therefore

$$\mu_{\text{eff}} = \frac{\mu_s}{1 + \mu_s V_{DS}/v_{\text{max}} L_{\text{eff}}} \qquad (4\text{-}100)$$

where μ_s is the surface mobility calculated from Eq. (4-91). In this model, the calculation of the saturation voltage of the short-channel MOST is modified by the parameter $\upsilon_{\max}$; the hypothesis is that the saturation voltage is reached when the carriers reach the limit speed $\upsilon_{\max}$. The saturation voltage can be expressed with an equation simpler than that of the LEVEL 2 model, i.e.,

$$V_{D,\text{sat}} = V_a + V_b - \sqrt{V_a^2 + V_b^2} \qquad (4\text{-}101)$$

where V_a is the saturation voltage if $\upsilon_{\max}$ is not included in the .MODEL card, and V_b modifies $V_{D,\text{sat}}$ if $\upsilon_{\max}$ is included. They are expressed as

$$V_a = \frac{V_{GS} - V_{TH}}{1 + F_B} \qquad (4\text{-}102)$$

$$V_b = \frac{\upsilon_{\max} L_{\text{eff}}}{\mu_s} \qquad (4\text{-}103)$$

The use of Eqs. (4-100) and (4-101) together with the model described in the following section allows the continuity of the derivative of the current I_{DS} with respect to V_{DS} at $V_{DS} = V_{D,\text{sat}}$ (see Fig. 4-39).

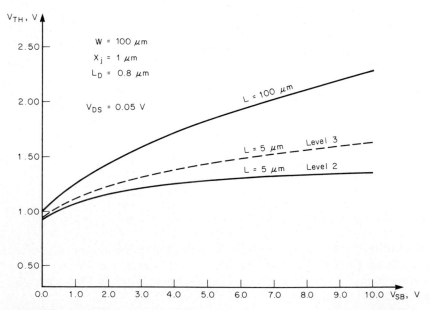

Figure 4-37 Comparison between the models of the body effect with the LEVEL 2 and LEVEL 3 models.

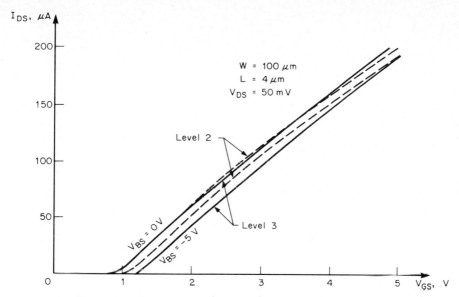

Figure 4-38 Comparison between the transfer characteristics simulated with (*a*) the LEVEL 2 model and (*b*) the LEVEL 3 model. The common parameters are $V_{T0} = 1$ V, $X_j = 1$ μm, $X_{jl} = 0.8$ μm; the parameters of the LEVEL 2 model are $\mu = 800$ cm/(V · s), $U_c = 5 \times 10^4$ V/cm, and $U_e = 0.15$; the parameters of the LEVEL 3 model are $\mu = 850$ cm(V · s), $\theta = 0.04$ V^{-1}.

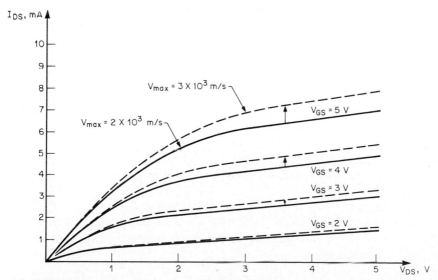

Figure 4-39 Variations of I_{DS} with v_{max}, LEVEL 3 model: output characteristics.

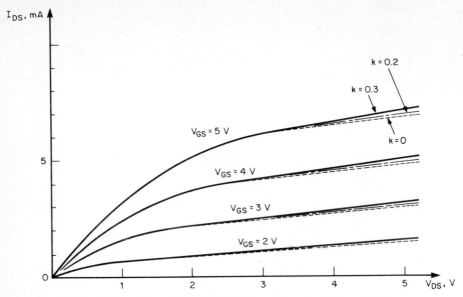

Figure 4-40 Variations of I_{DS} with K, LEVEL 3 model: output characteristics.

4.4.5 Variation of channel length in the saturation region

The equation of the LEVEL 3 model simulates the modulation of the channel length in saturation [19] as

$$L_{\text{eff}} - L' = \sqrt{\left(\frac{E_p X_D^2}{2}\right)^2 + K X_D^2 (V_{DS} - V_{D,\text{sat}})} - \frac{E_p X_D^2}{2} \quad (4\text{-}104)$$

where
$$E_p = \frac{I_{D,\text{sat}}}{G_{D,\text{sat}} L_{\text{eff}}} \quad (4\text{-}105)$$

$I_{D,\text{sat}}$ and $G_{D,\text{sat}}$ represent the current and the conductance for $V_{DS} = V_{D,\text{sat}}$, calculated with the equation derived for the linear region; K is an empirical fitting parameter, with a typical value of 1, whose effect is shown in Fig. 4-40.

With Eqs. (4-104) and (4-105), the derivative of I_{DS} vs. V_{DS} for $V_{DS} = V_{D,\text{sat}}$ in the saturation region (G_{DS}) should be equal to $G_{D,\text{sat}}$ calculated in the linear region; to obtain this continuity it is necessary to correct Eq. (4-105) according to

$$E_p = \frac{I_{D,\text{sat}} K}{G_{D,\text{sat}} L_{\text{eff}}} \quad (4\text{-}106)$$

4.4.6 Model of the gate capacitance

According to the user's guide for SPICE2 [6], it is possible with the
LEVEL 3 model to choose between the simple Meyer's model and the
charge-controlled Ward's model. Moreover, the equations of the LEVEL
3 model allow simpler equations also for Q_G and Q_B. Thus

$$Q_G = C'_{ox}WL_{eff}\left(V_{GS} - V_{BS} - 2\phi_p \right.$$

$$\left. + \sigma V_{DS} - \frac{V_{DS}}{2} + \frac{1 + F_B}{12F_i} V^2_{DS} \right) \qquad (4\text{-}107)$$

where
$$F_i = V_{GS} - V_{TH} - \frac{1 + F_B}{2} V_{DS} \qquad (4\text{-}108)$$

The charge in the substrate is

$$Q_B = C'_{ox}WL_{eff}\left[\gamma F_s \sqrt{2\phi_p - V_{BS}} + F_n(2\phi_p - V_{BS}) \right.$$

$$\left. + \frac{F_B}{2} V_{DS} \right]\left[-\frac{F_B(1 + F_B)}{12F_i} V^2_{DS} \right] \qquad (4\text{-}109)$$

The charge in the channel is

$$Q_{ch} = -Q_G - Q_B \qquad (4\text{-}110)$$

4.5 Comments on the Three Models

At this point, if someone needed to choose the best model for a circuit
simulation, it would be useful to reexamine the main differences between
the three models.

Usually the LEVEL 1 model is not sufficiently precise because the theory is too approximated and the number of fitting parameters too small;
its usefulness is in a quick and rough estimate of circuit performances.

The LEVEL 2 model can be used with differing complexity by adding
the parameters relating to the effects needed to simulate with this model.
However, if all the parameters are used, i.e., the greatest possible complexity is obtained, this model requires a great amount of CPU time for
the calculations, and it often causes trouble with the convergence of the
Newton-Raphson algorithm.

A comparison between the LEVEL 2 and LEVEL 3 models is interesting. In Fig. 4-41, the LEVEL 3 model shows an average quadratic error
just less than that obtained with LEVEL 2, but the CPU time for every
model evaluation is 25 percent less and the iteration number is lower.

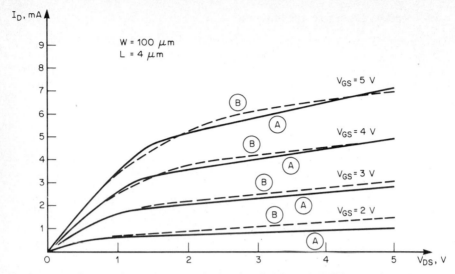

Figure 4-41 Output characteristics calculated with (*a*) the LEVEL 2 model and (*b*) the LEVEL 3 model. The parameters are the same as those specified in Fig. 4-38, with v_{max} = 5 × 10^4 m/s and K = 0.1 for the LEVEL 3 model.

In conclusion, the only disadvantage of the LEVEL 3 model is the complexity in the calculation of some of its parameters (see Chap. 6). It is best to use the LEVEL 3 model if possible, and the LEVEL 1 model if great precision is not required.

4.6   The Effect of Series Resistances

The presence of series resistances at the source and drain regions causes a degradation of the electrical characteristics of the MOST, because the effective voltages at the source and the drain are less than those applied at the external terminals (see Fig. 4-42).

In SPICE it is possible to insert in the equivalent circuit of the MOST two resistances R_S and R_D in series to the source and the drain. These resistances are specified in the .MODEL card, and they are equal for all MOSTs with the same model parameters. It is also possible to specify the resistance for the square of the diffused regions (R_{sh}); in this case the value of R_D and R_S is calculated for every MOST using the parameters N_{RD} and N_{RS}, which are the number of squares of diffused region in series to the drain and the source.

$$R_D = R_{sh}N_{RD} \qquad (4\text{-}111)$$

$$R_S = R_{sh}N_{RS} \qquad (4\text{-}112)$$

This method makes it possible to specify different values of series resistance for each MOST, and consequently it gives a better description of the circuit.

These models, however, do not make it possible to describe adequately the resistance contribution of the diffused region near the channel, where the resistance is linearly proportional to W^{-1} [29].

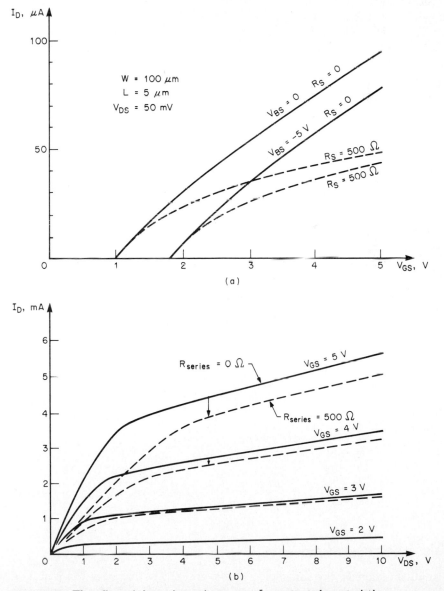

Figure 4-42   The effect of the series resistance on I_{DS}: output characteristics.

4.7   Small-Signal Model

SPICE uses differential parameters both for the small-signal equivalent circuit for ac analysis (Fig. 4-43) and for the linearized equivalent circuit for the Newton-Raphson method (Fig. 4-44). The linearized model includes capacitances, conductances, and voltage-controlled sources.

The precision required from the model in these two cases is not the same. In the *transient* and *dc analyses,* small errors in the linearized model do not influence the solution of the Newton-Raphson method; they influence only the speed of the convergence. In the *ac analysis,* a precise equivalent circuit is requested to simulate the behavior at low and high frequency. The models in SPICE have been developed for the transient analysis of digital circuits, and the precision of the parameters g_m, g_{mb}, and g_D is insufficient for the ac analysis, both for low and high frequency [46].

The best available solution is the old LEVEL 1 model, because the simple equations allow us to calculate easily the linearized circuit near the bias point of the nonlinear circuit.

For $V_{GS} > V_{TH}$ and $V_{DS} < V_{D,\text{sat}}$, it follows from the equations of the LEVEL 1 model that

$$g_D = \frac{dI_{DS}}{dV_{DS}} = \beta(V_{GS} - V_{TH} - V_{DS}) \qquad (4\text{-}113)$$

$$g_m = \frac{dI_{DS}}{dV_{GS}} = \beta V_{DS} \qquad (4\text{-}114)$$

$$g_{mb} = \frac{dI_{DS}}{dV_{BS}} = -\frac{g_m}{2\sqrt{2\phi_p - V_{BS}}} \qquad (4\text{-}115)$$

For $V_{DS} > V_{D,\text{sat}}$,

$$g_D = \frac{dI_{DS}}{dV_{DS}} = \lambda I_{DS} \qquad (4\text{-}116)$$

$$g_m = \frac{dI_{DS}}{dV_{GS}} = \frac{2I_{DS}}{V_{GS} - V_{TH}} \qquad (4\text{-}117)$$

$$g_{mb} = \frac{dI_{DS}}{dV_{BS}} = -\frac{g_m}{2\sqrt{2\phi_p - V_{BS}}} \qquad (4\text{-}118)$$

4.8   The Effect of Temperature

In all the models, the threshold voltage and the potential of the junctions $2\phi_p$ and ϕ_j depend on temperature, according to Eq. (4-9); the parameter

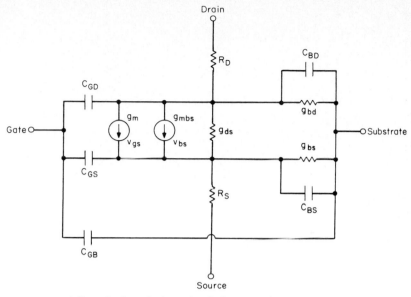

Figure 4-43 A linearized equivalent circuit for ac analysis.

$$I_0 = I_D - g_m V_{GS} - g_g V_{DS} - g_{mb} V_{BS}$$

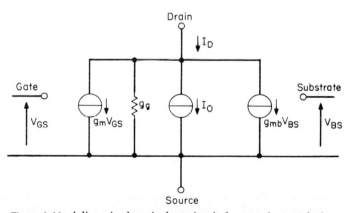

Figure 4-44 A linearized equivalent circuit for transient analysis.

n_i, the number of intrinsic carriers, and the energy gap E_g also depend on the temperature, according to the following empirical equations [30]:

$$n_i = 1.45 \times 10^{10} \left(\frac{T}{300}\right)^{1.5} e^{(q/2k)(1.16/300 - E_g/T)} \qquad (4\text{-}119)$$

$$E_g = 1.16 - \frac{7.02 \times 10^{-4} T^2}{T + 1108} \qquad (4\text{-}120)$$

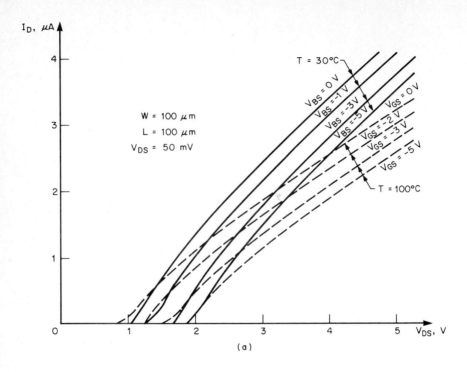

(a)

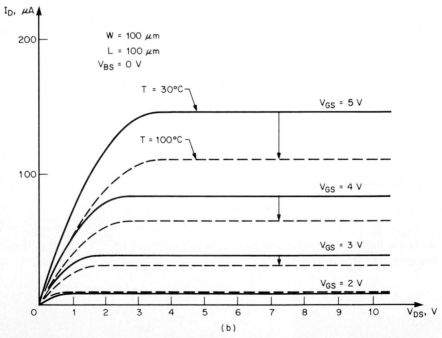

(b)

Figure 4-45 Temperature variation of I_{DS}: (a) transfer characteristics and (b) output characteristics.

The mobility, too, and therefore KP, varies through the following empirical equation:

$$\mu(T) = \mu(300)\left(\frac{300}{T}\right)^{1.5}$$

(4-121)

The values of n_i and E_g are used in Eqs. (4-9) to calculate $2\phi_p$, and then in Eqs. (4-11) to calculate V_{T0}; also the parameters of the diode model for the source and drain junctions (I_s, J_s, C_j, C_{jsw}, and ϕ_j) are affected by the temperature. The variations of the current I_{DS} (see Fig. 4-45), of the threshold voltage (see Fig. 4-46), and of the mobility (see Fig. 4-47) vs. the temperature are shown in the indicated figures. The current in weak inversion is also clearly affected by the temperature (see Fig. 4-48).

4.9 Problems to Be Solved

4.9.1 Effect of ion implantation

All the models for the MOST implemented in SPICE have been obtained with the hypothesis that the doping in the silicon is uniform. This is not true in most modern processes, because the surface doping concentration is increased by ion implantation [31]; the implanted doping concentration determines the value of the threshold voltage. If the implanted region is completely depleted, the value of the threshold voltage is simply increased

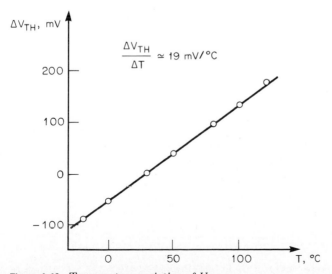

Figure 4-46 Temperature variation of V_{TH}.

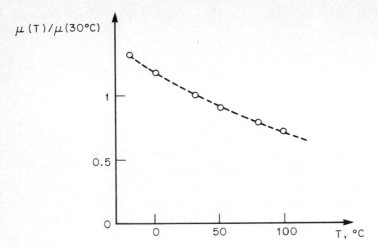

Figure 4-47 Temperature variation of μ.

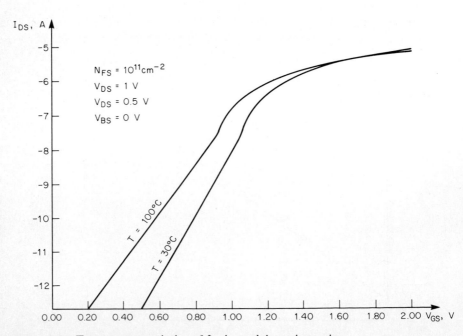

Figure 4-48 Temperature variation of I_{DS} in weak inversion region.

by an amount V_{sh}, where

$$V_{sh} = \frac{qN_{\text{imp}}}{C'_{\text{ox}}} \tag{4-122}$$

where N_{imp} is the number of implanted ions per square centimeter. This behavior can be simulated even with the present models, but SPICE calculates a voltage V_{FB} that is in reality $V_{FB} + V_{sh}$; this means that the capacitance C_{GB} for $V_{FB} < V_{GS} < V_{TH}$ (depleted surface) is not calculated correctly.

If the thickness of the depleted region X_B is lower than the thickness of the implanted region for low values of the body bias V_{BS}, the curve of V_{TH} vs. V_{BS} shows two different values of the body-effect parameter γ (Fig. 4-49); the only possible solution with the present models is to find an average value of γ in the range of V_{BS}.

Analytical models of the body effect with no uniform doping, applicable to circuit simulation, can be found in the literature; these models are based on approximations of box profiles [32] or Gaussian profiles [33]. It would be appropriate to introduce one of these models in LEVEL 3. The short-channel effects are also influenced by ion implantation [23].

4.9.2 Model of the depletion MOST

A model for the depletion MOST does not exist in the original version of SPICE, but it has been implemented in other versions [48].

The simulation of the depletion MOST with a model of the enhancement MOST is possible only for a limited range of V_{GS}, because the transfer characteristics are rather different. The results in Fig. 4-50 show that it is possible to simulate a depletion MOST used as a load device with the enhancement MOST model, because in this case V_{GS} is always equal to zero. However, this result cannot be obtained if V_{GS} is not constant (see Fig. 4-51).

Moreover, the simulated behavior of the gate capacitance is certainly incorrect. Several models in the literature can be applied to circuit simulation [35–38, 48].

4.9.3 Other problems

Some phenomena in the behavior of the MOST are not simulated by SPICE models. For example, models for breakdown and punch-through are missing; however, these phenomena occur outside the normal operating region of MOST's. It would, on the other hand, be more important to introduce a model for the carrier generation by ionization, which causes current leakage in the substrate [34]; this leakage is present even at rela-

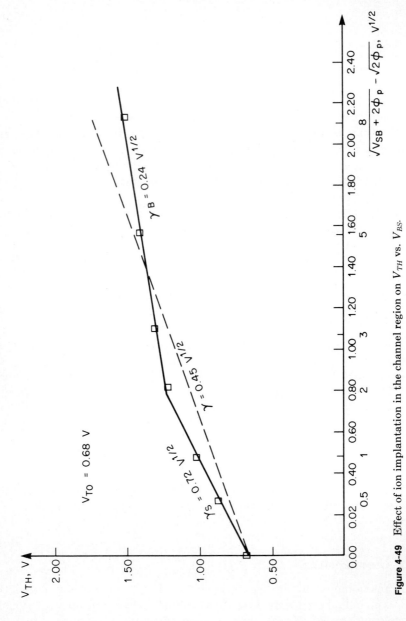

Figure 4-49 Effect of ion implantation in the channel region on V_{TH} vs. V_{BS}.

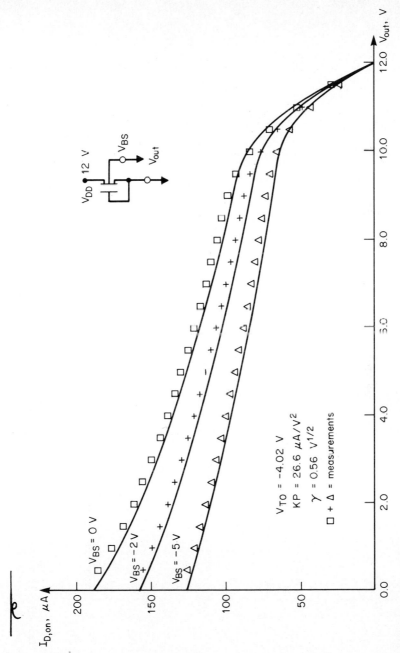

Figure 4-50 Load characteristics of a depletion MCST.

$V_{TO} = -4.02$ V
$KP = 26.6$ $\mu A/V^2$
$\gamma = 0.56$ V$^{1/2}$
□ + △ = measurements

$V_{BS} = 0$ V
$V_{BS} = -2$ V
$V_{BS} = -5$ V

V_{DD} 12 V

V_{BS}

V_{out}

$I_{D,on}$, μA

V_{out}, V

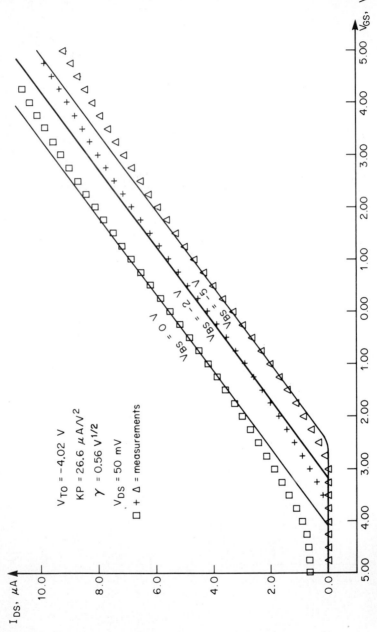

Figure 4-51 Transfer characteristics of a depletion MOST.

$V_{TO} = -4.02$ V

$KP = 26.6$ $\mu A/V^2$

$\gamma = 0.56$ $V^{1/2}$

$V_{DS} = 50$ mV

$\square$ + $\triangle$ = measurements

$V_{BS} = 0$ V

$V_{BS} = -2$ V

$V_{BS} = -5$ V

I_{DS}, μA

V_{GS}, V

tively low voltage. Lastly, models for the tunneling effects in the gate oxide, due to the presence of hot electrons, are also missing.

4.9.4 Possible future developments

In the last few years there have been many attempts to develop complete physical models of the MOST [39–42]. The purpose is to find an explicit equation of the current that is valid for all the working regions; in this way, a lot of the restrictive hypotheses made in Sec. 4.2.1 to obtain simple equations could be superseded.

The equations proposed are very complex, but they make it possible to avoid tests to identify the working region in the calculation of the model. The absence of discontinuity in the function of I_{DS} and its derivative is an advantage in the numerical solution of the nonlinear equations that represent the circuits.

Other models have been proposed where more simple and compact equations are preferred, with a lot of empirical parameters which are extractable from measurements in a fully automated fashion [47].

One possible solution for reducing the calculation time without reducing the model's precision is the use of look-up tables in the model evaluation [43].

Lastly, a complete physical model would make it possible to develop a precise and complete model to simulate the dynamic behavior of the MOST [44].

REFERENCES

1. A. S. Grove, *Physics and Technology of Semiconductor Devices,* Wiley, New York, 1977.
2. H. Shichman and D. A. Hodges, Modeling and Simulation of Insulated-Gate Field-Effect Transistor Switching Circuits, *IEEE J. Solid-State Circuits,* **SC-3,** 1968.
3. J. E. Meyer, MOS Models and Circuit Simulation, *RCA Rev.,* **32,** 1971.
4. D. Frohman-Bentchkowsky and A. S. Grove, On the Effect of Mobility Variation on MOS Device Characteristics, *Proc. IEEE,* **56,** 1968.
5. R. M. Swanson and J. D. Meindl, Ion-Implanted Complementary MOS Transistor in Low-Voltage Circuits, *IEEE J. Solid-State Circuits,* **14**(2), 1979.
6. A. Vladimirescu, A. R. Newton, and D. O. Pederson, *SPICE2 Version 2F.1 User's Guide,* University of California, Berkeley, 1979.
7. P. Antognetti, D. Caviglia, and E. Profumo, CAD Model for Threshold and Sub-threshold Conduction in MOSFETs, *IEEE J. Solid-State Circuits,* **17,** 1982.
8. D. Frohman-Bentchkowsky and A. S. Grove, Conductance of MOS Transistors in Saturation, *IEEE Trans. Electron Devices,* **ED-16,** 1969.
9. L. D. Yau, A Simple Theory to Predict the Threshold Voltage of Short-Channel IGFETs, *Solid-State Electron.,* **17,** 1974.
10. L. A. Akers, An Analytical Expression for the Threshold Voltage of a Small Geometry MOSFET, *Solid-State Electron.,* **24,** 1981.
11. O. Jantsch, A Geometrical Model of the Threshold of Short and Narrow Channel MOSFETs, *Solid-State Electron.,* **25**(1), 1982.

12. W. R. Bandy and D. P. Kokalis, A Simple Approach for Accurately Modeling the Threshold Voltage of Short-Channel MOSTs, *Solid-State Electron.*, **20,** 1977.

13. R. R. Troutman and A. G. Fortino, Simple Model for Threshold Voltage in a Short-Channel IGFET, *IEEE Trans. Electron Devices,* **10,** 1977.

14. Y. Omura and K. Ohwada, Threshold Voltage Theory for a Short-Channel MOSFET Using a Surface-Potential Distribution Model, *Solid-State Electron.,* **22,** 1979.

15. Y. Ohno, Short-Channel MOSFET V_{TH}-V_{DS} Characteristics Model Based on a Point Charge and Its Mirror Images, *IEEE Trans. Electron Devices,* **2,** 1982.

16. S. Selberherr, A. Schutz, and H. W. Potzl, MINIMOS—A Two-Dimensional MOS Transistor Analyzer, *IEEE Trans. Electron Devices,* **ED-27,** 1980.

17. J. A. Greenfield and R. W. Dutton, Nonplanar VLSI Device Analysis Using the Solution of Poisson's Equation, *IEEE Trans. Electron Devices,* **ED-27,** 1980.

18. E. D. Sun, A Short-Channel MOSFET Model for CAD, *Proceedings of the 12th Annual Asilomar Conference on Circuits, Systems and Computers,* 1978.

19. G. Baum and H. Beneking, Drift Velocity Saturation in MOS Transistors, *IEEE Trans. Electron Devices,* **17,** 1970.

20. K. O. Jeppson, Influence of the Channel Width on the Threshold Voltage Modulation in MOSFETs, *Electron. Lett.,* **11**(14), 1975.

21. D. E. Ward and R. W. Dutton, A Charge-Oriented Model for MOS Transistors Capacitances, *IEEE J. Solid-State Circuits,* **SC-13,** 1978.

22. S. Y. Oh, D. E. Ward, and R. W. Dutton, Transient Analysis of MOS Transistors, *IEEE Trans. Electron Devices,* **ED-27**(8), 1980.

23. L. M. Dang, A Simple Current Model for Short Channel IGFET and Its Application to Circuit Simulation, *IEEE J. Solid-State Circuits,* **14**(2), 1979.

24. G. Merkel, J. Borel, and N. Z. Cupcea, An Accurate Large Signal MOS Transistor Model for Use in Computer-Aided Design, *IEEE Trans. Electron Devices,* **ED-19,** 1972.

25. K. Y. Fu, Mobility Degradation Due to the Gate Field in the Inversion Layer of MOSFET's, *Electron. Lett.,* **EDL-3**(10), 1982.

26. T. Poorter and J. H. Satter, A D.C. Model for an MOS-Transistor in the Saturation Region, *Solid-State Electron.,* **23,** 1980.

27. R. R. Troutman, VLSI Limitations from Drain-Induced Barrier Lowering, *IEEE J. Solid-State Circuits,* **SC-14**(2), 1979.

28. H. Masuda, M. Nakai, and M. Kubo, Characteristics and Limitations of Scaled-Down MOSFET's Due to Two-Dimensional Field-Effects, *IEEE Trans. Electron Devices,* **ED-14**(2), 1979.

29. P. Antognetti, C. Lombardi, and D. Antoniadis, Use of Process and 2-D MOS Simulation in the Study of Doping Profile Influence on S/D Resistance in Short Channel MOSFET's, *IEDM,* Washington, 1981.

30. S. M. Sze, *Physics of Semiconductor Devices,* Wiley, New York, 1969.

31. J. Sansbury, Applications of Ion Implantation in Semiconductor Processing, *Solid-State Technol.,* 1976.

32. G. Doucet and F. Van De Wiele, Threshold Voltage in Nonuniformly Doped MOS Structures, *Solid-State Electron.,* **16,** 1973.

33. J. R. Brews, Threshold Shifts Due to Nonuniform Doping Profiles in Surface Channel MOSFET's, *IEEE Trans. Electron Devices,* **ED-26**(11), 1979.

34. J. Mar, S. Li, and S. Yu, Substrate Current Modeling for Circuit Simulation, *IEEE Trans. CAD,* **CAD-1**(4), 1982.

35. R. A. Haken, Analysis of the Deep Depletion MOSFET and the Use of DC Characteristics for Determining Bulk-Channel Charge Coupled Device Parameters, *Solid-State Electron.,* **21,** 1978.

36. T. E. Handrickson, A Simplified Model for Subpinchoff Conduction in Depletion-Mode IGFET's, *IEEE Trans. Electron Devices,* **25,** 1978.

37. T. E. Handrickson, Determination of Buried Channel Parameters from IV and CV Measurements, *Solid-State Electron.,* **22,** 1979.

38. Y. A. El-Mansy, Analysis and Characterization of the Depletion-Mode IGFET, *IEEE J. Solid-State Circuits,* **SC-15**(3), 1980.

39. Y. A. El-Mansy and A. R. Boothroyd, A New Approach to the Theory and Modelling

of Insulated-Gate Field-Effect Transistor, *IEEE Trans. Electron Devices,* **24**(3), 1977.

40. J. R. Brews, A Charge-Sheet Model of the MOSFET, *Solid-State Electron.,* **21,** 1978.
41. F. Van De Wiele, A Long-Channel MOSFET Model, *Solid-State Electron.,* **22,** 1979.
42. G. Baccarani, M. Rudan, and G. Spadini, Analytical IGFET Model Including Drift and Diffusion Currents, *IEEE Proc. Solid-State Electron Devices,* **2**(2), 1978.
43. T. Shima, T. Sugawara, S. Moriyama, and H. Yamada, Three-Dimensional Look-Up Table MOSFET Model for Precise Circuit Simulation, *IEEE J. Solid-State Circuits,* **SC-17**(3), 1982.
44. C. Turchetti, G. Masetti, and Y. Tsividis, On the Small-Signal Behavior of the MOS Transistor in Quasistatic Operation, *Solid-State Electron.,* **26,** 1983.
45. P. Yang, B. Epler, and P. Chatterjee, An Investigation of the Charge Conservation Problem for MOSFET Circuit Simulation, *IEEE J. Solid-State Circuits,* **18,** 1983.
46. Y. Tsividis and G. Masetti, Problems in Precision of the MOS Transistor for Analog Applications, *IEEE Trans. CAD,* **CAD-3**(1), 1983.
47. B. J. Sheu, D. L. Scharfetter, and H. C. Poon, Compact Short Channel IGFET Model (CSIM), Electronics Research Laboratory Rep. No. M84/20, University of California, Berkeley, 1984.
48. D. A. Divekar and R. Dowell, A Depletion Mode MOSFET Model for Circuit Simulation, *IEEE Trans. CAD,* **CAD-3**(1), 1984.

BJT Parameter Measurements

Giuseppe Massobrio

Department of Electronics (DIBE),
University of Genoa, Genoa, Italy

Accurate circuit simulation is possible only by specifying accurate and meaningful parameters for each model. This chapter discusses how the model input parameters can be extracted from the measurement data. The emphasis here is on the algorithms for extracting every parameter rather than on the detailed equations. The algorithms are written in a way that enables the user to extract the parameters manually or by writing computer programs.

These parameters are calculated from the forward I_B and I_C vs. V_{BE} measurements. Two different configurations can be used for this measurement. In the first configuration, the base current is varied by a current source while the base-emitter voltage and collector current are measured. This technique is easier, but because of the high impedance on the base

Note: The material contained in this chapter is based in part upon the book *Modeling the Bipolar Transistor,* by Ian E. Getreu, copyright © Tektronix, Inc., 1976. All rights reserved. Reproduced with permission.

terminal, the transistor may oscillate at the high currents, producing erroneous data. In the second case, the base-emitter voltage is swept with a voltage source while the base and collector currents are measured with two current meters. Since a low-impedance voltage source is now connected to the base terminal, the device is more stable, but the measurement needs two current meters. Also, the appropriate voltage range on the base terminal has to be measured first [1].

The knee points are found in order to divide the curves into low-current, midrange, and high-current regions. The knee points are where the slope of the curves change on the logarithmic scale [1].

The equations and the discussion in this chapter refer to the *SPICE Gummel-Poon* model. The model (and consequently its parameters) will automatically simplify to the simpler *Ebers-Moll* model (parameters) when certain parameters are not specified or certain restrictions are made (that is, $q_b = 1$, $C_2 = 0$, and so on).

The described measurement techniques are by no means meant to encompass all possible ones; there are other techniques, some of which may be just as good or even better.

5.1   Input and Model Parameters

To completely specify a transistor model, a program such as SPICE requires three types of information [2]: (1) fundamental physical constants, (2) operating conditions, and (3) model parameters.

The *fundamental physical constants,* such as Boltzmann's constant k and electronic charge q, are normally defined inside the program.

The *operating conditions* define the circumstances under which the model equations are to be used. In a nodal analysis program, for example, the operating conditions are normally the transistor's bias voltages, say V_{BE} and V_{BC}. These bias voltages are determined internally as the computer iterates to the solution. That is, the program assumes a set of bias voltages, solves the equations, and then selects new and better values until it converges to an adequate solution; this is all done internally.

In the previous chapters, it is has been assumed that the operating conditions consist of not only the bias voltages but also the temperature T at which the analysis is to be performed. The value of T is normally required as input, and it will be assumed throughout that T has been specified.

The third type of information required is the set of *model parameters* for each different device in the circuit. The measurement of the model parameters is the subject of this chapter.

The values of the model parameters must be supplied by the user in a manner predetermined by the program. Some programs are very flexible and allow some model parameters to be specified indirectly. For example,

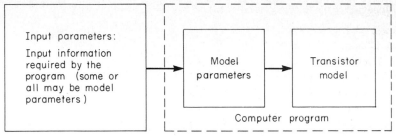

Figure 5-1 Distinction between input parameters and model parameters [2].

τ_F (the total transit time in the normal active region) is normally determined through the measurement of f_T (the unity-gain band width).

In some programs, the user can specify either τ_F or the f_T value at a bias condition. If the f_T value is provided, the program determines the value of τ_F internally, taking into account the effects of junction capacitances. It must be noted that in this particular case, τ_F is the model parameter, yet f_T is the input. Therefore, a distinction must be made between the model parameters and the program's *input parameters* [2]. This distinction is illustrated in Fig. 5-1.

The model parameters are those parameters used in the model equations to describe the device for a given set of operating conditions. The input parameters are the data required by the program to specify the model parameters. Some or all of the input parameters may be model parameters, depending on the program.

The formal distinction between model and input parameters, though appearing at first to be rather pedantic, is in fact very important; it helps to maintain the proper perspective.

In the preceding example, the fact that τ_F is the model parameter underscores its importance. If τ_F were measured directly (rather than via f_T), then there would be no need to measure f_T. The measurement of f_T arises *only* in that it is a means of determining τ_F.

Tables 5-1 and 5-2 list the parameters (which may be specified in the .MODEL card) required to model the BJT. Included in the tables are the keywords and the default values used in SPICE2 version G.

5.2 Parameter Measurements

5.2.1 Formulation of the Gummel-Poon model

The SPICE Gummel-Poon model was widely discussed in Chap. 2. The description of this model is given by the set of equations rewritten for convenience in the following, where $n_F = n_R = 1$:

$$I_C = \frac{I_S}{q_b}(e^{qV_{BE}/kT} - e^{qV_{BC}/kT}) - \frac{I_S}{\beta_{RM}}(e^{qV_{BC}/kT} - 1)$$

$$- C_4 I_S(e^{qV_{BC}/n_{CL}kT} - 1)$$

$$I_B = \frac{I_S}{\beta_{FM}}(e^{qV_{BE}/kT} - 1) + C_2 I_S(e^{qV_{BE}/n_{EL}kT} - 1)$$

$$\tag{5-1}$$

$$+ \frac{I_S}{\beta_{RM}}(e^{qV_{BC}/kT} - 1) + C_4 I_S(e^{qV_{BC}/n_{CL}kT} - 1)$$

The q_b term consists of two components, q_1 and q_2; that is,

$$q_b = \frac{q_1}{2} + \sqrt{\left(\frac{q_1}{2}\right)^2 + q_2} \tag{5-2}$$

Instead of representing q_1 as a complicated function of voltage, the junction capacitances are assumed to be constant (only for approximating changes in transport current). The charge associated with junction capacitances, therefore, will be directly proportional to the junction voltages. The parameters V_A and V_B are the constants of proportionality and give rise to the finite output conductance

$$q_1 = 1 + \frac{V_{BC}}{V_A} + \frac{V_{BE}}{V_B} \tag{5-3}$$

q_2 is the current-dependent charge contributed by the diffusion capacitances of the two junctions. It is assumed that there is no space charge in the collector region, which is equivalent to assuming that base pushout effects are neglected. The equation for q_2, therefore, can be written as

$$q_2 = \frac{I_S}{I_{KF}}(e^{qV_{BE}/kT} - 1) + \frac{I_S}{I_{KR}}(e^{qV_{BC}/kT} - 1) \tag{5-4}$$

The terms I_{KF} and I_{KR} determine the conditions for the onset of high-level effects; they are referred to as the *knee currents*. The coefficients n_{EL} and n_{CL} allow adjustment of the exponential slopes for the nonideal base current dependences, and C_2 and C_4 set the intercept coefficients. Equations (5-1) to (5-4) represent the BJT model used in SPICE. This model contains 11 parameters: I_S, β_{FM}, β_{RM}, C_2, C_4, n_{EL}, n_{CL}, I_{KF}, I_{KR}, V_A, and V_B. (Note that in the previous equations we have put $I_S = I_{SS}$ using SPICE convention.)

To illustrate the essential features of this implementation of the Gum-

TABLE 5-1   SPICE BJT Parameters: Ebers-Moll Model

Symbol	SPICE 2G keyword	Affected by area	Parameter name	Default value	Unit
\multicolumn{6}{c}{Static model parameters}					
I_S	IS	X	Saturation current	10^{-16}	A
β_F	BF		Ideal maximum forward current gain	100	
β_R	BR		Ideal maximum reverse current gain	1	
r_B	RB	X	Zero-bias base resistance	0	Ω
r_C	RC	X	Collector resistance	0	Ω
r_E	RE	X	Emitter resistance	0	Ω
V_A	VAF		Forward Early voltage	∞	V
E_g	EG		Energy gap	1.11	eV
\multicolumn{6}{c}{Dynamic model parameters}					
C_{JE}	CJE	X	Zero-bias base-emitter depletion capacitance	0	F
ϕ_E	VJE		Base-emitter built-in potential	0.75	V
C_{JC}	CJC	X	Zero-bias base-collector depletion capacitance	0	F
ϕ_C	VJC		Base-collector built-in potential	0.75	V
C_{JS}	CJS	X	Zero-bias collector-substrate capacitance	0	F
ϕ_S	VJS		Substrate-junction built-in potential	0.75	V
τ_F	TF		Ideal forward transit time	0	s
τ_R	TR		Ideal reverse transit time	0	s
\multicolumn{6}{c}{Noise parameters}					
k_f	KF		Flicker-noise coefficient	0	
a_f	AF		Flicker-noise exponent	1	

mel-Poon model, consider $\ln I_C$, $\ln I_B$ vs. V_{BE} characteristics for constant V_{BC} for a transistor, as shown in Fig. 5-2.

At very low forward bias, the component q_2 can be neglected. In this case, $q_b \simeq q_1$, and, for forward bias,

$$ I_C = \frac{I_S}{1 + V_{BC}/V_A + V_{BE}/V_B} \, e^{qV_{BE}/kT} \simeq I_S \, e^{qV_{BE}/kT} \left(1 - \frac{V_{BC}}{V_A} \right) \quad (5\text{-}5) $$

Thus, at low forward bias, collector current has an ideal component with a slope of q/kT and the finite output conductance is determined by the parameter V_A, the Early voltage. Measurements of I_C at several values of

TABLE 5-2 SPICE BJT Parameters: Gummel-Poon Model†

Symbol	SPICE 2G keyword	Affected by area	Parameter name	Default value	Unit
			Static model parameters		
C_2	ISE = $C_2 I_S$	X	Base-emitter leakage saturation current	0	A
C_4	ISC = $C_4 I_S$	X	Base-collector leakage saturation current	0	A
V_B	VAR		Reverse Early voltage	∞	V
I_{KF}	IKF	X	Corner for forward β high-current roll-off	∞	A
I_{KR}	IKR	X	Corner for reverse β high current roll-off	∞	A
n_{EL}	NE		Base-emitter leakage emission coefficient	1.5	
n_{CL}	NC		Base-collector leakage emission coefficient	2	
r_{BM}	RBM	X	Minimum base resistance at high currents	RB	Ω
I_{rB}	IRB	X	Current where base resistance falls halfway to its minimum value	∞	A
n_F	NF		Forward current emission coefficient	1	
n_R	NR		Reverse current emission coefficient	1	
			Dynamic model parameters		
m_E	MJE		Base-emitter junction grading coefficient	0.33	
m_C	MJC		Base-collector junction grading coefficient	0.33	
m_S	MJS		Substrate-junction exponential factor	0	
X_{CJC}	XCJC		Fraction of base-collector depletion capacitance connected to internal base node	1	
FC	FC		Coefficient for forward-bias depletion capacitance formula	0.5	
$X_{\tau F}$	XTF		Coefficient for bias dependence of TF	0	
$V_{\tau F}$	VTF		Voltage describing V_{BC} dependence of TF	∞	V
$I_{\tau F}$	ITF	X	High-current parameter for effect on TF	0	A
$P_{\tau F}$	PTF		Excess phase at $f = 1/2\pi\tau_F$	0	°
$X_{T\beta}$	XTB		Forward and reverse β temperature coefficient	0	
X_{TI}	XTI		Saturation current temperature exponent	3	

† Any combination of these parameters may be specified in addition to the Ebers-Moll model parameters.

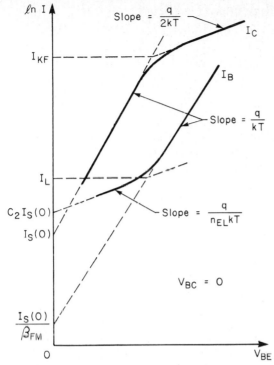

Figure 5-2 Plots of ln I_C and ln I_B vs. V_{BE} (for V_{BC} = 0).

V_{BC} are needed to determine V_A. At low forward bias, and hence at low currents, recombination in the space-charge layers is significant, and the curve for I_B has a nonideal slope of $q/n_{EL}kT$.

At high-current levels and for low values of V_{BC}, the component q_2 predominates and q_1 can be assumed equal to 1. Then [4],

$$q_b = \frac{q_1}{2}\left(1 + \sqrt{1 + 4\frac{q_2}{q_1^2}}\right) \simeq \tfrac{1}{2}\left(1 + \sqrt{1 + 4q_2}\right) \qquad (5\text{-}6)$$

In case of low-level injection,

$$\frac{4I_S}{I_{KF}}\,e^{qV_{BE}/kT} \ll 1$$

and the collector current follows the *ideal* law

$$I_C \simeq I_S\,e^{qV_{BE}/kT} \qquad (5\text{-}7)$$

For high-level injection,

$$\frac{4I_S}{I_{KF}}\,e^{qV_{BE}/kT} \gg 1$$

and the collector current becomes *nonideal*. Then

$$I_C = \frac{I_S}{q_b} e^{qV_{BE}/kT} \simeq \frac{I_S e^{qV_{BE}/kT}}{\sqrt{(I_S/I_{KF})e^{qV_{BE}/kT}}} = \sqrt{I_S I_{KF}} \, e^{qV_{BE}/2kT} \qquad (5\text{-}8)$$

At high-current levels, therefore, the curve for I_C has a slope of $q/2kT$. For the I_B curve, after a certain value of current, the ideal component dominates and the curve has a slope of q/kT. Again at high-current levels, the voltage drop in the extrinsic base resistance r_B starts to dominate and the curves for I_B and I_C deviate from their straight-line behavior on a semilogarithmic plot. I_B does not rise as steeply as before and, hence, the slope of the curve becomes less than q/kT [4].

5.2.2 Static parameter measurement techniques

Measurement techniques to determine SPICE forward dc parameters for a BJT are presented in this section. The reverse parameters are calculated from the reverse measurement using the same algorithm. In the reverse mode measurements, the emitter and the collector are interchanged and the transistor is handled as in the forward mode.

As previously explained ln I_C, ln I_B vs. V_{BE} curves (see Fig. 5-2) depict all the physical phenomena represented by the model. The parameters for the model can be determined from the curves, as described below. First, the curve for I_C is considered.

I_S. I_S is the transistor saturation current. With regard to the curve for I_C, by extrapolating the region with slope q/kT, the value of I_S is obtained. In this region, the collector current is given at low-current levels by

$$I_C = I_S \left(e^{qV_{BE}/kT} - 1\right) \qquad (5\text{-}9)$$

I_S is directly proportional to the emitter-base junction area and therefore can vary significantly from device to device. A typical value of I_S for an integrated circuit transistor is 10^{-16} A.

I_{SS}. I_{SS} is defined in the Gummel-Poon model, from considerations of the internal physics of the device (for an *npn* transistor), as [2]

$$I_{SS} \equiv \frac{qD_n n_i^2 A_J}{\int_{x_{E0}}^{x_{C0}} p_0(x) \, dx} \qquad (5\text{-}10)$$

In terms of terminal measurements, I_{SS} is equal to $I_S(0)$, the value of I_S obtained from extrapolation of the I_C vs. V_{BE} characteristics at $V_{BE} = 0$ (and with $V_{BC} = 0$).

Since $I_{SS} = I_S(0)$, it has the same typical value: on the order of 10^{-16} A for an integrated circuit transistor. As seen from Eq. (5-10), I_{SS} is proportional to the emitter area A_J.

I_{SS}, like I_S, is obtained from a plot of ln I_C as a function of qV_{BE}/kT with $V_{BC} = 0$. The value of I_{SS} is obtained by extrapolation of the straight line at low currents at $V_{BE} = 0$. Alternatively, I_{SS} could be obtained from several points on the graph by fitting these points to the expression

$$I_C = \frac{I_{SS}}{1 + V_{BE}/V_B} (e^{qV_{BE}/kT} - 1) \tag{5-11}$$

This may be necessary when V_B is small ($< {\sim}5$ V).

I_{KF}. I_{KF} is the knee current; it models the drop in β_F at high collector currents due to high-level injection. As already seen, in regions where high-level injection dominates, the curve has a slope of $q/2kT$. If it were possible to assume that low-level and high-level effects could be distinguished by regions, a value of I_{KF} could be obtained by the intersection of the two asymptotes with slopes q/kT and $q/2kT$. In most transistors, however, the two effects overlap and some trial and error is required to produce the correct parameter [4]. I_{KF} typically ranges from approximately 0.1 to 10 mA.

C_2 and n_{EL}. With regard to the curve for I_B, by extrapolating the region with a slope of $q/n_{EL}kT$ at low-current levels, the value of C_2 can be determined and, by determining this slope, n_{EL} can be found.

C_2 (magnitude) and n_{EL} (variation with V_{BE}) describe the nonideal component of I_B, which is dominant at low currents. Then,

$$I_{B,\text{nonideal}} = C_2 I_S(0)(e^{qV_{BE}/n_{EL}kT} - 1) \tag{5-12}$$

This component of I_B is responsible for the drop in β_F at low currents.

The value of C_2 is typically on the order of 100 to 1000; n_{EL} is typically between 2 and 4 (most often 2).

β_{FM}. β_{FM} is the maximum value of β_F, which is the ratio of I_C to I_B when the transistor is in the normal active region. β_{FM} is determined by the regions of the two curves when they are parallel and have a slope of q/kT. In this region, both I_C and I_B have dominant ideal components, and the ratio of I_C and I_B in this region gives β_{FM}, which can be determined from the I_B curve by extrapolating the region with a slope of q/kT, as shown in Fig. 5-2.

Thus β_{FM} defines the magnitude of the ideal component of I_B at a given value of V_{BE}.

$$I_{B,\text{ideal}} = \frac{I_S(0)}{\beta_{FM}} (e^{qV_{BE}/kT} - 1) \tag{5-13}$$

Typical values of β_{FM} range from the order of 10 for an IC lateral *pnp* or a power transistor, to about 100 for a small-signal *npn* transistor, to the order of 1000 for a super-β *npn* transistor.

V_A *and* V_B. V_A is the Early voltage that models the effect on the transistor characteristics of base-width modulation due to variations in the collector-base space-charge layer. It is always a positive number.

Typical values of V_A are 50 to 70 V for *npn* transistors, 100 V for lateral *pnp* transistors, and on the order of 1 to 10 V for devices with extremely thin base widths (such as punch-through devices or super-β transistors).

The simplest method of obtaining V_A is from the slope of the I_C vs. V_{CE} curve in the linear region. The base voltage should be constant and the transistor should be biased to its normal operating point. In the forward region, the collector current can be simplified as

$$I_C = \frac{I_S}{q_b} e^{qV_{BE}/kT} \tag{5-14}$$

By keeping the base voltage constant, the collector current would be proportional to $1/q_b$ and $1/q_1$. Therefore, in the linear region, the collector current is proportional to $1 - V_{BC}/V_A - V_{BE}/V_B$. If V_{CE} is much larger than V_{BE}, the above term simplifies to $1 + V_{CE}/V_A$. The parameter V_A can now be calculated as the intersection of the extrapolated I_C curve with the V_{CE} axis, as shown in Fig. 5-3.

It is best to repeat the extraction at few base-emitter voltages and average the V_A [1].

The extrapolation can be performed graphically directly from the curve-tracer display. However, the experimental error associated with this approach can be very high.

A more accurate technique is the use of ln I_C vs. V_{BE} curves. This technique [2] for finding V_A is the determination of the ln I_C vs. V_{BE} characteristic at two different values of V_{BC}; this theoretically results in two parallel lines. The Early voltage is then determined from the ratio of the

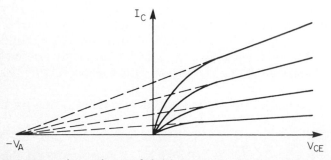

Figure 5-3 Approximate definition of V_A from I_C vs. V_{CE} characteristics.

(extrapolated) I_S values or the ratio of the value of two I_C values at the same V_{BE}.

$$\frac{I_S(V_{BC1})}{I_S(V_{BC2})} = \frac{I_C(V_{BC1})}{I_C(V_{BC2})} = \frac{1 + V_{BC2}/V_A}{1 + V_{BC1}/V_A} \tag{5-15}$$

Equation (5-15), which is used to find V_A, can be simplified if one of the V_{BC} values is zero. To simplify or eliminate the need for correcting for finite r_B, r_E, and r_C, the curves should be measured at currents as low as possible and with V_{BC} as close to zero as practical [2].

When V_A is large, all of the preceding techniques for finding V_A become very inaccurate. However, this is acceptable because the larger V_A is, the less important the effect of base-width modulation becomes and, therefore, the need to accurately determine V_A diminishes.

V_B is the reverse Early voltage that models the effect of base-width modulation due to variations in the width of the emitter-base space-charge layer. It is always a positive number. V_B is typically on the order of 10 to 50 V.

There is a complicating factor in measuring V_B in an analogous way to that used for measuring V_A. The measurement of V_A assumed that the variation in the width of the emitter-base space-charge layer had a negligible effect on the transistor's characteristics in the normal active region, that is, that V_{BE}/V_B is much less than unity. The equivalent assumption that would be necessary if the schemes for measuring V_A were to be used directly for measuring V_B is that V_{BC}/V_A is much less than unity. However, the technique for measuring V_B given in the following assumes that V_{BC}/V_A is *not* negligible compared with unity [2].

From the $\ln I_C$ vs. V_{BE} characteristics in the active region and from the $\ln I_E$ vs. V_{BC} characteristics in the inverse region, it follows that

$$I_C = \frac{I_S(e^{qV_{BE}/kT} - 1)}{1 + V_{BE}V/V_B + V_{BC}/V_A} \rightarrow \frac{I_C(0)}{I_C(V_{BC1})}$$

$$= \frac{1 + V_{BE1}/V_B + V_{BC1}/V_A}{1 + V_{BC1}/V_B}$$

$$I_E = \frac{I_S(e^{qV_{BC}/kT} - 1)}{1 + V_{BE}/V_B + V_{BC}/V_A} \rightarrow \frac{I_E(0)}{I_E(V_{BE2})} \tag{5-16}$$

$$= \frac{1 + V_{BE2}/V_B + V_{BC2}/V_A}{1 + V_{BC2}/V_A}$$

Equations (5-16) can be solved for V_A and V_B.

A determination of the importance of V_B in the normal active region can be obtained by measuring the slope of the $\ln I_C$ vs. qV_{BE}/kT curve (with $V_{BC} = 0$). If this slope departs significantly from unity, V_B can be important and may need better modeling. The slope of the $\ln I_C$ vs. qV_{BE}/kT curve in the normal region at low currents is given approximately as [2]

$$\frac{1}{n_{EL}} = 1 - \frac{kT/q}{V_B + V_{BE}} \simeq 1 - \frac{kT}{qV_B} \tag{5-17}$$

The measurement of this slope can be used to determine the appropriate value of V_B for the normal active region, although the measurement could be inaccurate if the slope is near unity [2].

E_g. E_g is the effective energy gap of the semiconductor material. E_g is typically

1.11 eV for Si

0.67 eV for Ge

0.69 eV for SBD

E_g is used to model the variation of I_S with temperature and is therefore obtained by curve-fitting the model equation to the I_S vs. T curve. The I_S vs. T curve is determined from measurements of I_S with the device in a controlled-temperature environment, for example, an oven (in making these measurements care must be taken that the power dissipated by the device does not raise the junction temperature significantly above the ambient temperature).

The model equation given in Sec. 2.8.1 [see Eq. (2-143)], assumes that I_S is proportional to $T^{X_{TI}}e^{-E_g/kT}$. When the T-term power before the exponential term is allowed to vary, typical values generated by curve-fitting for Si are 1 to 4 for the power of T, while for E_g (the extrapolated, zero-temperature energy gap), typical values range from 1.206 eV for low doping levels to approximately 1.11 eV for high doping levels [2].

r_E. r_E is a constant resistor that models the resistance between the emitter region and the emitter terminal. r_E is extracted by measuring the external collector-emitter voltage at different base currents when the collector is open ($I_C = 0$). The internal V_{CE} is then simulated using the forward and reverse parameters, and r_E is calculated as the difference divided by $I_B = I_E$ [see Eq. (5-18)].

This technique is more accurate than that of just looking at the slope of V_{CE} vs. I_B. However, due to the external base-collector diode in the transistor, the r_E is still overestimated to some extent. The cable resistances must also be subtracted from this value. It must be noticed that the resistances on the base and collector terminals do not affect the measurements, and only the resistance of the emitter terminal to ground should be subtracted.

For simulating the internal collector-emitter voltage, V_{BE} is varied and I_B and V_{CE} are calculated so that $I_C = 0$. This is done by the following algorithm. At every V_{BE}, a starting value for V_{CE} is assumed and V_{BC} is calculated as $V_{BC} = V_{BE} - V_{CE}$. I_B and I_C are then calculated by the model using the extracted parameters from the forward and reverse regions. A Newton-Raphson iteration is then used to find V_{CE} so that I_C becomes equal to zero. Now by measuring external V_{CE} at the same I_B range, the emitter resistance can be calculated as the average value of [1]

$$r_E = \frac{V_{CE,\text{ext}} - V_{CE,\text{int}}}{I_B} \tag{5-18}$$

A typical value of r_E is approximately 1 Ω, of which the metal contact resistance is normally a significant portion.

r_C. r_C models the resistance between the transistor's collector region and its collector terminal. r_C is calculated from the slope of the V_{CE} vs. I_C curve when β_F is forced to 1 in the measurement. Forced β measurement can be done by forcing the base-emitter voltage and the collector current equal to the measured base current.

The external V_{CE} is equal to the internal V_{CE} plus the voltage drop on r_C and r_E [1]:

$$V_{CE,\text{ext}} = I_C \left[r_C + r_E \left(1 + \frac{1}{\beta_F} \right) \right] + V_{CE,\text{int}} \tag{5-19}$$

After forcing β_F to 1 ($I_C = I_B$), the r_E term in the preceding equation would be constant, and r_C can be calculated from the slope of the V_{CE} vs. I_C curve.

Parameter r_C, which is calculated by this technique, is the collector resistance of the transistor in saturation mode $r_{C,\text{sat}}$. This value should not be confused with $r_{C,\text{lin}}$ found by measuring the slope of the I_C vs. V_{CE} curve in the linear region, which is usually higher. For switching circuits, such as emitter-coupled logic (ECL), $r_{C,\text{sat}}$ can be used. But for linear applications, the higher value of $r_{C,\text{lin}}$ will be more appropriate [1]. Figure 5-4 shows these two limiting values of r_C.

For a given transistor, r_C actually varies with current level, but for SPICE it is considered constant. The value of r_C can vary significantly from a few ohms for discrete and deep-collector integrated devices to hundreds of ohms for standard integrated devices.

r_B. r_B models the resistance between the base region and the base terminal. Values of r_B can range from about 10 Ω (for microwave devices) to several kilohms (for low-frequency devices).

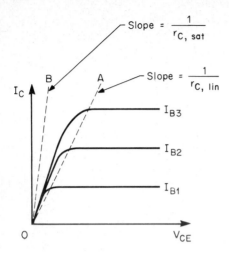

Figure 5-4 I_C vs. V_{CE} characteristics at constant I_B, showing the two limiting values of r_C [2].

The value obtained for r_B depends strongly on the measurement technique used as well as the transistor's operating conditions.

r_B should be determined by the method closest to the operating conditions under study. If r_B is being measured to ascertain its effect on noise performance, the noise measurement technique [2] should be used. Similarly, if the transistor is to be used in a switching application, the pulse measurement techniques [2] may provide the most appropriate value. For small-signal analyses, four measurement techniques can be used: the input impedance circle method, the phase-cancellation technique, the two-port network method, and the h-y ratio technique [2, 4].

Here, a technique involving all the parameters affecting the base resistance is presented [1].

At I_{rB} (a Gummel-Poon model parameter), the base resistance is halfway between minimum and maximum; that is,

$$r_{BB'} = \frac{r_B + r_{BM}}{2} \tag{5-20a}$$

where r_{BM} is the minimum base resistance at high currents.

Assuming r_{BM} and I_{rB} are known, $r_{BB'}$ at I_{rB} can be found from the measured data. Now r_B can be calculated as

$$r_B = 2r_{BB'} - r_{BM} \tag{5-20b}$$

At this point, the base resistance can be simulated over the I_B range and the error in the measured $r_{BB'}$ is calculated. Sweeping the r_{BM} and I_{rB}, the point where the error is minimum is the optimum point for the parameters.

In addition to determining the parameters for the SPICE model, values of I_B with I_C as a parameter can be obtained. If the ratio of respective values of I_C and I_B against I_C are plotted, the β_F vs. I_C curve for that particular device can be obtained. Typically, the curve is in the form displayed in Fig. 5-5.

Figure 5-2 illustrated the typical I_C and I_B dependence for a bipolar transistor for active forward-bias conditions. It has been seen that the Gummel-Poon parameters can be determined from these curves. However, to obtain a good estimation of these parameters, care and accuracy are required. The parameters I_S and C_2 are determined by extrapolating the curves at lower currents. In order to have an accurate estimate of the intercepts, it is necessary to take measurements at the lowest currents possible. On the other hand, a good estimate of I_{KF} and r_B requires that measurements be taken at high-current values. Thus, it is necessary to take measurements over a wide range of current values.

5.2.3 Dynamic parameter measurement techniques

The parameters for each junction capacitance and the algorithm for extracting the parameters associated with τ_F are described in this section.

C_{J0}, ϕ, and m.   These three parameters describe the junction capacitance due to the fixed charge in a junction depletion region.

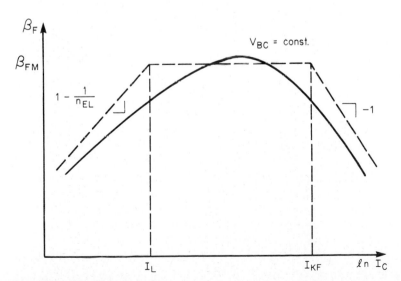

Figure 5-5 Plot of β_F vs. $\ln I_C$.

In the Gummel-Poon model, when the appropriate junction voltage V is less than or equal to $\phi/2$, the junction capacitance is modeled by

$$C_J(V) = \frac{C_{J0}}{(1 - V/\phi)^m} \qquad (5\text{-}21)$$

where C_{J0} is the value of C_J at $V = 0$, ϕ is the built-in potential, and m is the capacitance gradient factor. For the emitter-base junction, the subscript E is added. The junction capacitance C_{JE} is a function of the internal base-emitter voltage V_{BE}, and the parameters are C_{JE0}, ϕ_E, and m_E. Similarly, for the collector-base junction, C_{JC} is expressed in terms of C_{JC0}, ϕ_C, and m_C.

In SPICE, regardless of the type of device (npn or pnp), V is positive if the junction is forward-biased and negative if the junction is reverse-biased.

Both junction capacitances can be obtained as a function of voltage by means of a bridge. The two junction contacts are connected to the bridge and the third contact is left open. For example, for C_{JE}, the emitter and base leads are connected to the bridge and the collector contact is left open. The measurement frequency is normally low enough that it can be assumed that the ohmic resistances have a negligible effect.

A complicating factor is the extra capacitance C_k, caused mainly by pin capacitance, stray capacitance, and pad capacitance. C_k is normally assumed to be constant. The capacitance measured by the bridge is

$$C_{\text{meas}} = \frac{C_{J0}}{(1 - V/\phi)^m} + C_k \qquad (5\text{-}22)$$

C_k can be determined in four ways: by an estimate (approximately 0.4 to 0.7 pF), by measurement with a dummy can, by a computer parameter optimization procedure, or by graphical techniques [2].

The dummy-can technique is the most accurate method. It requires an identical device can with its metal run disconnected (scratched off) at the emitter or collector (but not at the base). Either this dummy package can be used to zero the capacitance bridge or its capacitance can be measured separately and the measured value subtracted from the bridge measurements.

Using an optimization algorithm on a computer or calculator is fast and convenient once the algorithm has been written and tested. However, as with the graphical techniques described subsequently, the solution is often not unique; several sets of solutions can be obtained depending on the initial estimates and the methods used. Since the parameters C_{J0}, ϕ, and m are used in the programs *only* to recreate the junction capacitances, any set of positive values for these parameters is acceptable.

A method of reducing the data by graphical techniques involves making an initial guess for ϕ and C_k and then plotting the resultant value of C_{meas} — C_k as a function of ϕ — V on ln-ln graph paper [2]. If a straight line (with a slope between 0.5 and 0.333) results, the chosen values are assumed correct. If the plotted line is not straight, a second guess is made for C_k and/or ϕ and the plot redone. This process continues until the appropriate straight line is obtained. Since the slope of the straight line is equal to $-m$, the values of ϕ, m, C_k, and C_{J0} can be determined from this plot.

An alternative graphical technique is to plot $(C_{\text{meas}} - C_k)^{-1/m}$ as a function of V. When a straight line is obtained, ϕ is determined by extrapolating the line to the V axis.

Typically C_J varies with V as shown in Fig. 5-6 for V less than or equal to $\phi/2$. Built-in potential ϕ is usually about 0.5 to 0.7 V. Gradient factor m is assumed to be between 0.333 and 0.5, the graded-junction and abrupt-junction values, respectively.

C_{JS}. C_{JS} is the epitaxial-layer–substrate capacitance (only in the case of an integrated circuit). It can be measured directly on a capacitance bridge at the bias voltage to be used in the analysis. If the bias voltage will change drastically, an averaging process should be used. Alternatively, a separate reverse-biased diode (with its built-in junction capacitor) or a parasitic transistor can be added to the circuit description, as shown in Fig. 5-7. If the parasitic transistor is used, care must be taken with the junction saturation currents. For example, for the *npn* transistor of Fig. 5-7, the collector-base junction of T_A and the emitter-base junction of T_B are the *same* junction. It should not be modeled twice.

C_{JS} is mainly important for integrated *npn* transistors and lateral *pnp* transistors. For *npn* devices, C_{JS} is represented as a constant capacitance, typically 1 to 2 pF, from the collector terminal to ground. Ideally, C_{JS} should be modeled by a junction capacitance distributed across r_C (with its dependence on the epitaxial-layer–substrate voltage) [2].

X_{CJC}. X_{CJC} models the split of the base-collector junction capacitance C_{JC} across the base resistance r_B. The capacitance ($X_{CJC}C_{CJ}$) is placed between the internal base node and the collector. $(1 - X_{CJC})C_{JC}$ is the

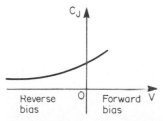

Figure 5-6   Variation of C_J as a function of V [2].

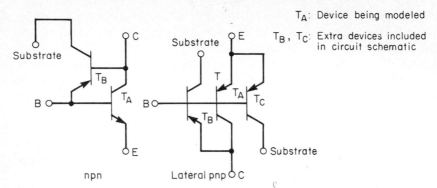

Figure 5-7 Parasitic devices for *npn* and lateral *pnp* IC transistor for determining C_{JS} [2].

capacitance from external base to collector, C_{JC} being the total base-collector capacitance. This parameter is usually important only at very high frequencies [1, 2].

X_{CJC} is a difficult parameter to determine from terminal measurements. It can be found with relative ease if the geometry of the device is known, since it is equal to $1 - A_E/A_B$, where A_E is the area of the emitter and A_B is the total area of the base, including the emitter area.

It is possible to obtain X_{CJC} via terminal measurements, too, by measuring the product $r_B(1 - X_{CJC})C_{JC}$, [2]. X_{CJC} must lie between 0 and 1. It is typically on the order of 0.8.

τ_F. τ_F, the total forward transit time, is used for modeling the excess charge stored in the transistor when its emitter-base junction is forward-biased and $V_{BC} = 0$. It is needed to calculate the transistor's emitter diffusion capacitance.

Generally, τ_F is determined from f_T, the transistor's unity-gain bandwidth, which is defined as the frequency at which the common-emitter, zero-load, small-signal current gain extrapolates to unity.

f_T also varies with the operating point, as well as from device to device. A typical variation of f_T with $\ln I_C$ is sketched in Fig. 5-8.

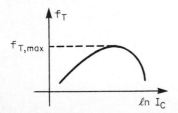

Figure 5-8   Variation of f_T with I_C [2].

For discrete devices and integrated *npn* transistors, the peak f_T is generally on the order of 600 MHz to 2 GHz. For integrated *pnp* transistors, the peak f_T is usually 10 MHz for a substrate *pnp* device and 1 MHz for a lateral *pnp* device. The drop in f_T at high currents is caused by the increase in τ_F at high currents. The drop in f_T at low currents is caused by junction capacitances C_{JE} and C_{JC}. Since these two capacitances are modeled separately, the drop in f_T at low currents is included inherently in the Gummel-Poon model.

In the region where f_T is constant, τ_F is given by

$$\tau_F = \frac{1}{2\pi f_{T,\text{max}}} - C_{JC}r_C \tag{5-23}$$

where $f_{T,\text{max}}$ is the peak value of f_T and r_C is the $r_{C,\text{lin}}$ value (see r_C measurement). When there is no constant f_T region, τ_F is obtained by plotting $1/f_T$ as a function of $1/I_C$. The resultant curve can then be extrapolated to obtain τ_F. The intercept (noted by $1/f_A$) of the extrapolated straight line at $1/I_C = 0$ is related to τ_F by

$$\tau_F = \frac{1}{2\pi}\frac{1}{f_A} - C_{JC}r_C \tag{5-24}$$

The unity-gain bandwidth can be measured with a small-signal measurement setup or an *s*-parameter measurement system [2].

Typically, τ_F varies with I_C as shown in Fig. 5-9. However, for the Ebers-Moll model, τ_F is assumed to be constant. Values of τ_F generally range from 0.3 ns for a standard *npn* transistor to 80 ps for a high frequency device ($f_T \simeq 2$ GHz).

$X_{\tau F}$, $I_{\tau F}$, $V_{\tau F}$, and $P_{\tau F}$.   The four parameters associated with the forward transit time τ_F should be calculated from the f_T measurement. A method to determine the parameters $X_{\tau F}$, $I_{\tau F}$, $V_{\tau F}$, and $P_{\tau F}$ is described next; this method finds the approximate values for the parameters.

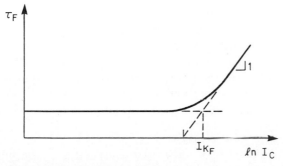

Figure 5-9   Variation of τ_F with I_C.

Because of the interaction with other model parameters, a good fit to the measured data can be made only by trial and error.

At a high value of V_{CE}, the peak value of f_T is first found ($f_{T,\max}$). τ_F is then calculated. At a low value of V_{CE} such that $V_{BC} = 0$, I_C is set to a large value such that f_T goes to a minimum. The f_T at this point is called $f_{T,\min}$. $X_{\tau F}$ is then calculated as [1]

$$X_{\tau F} = \frac{f_{T,\max}}{f_{T,\min}} - 1 \tag{5-25}$$

With V_{BC} set to almost zero, I_C can be found at the peak of the f_T curve. Assuming the peak value is f_T^*, $I_{\tau F}$ can be calculated as [1]

$$I_{\tau F} = \frac{I_C}{\sqrt{X_{\tau F}(f_{T,\max}/f_T^* - 1)}} - I_C \tag{5-26}$$

If $f_T(f_{TX})$ is measured at a high $V_{CE}(V_{CEX})$ and a high I_C and also if $f_T(f_{TY})$ is measured at a low $V_{CE}(V_{CEY})$ and at the same I_C, $V_{\tau F}$ can be calculated as [1]

$$V_{\tau F} = \frac{V_{CEX} - V_{CEY}}{1.44 \ln \left[(f_{TY}/f_{TX})(f_{T,\max} - f_{TX})/(f_{T,\max} - f_{TY}) \right]} \tag{5-27}$$

After the parameters have been approximated, they should be checked with an actual simulation. The parameters can then be trimmed to find a better match.

The excess phase parameter $P_{\tau F}$ is found by measuring the phase at unity-gain bandwidth of the transistor. $P_{\tau F}$ is the difference between the actual phase at f_T and $90°$.

TNOM, T_{C1}, T_{C2}, X_{TI}, and $X_{T\beta}$. TNOM is the temperature at which all the model parameters are obtained. The simplest technique for measuring TNOM is by means of a thermometer placed near the transistor. As long as the power dissipation of the device is low enough to cause a negligible increase in the junction temperature, then the junction temperature is approximately room temperature. TNOM is considered to be room temperature, about $27°C$, or 300 K.

T_{C1} and T_{C2} are the first- and second-order temperature coefficients and are used to model the effect of temperature on resistors (and consequently on r_B, r_C, and r_E if they are specified as *external* resistances and not as model parameters). Thus,

$$r(T) = r(\text{TNOM})[1 + T_{C1}(T - \text{TNOM})$$
$$+ T_{C2}(T - \text{TNOM})^2] \tag{5-28}$$

where r is the resistor being considered, TNOM is the nominal temperature at which the parameter was measured, and T is the temperature at which the analysis is to be performed [2].

The temperature variation of each parameter is obtained by placing the device in a controlled-temperature environment and making the appropriate measurements. The observed temperature variation is then fitted to Eq. (5-28). T_{C1} represents a linear variation with temperature, and T_{C2} represents a nonlinear (square-law) variation with temperature.

Values of T_{C1} and T_{C2} vary from one parameter to the other. Typical values of T_{C1} and T_{C2} for r_C and r_B are 2×10^{-3} K^{-1} and 8×10^{-6} K^{-2}, respectively.

X_{TI} is the saturation current temperature exponent (usually equal to 3) that appears explicitly in the temperature dependence of the saturation current formula:

$$I_S(T_2) = I_S(T_1) \left(\frac{T_2}{T_1} \right)^{X_{TI}} e^{[-qE_g(300)/kT_2](1 - T_2/T_1)} \qquad (5\text{-}29)$$

$X_{T\beta}$ is the forward and reverse β temperature coefficient, according to the formula

$$\beta_{F,R}(T_2) = \beta_{F,R}(T_1) \left(\frac{T_2}{T_1} \right)^{X_{T\beta}} \qquad (5\text{-}30)$$

REFERENCES

1. E. Khalily, Hewlett-Packard Co., private communication.
2. I. E. Getreu, *Modeling the Bipolar Transistor*, Tektronix, Inc., Beaverton, Ore., 1976.
3. H. K. Gummel and H. C. Poon, An Integral Charge Control Model of Bipolar Transistors, *Bell Syst. Tech. J.*, **49**, 1970.
4. F. Van De Wiele, W. L. Engl, and P. G. Jespers (eds.), *Process and Device Modeling for Integrated Circuit Design*, Nijhoff, The Hague, 1977.
5. W. D. Mack and M. Harowitz, Measurement of Series Collector Resistance in Bipolar Transistors, *IEEE J. Solid-State Circuits*, **SC-17**, 1982.
6. R. D. Thornton, J. G. Linvill, E. R. Chenette, H. L. Ablin, J. N. Harris, A. R. Boothroyd, J. Willis, and C. L. Searle, *Handbook of Basic Transistor Circuits and Measurements*, vol. 7, Semiconductor Electronics Education Committee, Wiley, New York, 1965.
7. W. M. C. Sansen and G. Meyer, Characterization and Measurement of the Base and Emitter Resistances of Bipolar Transistors, *IEEE J. Solid-State Circuits*, **SC-7**, 1972.

Chapter 6

MOS Parameter Measurements

Enrico Profumo

SGS Microelettronica, Milan, Italy

This chapter considers the methods of calculating and measuring the parameters of the three existing levels of the MOS model implemented in SPICE. For each parameter, the values necessary to simulate n-channel MOS transistors will be obtained as examples, with different channel lengths and widths, starting from a material with a doping value of $N_A = 7 \times 10^{14}$ cm^{-3}, with source and drain junctions made with phosphorus diffusions 1 μm deep, and with a gate oxide thickness $t_{ox} = 80$ nm. This example allows us to evaluate the degree of precision reached by the three models by comparing the results of the simulations with the measurements.

6.1 Calculation of the Parameters of the LEVEL 1 Model

The equations of the LEVEL 1 model were given in Sec. 4.2. As an example of the calculation of the parameters for the LEVEL 1 model, a MOS with a very large and wide channel will be used, with $W_{nom} = L_{nom} = 100$ μm. This choice allows us to neglect possible differences between the nominal dimensions and the real dimensions W_{eff} and L_{eff}; moreover, this transistor does not present short- and narrow-channel effects, which cannot be simulated with this simple model.

6.1.1 Measurements of V_{T0} and *KP* in the linear region

The parameters V_{T0} and KP can be calculated from the transfer characteristics measured in the linear region or in the saturation region. Figure 6-1*a* shows the setup for the measurement in the linear region.

In order to account for the condition that the transistor is in the linear region, that is, $V_{DS} < V_{GS} - V_{TH}$, it is necessary to carry out the measurements with a small V_{DS}. In the example in Fig. 6-2, the voltage on the channel is $V_{DS} = 50$ mV; in this condition, all the measurements are in the linear region except those with $V_{TH} < V_{GS} < V_{TH} + V_{DS}$.

This means that the measurements of the example, carried out with $V_{GS} - V_{TH} < 50$ mV, are in the saturation region and not in the linear region (see Fig. 4-4).

With this only exception, the current is expressed by Eq. (4-23), in which the term $V_{DS}/2$ can also be eliminated (the term $V_{DS}/2$ is negligible with respect to V_{GS}).

$$I_{DS} \simeq \beta(V_{GS} - V_{TH})V_{DS} \qquad (6\text{-}1)$$

where
$$\beta = KP\,\frac{W_{\text{eff}}}{L_{\text{eff}}}$$

The term $W_{\text{eff}}/L_{\text{eff}}$ is the ratio of the effective dimensions of the channel; its method of calculation will be shown later (see Sec. 6.3.1). In this example, the ratio is equal to 1.

Equation (6-1) expresses a linear relationship between I_{DS}, V_{DS}, and $V_{GS} - V_{TH}$, and suggests the calculation of β and V_{T0} by linear extrapolation, as shown in Fig. 6-2.

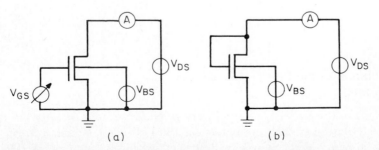

Figure 6-1 Circuit plans for the measurement of the threshold voltage: (*a*) measurement with V_{DS} constant and (*b*) measurement with $V_{DS} = V_{GS}$.

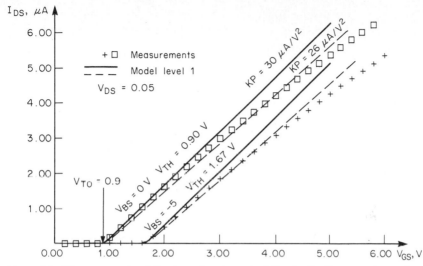

Figure 6-2 Measurements of the transfer characteristics in the linear region; calculation by extrapolation to $I_D = 0$ V of V_{TH} and of KP. The value of KP calculated using the lowest current measurements is $KP = 30$ $\mu A/V^2$ (continuous lines), and the average value is $KP = 26$ $\mu A/V^2$ (dashed lines).

This model is sufficiently accurate only for a short range of the transfer characteristic. This is due to the fact that measurements do not show the line to improve the precision of the model by accurate evaluation of the parameters V_{T0} and β_0. One way to evaluate these parameters is first to find V_{T0} by extrapolation of the measurements at low current and then to find an average value of β_0 in the range of measured V_{GS} voltages.

This extraction can be performed graphically or with the following equation, obtained from Eq. (6-1) applied to the voltages $V_{GS} = V_{GS1}$ and $V_{GS} = V_{GS2}$. From this relationship it is possible to obtain the threshold voltage from the two measurements I_{DS1} and I_{DS2}:

$$V_{TH} = \frac{V_{GS1} - (I_{DS1}/I_{DS2})V_{GS2}}{1 - I_{DS1}/I_{DS2}} \tag{6-2}$$

The input parameter for SPICE, V_{T0}, is the threshold voltage V_{TH} with zero voltage between source and substrate ($V_{BS} = 0$ V). In order to calculate the input parameter KP from β_0, it is sufficient to substitute the values obtained from Eq. (6-1). If the currents I_{DS1} and I_{DS2} are related in such a way that $I_{DS2} = 2I_{DS1}$, Eq. (6-2) becomes simply

$$V_{TH} = 2V_{GS1} - V_{GS2} \tag{6-3}$$

The currents I_{DS1} and I_{DS2} are to be chosen so as to avoid both the weak inversion region and the variable transconductance region. In the example in Fig. 6-2, $I_{DS1} = 1$ μA and $I_{DS2} = 2$ μA have been used.

This equation is very convenient for calculating the threshold voltage with only two measurements, for which a curve tracer can be used. The values calculated from the measurements in Fig. 6-2 are:

1. Using Eq. (6-3) and the experimental points $I_{DS1} = 1$ μA and $I_{DS2} = 2$ μA, the result is $V_{T0} = 0.9$ V and $KP = 30$ μA/V^2.
2. Using Eq. (6-2) and $V_{GS1} = 1$ V and $V_{GS2} = 4$ V, the result is $V_{T0} = 0.85$ V and $KP = 26$ μA/V.

The best solution is choosing the voltage V_{T0} calculated at the lowest currents, that is, $V_{T0} = 0.9$ V, and the value of KP calculated at the highest voltages, that is, $KP = 26$ μA/V^2.

Finally, to make the best choice of parameters, it is convenient to use a larger number of measurements and calculate V_{TH} and β using a simple algorithm with a linear regression.

6.1.2 Measurements of V_{T0} and KP in the saturation region

As an alternative to the method presented in the preceding section, it is possible to measure the same parameters in the saturation region by referring to Eq. (4-26) and using the setup of Fig. 6-1b. This setup has the advantage of using only two adjustable voltage sources.

Moreover, if the threshold voltage is positive (or negative for p channel), all the measurements are certainly in the saturation region; this method cannot be applied to MOS depletion transistors, because these transistors are in the linear region for $V_{GS} = V_{DS}$.

In order to again obtain a linear relationship similar to Eq. (4-26), the current in the saturation region can be described in the following way:

$$\sqrt{I_{DS}} = \sqrt{\frac{\beta}{2}} (V_{GS} - V_{TH}) \tag{6-4}$$

In this case a graphical calculation can be carried out (where the square root of I_{DS} is on the y axis, Fig. 6-3) before proceeding as in Sec. 6.1.1.

In the same way as in Eq. (6-2), the following equation calculates V_{TH} in an analytical way using only two measurements:

$$V_{TH} = \frac{V_{GS1} - \sqrt{I_{DS1}/I_{GS2}} \; V_{GS2}}{1 - \sqrt{I_{DS1}/I_{DS2}}} \tag{6-5}$$

In order to use Eq. (6-3) it is sufficient to make $I_{DS2} = 4I_{DS1}$. From the measurements shown in Fig. 6-3, the following values can be calculated:

$$V_{TH} = 0.88 \text{ V} \qquad KP = 22.5 \ \mu\text{A/V}^2$$

Another possible circuit setup for this measurement is shown in Fig. 6-1a, in which V_{DS} is maintained constant at a value sufficiently high to make the condition $V_{GS} < V_{DS} + V_{TH}$ valid; if this condition is not verified, the measurements of the transfer characteristics with $V_{GS} > V_{DS} + V_{TH}$ are in the linear region and not in the saturation region (see Fig. 4-4). This method is not as straightforward as the preceding one; however, it is useful afterward to find the law of variation of V_{TH} with V_{DS} in the short-channel MOS transistors.

It is worth nothing that the calculation of the same parameters with the two preceding methods does not give equal results. The short transition into the saturation region of the transfer characteristics with $V_{DS} = 50$ mV is the principal reason for the small difference in threshold voltages when measured with the two methods; it is easy to demonstrate that the measurements in the linear region exceed by $V_{DS}/2$ the measurements in the saturation region; furthermore, the latter method yields the correct result.

It is now necessary to examine the reason for the much clearer difference in the values calculated for KP. The result in Fig. 6-4, showing the measurements and simulations of the output characterisitcs, allows us to attribute this difference to the imprecision of the model; in fact, the simulations with the parameters calculated in the linear region are successful only in the linear region with low V_{DS}, while those with the parameters calculated in the saturation region successfully simulate the characteristics of the saturation region, as shown in Fig. 6-4.

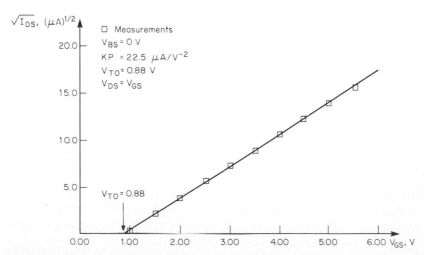

Figure 6-3 Measurements of the transfer characteristics in the saturation region; calculation by extrapolation to $I_D = 0$ of V_{TH} and KP.

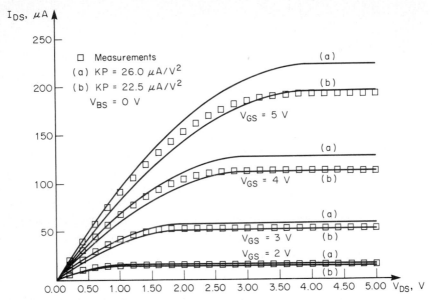

Figure 6-4 Measurements of output characteristics and simulations with parameters calculated in Figs. 6-2 and 6-3.

Given that a good agreement between measurement and simulation is possible only near the region where the parameters have been evaluated, the choice between one or another type of measurement should be based on which working range has to be simulated with greater precision.

6.1.3 Calculation of the body effect

From Eq. (4-28) we can easily calculate the parameter γ, which links the value of V_{TH} to the voltage between source and substrate (V_{BS}), using at least two measurements of V_{TH} at different substrate voltages:

$$\gamma = \frac{V_{TH2} - V_{TH1}}{\sqrt{2\phi_p - V_{BS2}} - \sqrt{2\phi_p - V_{BS1}}} \tag{6-6}$$

If more measurements are available, we can resort to a linear interpolation of the measurements of the threshold voltage V_{TH} against ($\sqrt{2\phi_p - V_{BS}} - \sqrt{2\phi_p}$) (see Fig. 6-5).

The value of $2\phi_p$, at this point of the calculations, is not yet known. It is possible to calculate this parameter from the value of the substrate doping, which is obtained from the resistivity of the starting material, or it can be taken to be equal to 0.6 V as a first estimate.

Referring to the example, we find that, with a doping of $N_A = 7 \times 10^{14}$ cm^{-3}, $2\phi_p = 0.56$ V with Eq. (4-33).

Calculating V_{TH1} and V_{TH2} with Eq. (6-6) from the curves in Fig. 6-2, we find that $\gamma = 0.484$ $V^{1/2}$, while with a fitting for all the measurements in Fig. 6-5, we find that $\gamma = 0.465$ $V^{1/2}$; the latter value will be used in future examples.

6.2 Calculation of the Parameters of the LEVEL 2 Model (Long Channel)

In this and the following sections the methods of calculating the parameters for the equations of the LEVEL 2 model seen in Sec. 4.3 will be presented. The topic has been divided in two sections to distinguish two types of parameters: those of the basic model, for which the example from the preceding chapter will be used (i.e., a long-channel MOS transistor), and the parameters relative to the second-order effects, not included in the basic model. The measurements for this second group of parameters concern almost all short-channel MOS transistors.

6.2.1 Measurements in the linear region

The parameters V_{T0}, KP, γ, and $2\phi_p$ have the same significance as in the previous model and are calculated in the same way. It is useful to observe that Eq. (4-23) of the current of the simple model with low V_{DS} supplies practically the same values as Eq. (4-40), which relates to the LEVEL 2 model; however, the method proposed in Sec. 6.1.1 is still applicable, while

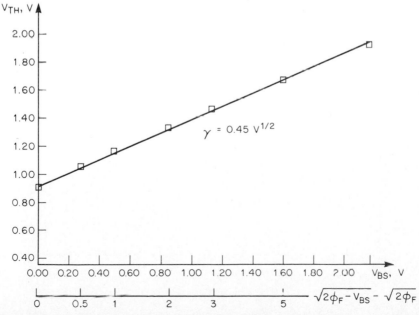

Figure 6-5 Graph calculation of the body-effect constant.

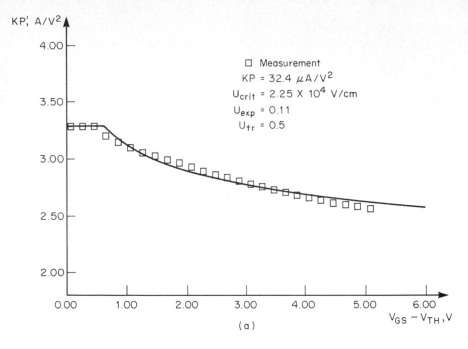

(a)

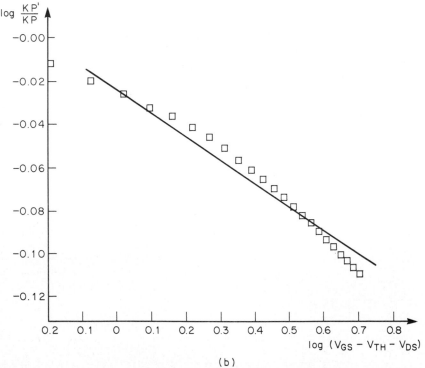

(b)

Figure 6-6 (*a*) Variation of β' vs. $V_{GS} - V_{TH}$ and (*b*) graphic calculation of U_c and U_e.

Eq. (4-26) for the current in the saturation region used in Sec. 6.1.2 is very different from that of the model now being examined. It is not possible, therefore, to calculate KP in the saturation region with Eq. (6-5) and use the same value in Eq. (4-41); we can, on the other hand, calculate the same parameter with Eq. (6-1) and use it in Eq. (4-40) without introducing large errors.

Moreover, if the present model is used to simulate a long-channel MOS transistor, the parameters calculated in the linear region are sufficient to simulate quite accurately the current value in the saturation region, too. This good correlation becomes more evident if we use the model indicated in Sec. 4.3.3 instead of using an average constant value for KP. The parameters of Eq. (4-45)—U_c, U_e, and U_t—can be calculated by a graphical method [1]. We can, in fact, rewrite the model in the following way:

$$ \log \frac{\beta'}{\beta} = U_e \left[\log \frac{t_{ox}\epsilon_{Si}U_c}{K_{ox}} - \log (V_{GS} - V_{TH} - U_t V_{DS}) \right] \quad (6-7) $$

where β' is the value of β calculated with Eq. (6-1) for the measurements in Fig. 6-2, with a different V_{GS} and a constant V_{DS}; V_{TH} and β are evaluated by extrapolation at the voltages nearest to the threshold, i.e., in the region where the dependence of I_{DS} on V_{GS} is at a maximum, and then are used in Eq. (6-7).

The values relating to the example are shown in Fig. 6-6a, as KP vs. $V_{GS} - V_{TH}$, and in Fig. 6-6b as $\log (\beta'/\beta)$ vs. $\log (V_{GS} - V_{TH} - V_{DS}U_t)$; in this way we can calculate U_e and U_c with Eq. (6-7). The value of the parameter U_t cannot be calculated with this method. However, this parameter does not generally have a great influence on the current calculated from the model; it is best therefore to set $U_t = 0.5$ and to give up the idea of a more precise calculation.

By doing so, we obtain the following values:

$V_{TH} = 0.9$ V

$KP = 32 \ \mu A/V^2$

$U_c = 2.25 \times 10^4$ V/cm

$U_e = 0.11$

With a suitable computer, it is best to use a linear regression method which supplies the parameters of the linear equation $y = ax + b$, where $y = \log (\beta'/\beta)$ and $x = \log (V_{GS} - V_{TH} - V_{DS}/2)$. From this expression we obtain U_e and U_c:

$$ U_e = -a \quad (6-8) $$

$$ U_c = \frac{K_{ox}}{t_{ox}K_{Si}} 10^{-b/a} \quad (6-9) $$

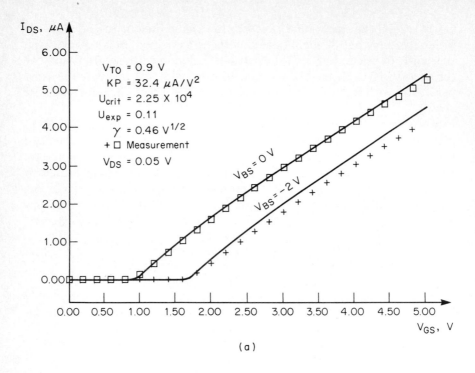

(a)

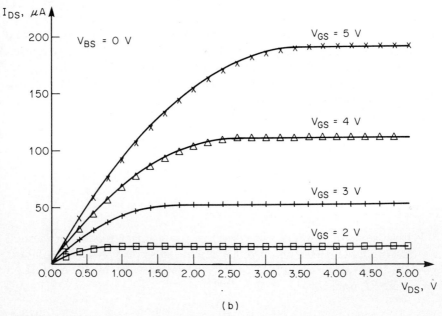

(b)

Figure 6-7 Measurements and simulations with the LEVEL 2 model of (a) transfer characteristics and (b) output characteristics with a long channel.

The value of γ can be calculated in the way shown in Sec. 6.1.3.

Figure 6-7a and b shows the simulations of the transfer characteristics and the output characteristics with this model; the improvement with respect to the simple model (see Fig. 6-4) is clear.

6.2.2 Calculation of physical parameters

The parameters V_{T0}, KP, $2\phi_F$, and γ are of an *electrical* type and are related to the parameters of a *physical* type—mobility (μ_0), substrate doping (N_A), gate oxide thickness (t_{ox}), and number of surface states (N_{SS}). If the value of t_{ox} is not already known from direct measurements, it can be calculated from a measurement of the gate capacitance, as described in Sec. 6.5.1. Once t_{ox} and C'_{ox} are known, the calculation of the mobility μ_0 and the doping N_A can be obtained by inverting Eqs. (4-31) and (4-32):

$$\mu_0 = \frac{KP}{C'_{ox}} \tag{6-10}$$

$$N_A = \frac{C'_{ox}{}^2 \gamma^2}{2q\epsilon_0 K_{Si}} \tag{6-11}$$

Once N_A is known, we can recalculate $2\phi_p$ with Eq. (4-33); if the value found is different from that used to calculate γ (see Sec. 6.1.3), it is better to repeat the calculation with Eq. (6-6).

With the parameters already calculated as in the example and with t_{ox} = 80 nm, we obtain

C'_{ox} = 4.32 $\times$ 10^{-8} F/cm^2

μ_0 = 750 cm^2/(V·s)

N_A = 1.2 $\times$ 10^{15} cm^{-3}

Recalculating $2\phi_p$ from N_A we obtain

$$2\phi_p = 2\,\frac{kT}{q}\ln\frac{N_A}{n_i} = 0.58 \text{ V}$$

Lastly, we can substitute the new value of $2\phi_p$ in Eq. (6-6) and recalculate the body-effect constant γ = 0.463 V$^{1/2}$. This value does not differ significantly from that calculated in Sec. 6.1.3; it is therefore superfluous to calculate the doping N_A from Eq. (6-11) again.

The threshold voltage V_{T0} can also be calculated from physical-type parameters, by using Eqs. (4-38) and (4-39).

In our example, the polysilicon gate has a different doping than the substrate (this situation is indicated in the SPICE .MODEL card with the parameter $T_{PG} = 1$); since the equations of the model calculate the same value of threshold voltage measured, that is, $V_{T0} = 0.9$ V, we need a number of surface states $N_{SS} = -2.5 \times 10^{10}$ cm^{-2}. This value is physically absurd, due to the fact that in the real transistor the threshold voltage has been obtained with ion implantation; the effect of a doping region different from that of the substrate is not included in the model (see Sec. 4.9.1). Equation (4-38) can therefore be used in the absence of ion implantation.

The use of this group of parameters, as an alternative to the electrical ones, is convenient in order to analyze the sensitivity with respect to the process parameters. As an example, it is possible to predict with reasonable accuracy the currents of a MOS transistor with a gate oxide $t_{ox} = 75$ nm, starting from physical-type parameters calculated with a gate oxide of $t_{ox} = 80$ nm, only by changing the t_{ox} parameter. This is not possible if the values of KP and γ have been given in the SPICE .MODEL card, because these two parameters depend in reality on t_{ox} [see Eqs. (4-31) and (4-32)].

6.2.3 Measurements in the weak inversion region

The parameter to be calculated is the number of fast surface states N_{FS}. The presence of this parameter makes possible the calculation of the subthreshold current (see Sec. 4.3.4), and it is determined by the exponential law that links I_{DS} to V_{GS}, for $V_{GS} < V_{on}$, where V_{on} is the boundary voltage between the regions of weak and strong inversion [see Eq. (4-43)].

The value of N_{FS} can therefore be calculated by inverting Eq. (4-44):

$$N_{FS} = \left(\frac{\Delta V_{GS}}{\Delta \log I_{DS}} \frac{q}{kT} - 1 \right) \frac{C'_{ox}}{q} \tag{6-12}$$

In Fig. 6-8, the measurements of the transfer characteristics are in the weak inversion region for $V_{GS} < 0.85$ V, and the exponential coefficient is

$$\frac{\Delta V_{GS}}{\Delta \log I_{DS}} = 116 \text{ mV per decade}$$

From the measurements in Fig. 6-8 we obtain $N_{FS} = 9.3 \times 10^{11}$ cm^{-2}. Unfortunately, the model does not ensure a good correlation with the measurements, even if the slope of the region exponentially growing is calculated correctly; this happens because it is not possible to vary the transition voltage V_{on}.

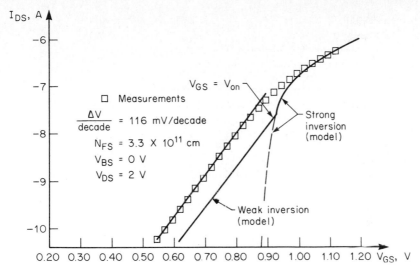

Figure 6-8   Measurements, calculation of N_{FS}, and simulation in weak inversion region.

6.3   Calculation of the Parameters of the LEVEL 2 Model (Short Channel)

In the preceding section the basic parameters were found for any channel length or width. In this section the parameters not provided by the basic model, which relate to short- and narrow-channel effects, will be calculated. We will therefore use as examples MOS transistors with short and narrow channels.

6.3.1   Calculation of the effective dimensions and of the series resistance

All the parameters still to be calculated depend on the length and width of the channel. It is therefore important to determine as precisely as possible the effective values of the length L_{eff} and the width W_{eff}; these values differ from the nominal ones because of constant contributions that are due to the process: the lateral diffusion X_{jl} for the length L_{nom} (Fig. 6-9a) and the step of the thick oxide W_{ox} for the width W_{nom} (Fig. 6-9b). In addition, there are random contributions due to imprecisions of the process in the definition of the gate (L_{err} and W_{err}). Therefore,

$$\Delta L = L_{nom} - L_{eff} = 2X_{jl} + 2L_{err} \qquad (6\text{-}13)$$

$$\Delta W = W_{nom} - W_{eff} = 2W_{ox} + 2W_{err} \qquad (6\text{-}14)$$

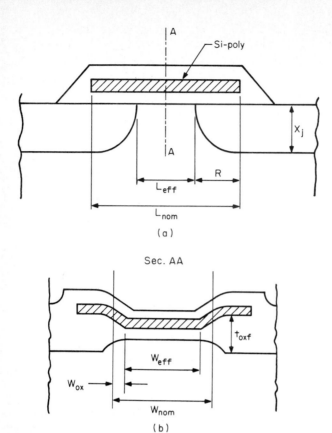

Figure 6-9 Definitions of effective length and width of a MOS transistor.

The definition of the value of the parameters W and L in the input description for SPICE of the single-transistor MOS and of X_{jl} in the model must also take into account the value of the parameters L_{err}, W_{ox}, and W_{err}. Moreover, we will see that we also need to know the parasitic series resistance in order to measure L_{eff} precisely.

The effective length and width can be calculated from the electrical measurements, using the formulas relating L_{eff} and W_{eff} to the resistance and to the equivalent channel conductance G_{on} in the linear region, obtained from Eq. (6-1):

$$R_{on} = \frac{V_{DS}}{I_{DS}} = \frac{1}{\beta(V_{GS} - V_{TH})} = \frac{L_{eff}}{KPW_{eff}(V_{GS} - V_{TH})} \tag{6-15}$$

$$G_{on} = \frac{I_{DS}}{V_{DS}} = KP\frac{W_{eff}}{L_{eff}}(V_{GS} - V_{TH}) \tag{6-16}$$

Repeating the measurements of R_{on} with the same voltages V_{GS} and V_{DS} on MOS transistors of equal width and of scaled lengths, we obtain the graph in Fig. 6-10a. With a linear extrapolation to $R_{on} = 0$, we obtain the total difference between the nominal length L_{nom} and the effective length L_{eff}, in the hypothesis that this difference is the same for all the MOS transistors measured; if all the MOS transistors come from the same chip, this assumption is certainly realistic. Similarly, the graph of G_{on} vs. the nominal width W_{nom} in Fig. 6-10b gives ΔW.

To obtain greater precision, we can repeat the measurements with different values of V_{GS} and find the average values of ΔL and ΔW. If the

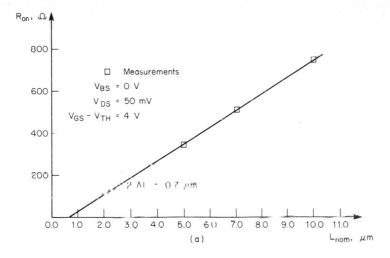

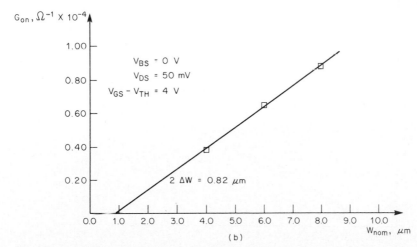

Figure 6-10 Electrical measurement of effective length and width of a MOS transistor, without the effect of series resistance.

threshold voltage is not the same on all transistors with different W or L—that is, short- and narrow-channel effects are noticeable—it is better to make the measurements at constant $V_{GS} - V_{TH}$ instead of at constant V_{GS}. The values obtained from the example in Fig. 6-10 are $L_{nom} - L_{eff} = 0.70\ \mu m$ and $W_{nom} - W_{eff} = 0.82\ \mu m$; in order to separate the contributions indicated in Eq. (6-14) optical measurements are necessary.

If the channel length is small, R_{on} becomes comparable to the parasitic series resistance external to the MOS transistor, which disturbs the measurement of the effective length. Rather complex methods of measurement exist to evaluate at the same time the total resistance series $R_t = R_D = R_S$ and the effective length L_{eff} [2, 3]. This explanation will be limited to the simplest case, following the hypothesis that the series resistance is constant and equal for all the MOS measured; moreover, the effect of the voltage drop on R_S on the voltage V_{BS} will be neglected, and therefore its influence on the value of V_{TH} will be neglected, too.

In this case, adding the contribution of the total parasitic resistance R_t in Eq. (6-15), we obtain

$$R_{on} + R_t = \frac{L_{nom} - \Delta L}{W_{nom} - \Delta W} \frac{1}{KP(V_{GS} - V_{TH})} \qquad (6\text{-}17)$$

This means that the linear interpolation of R_{on} against L_{nom}, with V_{GS} constant, intersects not on the axis $R_{on} = 0$ but at the point $R = R_t$ and $L_{nom} = \Delta L$ (see Fig. 6-11), where R_t is the sum of all the parasitic series

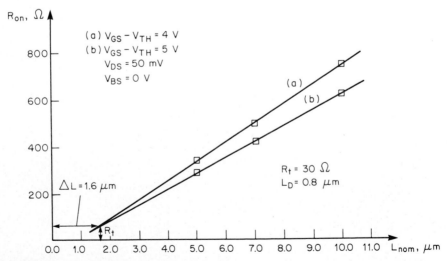

Figure 6-11 Electrical measurement of the effective length and of the resistance series of a MOS transistor.

resistances at the source and at the drain and ΔL is the difference between L_{nom} and L_{eff}.

From the example in Fig. 6-11 we find that the value of the effective length for a transistor with a nominal length of $L_{\text{nom}} = 5$ μm is $L_{\text{eff}} = 3.4$ μm, which is significantly different from that found in Fig. 6-10a, neglecting the series resistance. We therefore obtain $\Delta L = 1.6$ μm, equivalent to the double of the process value of X_{jl}; substituting the values of L_{eff} found in this way and the process value of the lateral diffusion X_{jl} in Eq. (6-13), we find that the error L_{err}, due to the real length of the gate, is very small.

The series resistance calculated from the example measurements in Fig. 6-11 is $R_t = 30$ Ω. To insert this value in SPICE, we can use the parameters for the resistance of the source and of the drain ($R_S = R_D = 15$ Ω) or the parameter for the sheet resistance (R_{SH}), which is multiplied by the number of square in series, i.e., by the parameters NRD and NRS in the description of the MOS transistor. Therefore, it is best to use R_D and R_S if the series resistances of the transistors in the simulation all have the same width and to use R_{SH} in all other cases.

6.3.2   Parameters for the model of the conductance in the saturation region

Once the effective dimensions are known, we can calculate the dimension-dependent parameters, beginning at the simplest equations of the conductance in the saturation region available on the LEVEL 2 model; the parameter used by this model is λ, which relates to the average conductance for $V_{DS} > V_{D,\text{sat}}$ as follows:

$$G_{D,\text{sat}} = \frac{I_{D,\text{sat}}\lambda}{(1 - \lambda V_{DS})^2} \qquad (6\text{-}18)$$

The value of λ that we obtain from Eq. (6-18) can be calculated approximately, neglecting the term λV_{DS}, from the following equation:

$$\lambda = \frac{G_{D,\text{sat}}}{I_{D,\text{sat}}} \qquad (6\text{-}19)$$

The value of λ calculated in Fig. 6-12 is 0.053 V^{-1}, while the value computed by SPICE with Eq. (4-50), starting from the substrate doping, is 0.103 V^{-1} for $V_{GS} = V_{DS} = 5$ V.

Remember that the corrective term of the current in the saturation region [see Eq. (4-41)], to avoid discontinuity between the two regions, acts in the linear region as well [see Eq. (4-40)]. This term, however, gives a false reading for the current in the linear region, with respect to that calculated with the model in Sec. 6.2.1; to solve the problem, reduce the value of KP, which has already been calculated, by a term dependent on λ and on a value of V_{DS} to be chosen:

$$KP' = KP(1 - \lambda V_{DS}) \tag{6-20}$$

The conductance in the saturation region can also be simulated with a more complex model (see Sec. 4.3.7); this model will be described near the end of this chapter because the parameters X_j and v_{max} must be used but have not yet been calculated.

6.3.3 Parameters for the short-channel effect on the threshold voltage

The equations in Sec. 4.3.5 relate the threshold voltage to the channel length, if this is small enough to be compared with the depth of junction X_j or with the thickness of the depleted regions.

The parameter used in SPICE to model the effect of the short channel on the threshold voltage is the depth of the source and drain junctions, that is, X_j. We are dealing therefore with a parameter with a precise physical significance; however, due to the limited precision of the model, if we

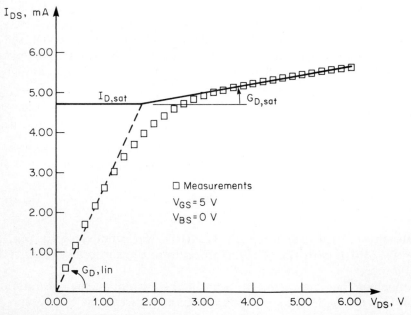

Figure 6-12 Determination of the average conductance in saturation and calculation of the parameter λ.

calculate X_j starting from the measurements of the threshold voltages for the short channel, the value obtained is not closely related to the real one.

The model uses the body-effect constant for the long channel γ, whose value is modified for short-channel MOS transistors with Eq. (4-53). The value of the depth of the junction X_j is therefore calculated by solving Eq. (4-53), with the hypothesis that the depleted regions under the source and drain are equal; that is, with V_{DS} very small, X_j can be calculated with the following explicit equation:

$$X_j = \frac{L_{\text{eff}}^2(1 - \gamma'/\gamma)^2}{2[W_S - L_{\text{eff}}(1 - \gamma'/\gamma)]} \tag{6-21}$$

where γ' is the body-effect constant calculated for the voltage V_{BS}, and W_S is the thickness of the depleted regions under the source, the drain, and the channel:

$$W_S = \sqrt{\frac{2\epsilon_s}{qN_A}(V_{BS} + 2\phi_p)} \tag{6-22}$$

Another possibility is to calculate a new value for the substrate doping N_A from the measurements of threshold voltage on a short-channel MOS transistor; this is done by using in the calculations and in the SPICE .MODEL card the real value for X_j. In this case, however, it is necessary to indicate explicitly in the SPICE .MODEL card all the parameters that the program would otherwise calculate from N_A: γ, $2\phi_p$, and ϕ_J. The value of N_A that solves Eq. (4-53) is that for which the thickness of the depleted regions W_S is equal to

$$W_S = \frac{L_{\text{eff}}^2(1 - \gamma'/\gamma)^2}{2X_j} + L_{\text{eff}}\left(1 - \frac{\gamma'}{\gamma}\right) \tag{6-23}$$

from which

$$N_A = \frac{2\epsilon_s}{qW_S^2}(V_{BS} + 2\phi_p) \tag{6-24}$$

The measurements in Fig. 6-13 have been carried out in the linear region of the short-channel MOS transistor, which was chosen as an example.

The value for the short-channel body effect calculated with $V_{BS} = -5$ V is $\gamma' = 0.269$ $V^{1/2}$. In the example, both methods have been tried: first the value of X_j has been calculated from Eq. (6-21) using the value of N_A calculated in Sec. 6.2.2; then N_A has been calculated from Eqs. (6-23) and

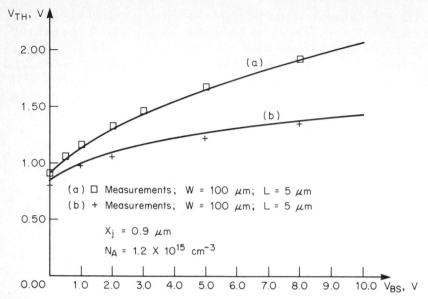

Figure 6-13 Measurements and simulations with the Yau model of the body effect of a short-channel MOS transistor.

(6-24) using the real value of X_j. The results obtained with the two procedures are as follows:

1. From Eq. (6-21):

$$N_A = 1.2 \times 10^{15} \text{ cm}^{-3} \qquad X_j = 1.1 \ \mu\text{m}$$

2. From Eq. (6-23):

$$N_A = 1.1 \times 10^{15} \text{ cm}^{-3} \qquad X_j = 1.0 \ \mu\text{m}$$

Verification made with a "best-fitting" program, solving Eq. (4-53) without any simplifying hypothesis and using all the measurements in Fig. 6-13, has given

$$N_A = 1.2 \times 10^{15} \text{ cm}^{-3} \qquad X_j = 0.9 \ \mu\text{m}$$

The latter two values provide the most precise results.

6.3.4   Parameters for the short-channel effect on mobility

As seen in Sec. 4.3.7, the reduction of mobility with the transverse electric field E_y is represented in this model by a reduction in the saturation voltage related to the parameter v_{max}.

This same parameter is used, together with X_j and N_{eff}, in the calculation of the conductance in the saturation region with a model different from that seen in Sec. 6.3.2.

Given the complexity of the models, there are no methods that are simple and sufficiently precise to evaluate v_{max} and N_{eff}. Thus, we can only try varying v_{max} in order to find the right saturation current (by increasing v_{max}, $I_{D,sat}$ increases) and varying N_{eff} in order to find the conductance (by increasing N_{eff}, $G_{D,sat}$ decreases). Figure 6-14 represents the best result obtainable with the LEVEL 2 model in the simulation of the short-channel transistor used as an example, with the parameters $v_{max} = 3 \times 10^4$ V/cm and $N_{eff} = 10$.

6.3.5 Parameters for the narrow-channel effect

As seen in Sec. 4.3.8, the narrowness of the channel influences the threshold voltage as well. The model uses parameter δ in order to calculate the value of the threshold voltage in relation to the width W. The value of δ can be calculated from the fitting of the body-effect curve of a narrow-channel MOS transistor; in our example, $W_{eff} = 5.2$ μm. We can find δ with a trial-and-error procedure similar to that used to calculate X_j (see

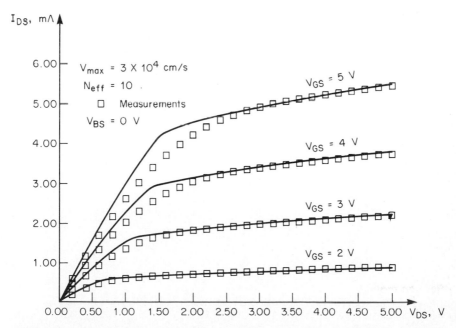

Figure 6-14 Measurements and simulations of the output characteristics of a LEVEL 2 model of a short-channel MOS transistor.

Fig. 6-15) or with a graph [1] evaluating the dependence of the measurements of V_{TH} on $1/W$ by using the formula

$$\delta = \frac{\Delta V_{TH}}{\Delta(1/W)} \frac{4C'_{ox}}{\pi X_D^2} \tag{6-25}$$

where

$$X_D = \sqrt{\frac{2\epsilon_s}{qN_A}}$$

To use this method, however, we must have a sufficient number of MOS transistors with different widths, although what we usually need to know is the dependence of the threshold voltage on the substrate voltage V_{BS}—this for the typical minimum width of the MOS transistors permitted by the process.

The MOS transistors that are at the same time "short" and "narrow" are worthy of special note. These cannot always be simulated by simply overlapping the short-channel and narrow-channel effects, but often require that we choose the values of the parameters X_j, δ, and v_{max} which are specific for them; that is due also to the difficulty of calculating the effective dimensions of these devices with the methods seen in Sec. 6.3.1.

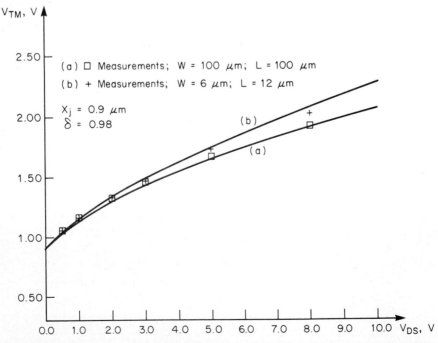

Figure 6-15   Measurements and simulations of the body effect of a narrow-channel MOS transistor.

6.4   Calculation of the Parameters of the LEVEL 3 Model

The equations for the LEVEL 3 model have been given in Sec. 4.4; even if the model has been explicitly developed for short-channel MOS transistors, the fundamental parameters V_{T0}, KP (or μ), and γ (or N_A) are those of a long-channel MOS and can be calculated as for the other models.

This model is applicable to MOS transistors with L_{nom} of 2 μm. However, in the example, we will again employ the MOS transistor used to characterize the LEVEL 2 model ($L_{nom} = 5$ μm), so as to be able to make a comparison between the results obtained by simulating the same MOS transistor with the two models.

6.4.1   Measurements in the linear region

For this model, too, we begin with the transfer characteristics in the linear region, using the equation for the simplest model in the calculations; it is still true that, for a small V_{DS}, the more complex models and the simplest model supply the same current values. Therefore, once V_{T0}, γ, μ_0, N_A, and $2\phi_p$ have been found (in the ways already explained), we must find the parameters for the model of the mobility variation with V_{GS} (see Sec. 4.4.1); this is different from what we saw in Sec. 6.2.1. The simplest method consists of calculating β' with at least two different values for V_{GS} and then finding the values of KP (or μ_0) and of θ to be inserted in the model by solving a system of two equations and two unknowns starting from Eq. (4-82):

$$\theta = \frac{\beta_1'/\beta_2' - 1}{V_{GS2} - V_{T0} - (\beta_1'/\beta_2')(V_{GS1} - V_{T0})} \tag{6-26}$$

$$KP = \beta_1' \frac{L}{W_{eff}} [1 + \theta(V_{GS1} - V_{T0})] \tag{6-27}$$

We can obtain a better result with an easily applicable method, with the help of a calculator if necessary [4].

Using an iterative method, we must calculate V_{T0} and θ, starting from a value of V_{T0} linearly extrapolated from the measurements with V_{GS} nearer to the threshold voltage V_{TH}, provided that it is in a region of strong inversion. Then we have to calculate β and θ with a linear regression method from the equation

$$\frac{\beta}{\beta_i'} = 1 + \theta(V_{GSi} - V_{T0}) \tag{6-28}$$

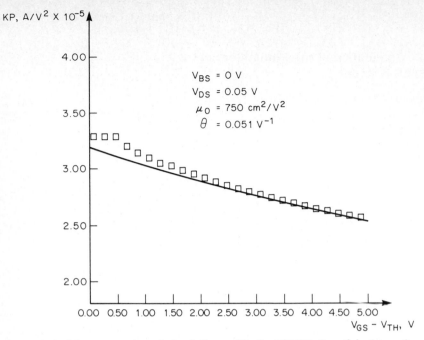

Figure 6-16 Measurements and simulations with the LEVEL 3 model of transfer characteristics, calculation of β and θ.

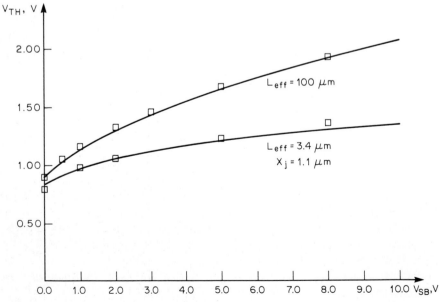

Figure 6-17 Measurements and simulations with the Dang model of body effect of a short-channel MOS transistor.

where β'_i is the value of β' calculated for $V_{GS} = V_{GSi}$. Once β and θ have been obtained, we can recalculate V_{T0} from Eq. (6-28) for each value of V_{GSi} and β'_i and repeat the procedure up to the convergence, which is usually reached very quickly. For the example in Fig. 6-16, the same measurements as in Fig. 6-2 have been used; the values calculated previously (Sec. 6.2) are

$$V_{T0} = 0.90 \text{ V}$$

$$\gamma = 0.46 \text{ V}^{1/2}$$

$$N_A = 1.2 \times 10^{15} \text{ cm}^{-3}$$

With the simple method, and $V_{GS1} = 2 \text{ V}$, $V_{GS2} = 5 \text{ V}$:

$$KP = 30.9 \ \mu\text{A/V}^2$$

$$\theta = 0.051 \text{ V}^{-1}$$

$$\mu_0 = 715 \text{ cm}^2/(\text{V}\cdot\text{s})$$

With the iterative procedure:

$$KP = 32.4 \ \mu\text{A/V}^2$$

$$\theta = 0.052 \text{ V}^{-1}$$

$$\mu_0 = 750 \text{ cm}^2/(\text{V}\cdot\text{s})$$

These latter values provide greater precision; they have been obtained, minimizing the error, by using a least-squares fit starting from a greater number of measurements than in the first method.

6.4.2 Parameters for the short-channel effect on the threshold voltage

The model under examination (see Sec. 4.4.2) uses the two parameters for the effect of the short channel on the threshold voltage: X_j and η. The parameter X_j is used in a model of the body effect similar to, but more accurate than, that used in the equations of the LEVEL 2 model.

The value of X_j calculated with Eq. (6-21) can be used only as an initial value for a trial-and-error procedure (see Fig. 6-17).

The parameter η, proportional to the variation of the threshold voltage with V_{DS} (see Sec. 4.4.2), is easily calculable from the threshold measurements in the saturation region and with V_{DS} constant (see Fig. 6-17). We can use the second method in Sec. 6.1.2.

The parameter η can be obtained from the linear dependence of V_{TH} on V_{DS} by inverting Eq. (4-84):

$$\eta = \frac{\Delta V_{TH}}{\Delta V_{DS}} \frac{C'_{ox} L_{eff}^3}{8.15 \times 10^{-22}} \qquad (6\text{-}29)$$

From the example in Figs. 6-17 and 6-18 we can calculate

$$X_j = 1.1 \ \mu\text{m} \qquad \eta = 0.98$$

6.4.3 Parameters for the short-channel effect on mobility

This model also uses the parameter v_{max}, but in a way different from that seen in Sec. 6.3.4. In fact, as well as being part of the calculation of $V_{D,sat}$, the parameter v_{max} reduces the effective mobility μ_{eff} in relation to the average electrical field in the channel $E_y = V_{DS}/L_{eff}$.

The method suggested in Ref. [1] consists of calculating the mobility with a small V_{DS}, using the method of Sec. 6.4.1, and then calculating, with an increasing V_{DS} and a constant V_{GS}, the values of the effective mobility μ_{eff} from the measured current.

We can then calculate v_{max} also by using a graph based on the following equation, where μ_s is the mobility calculated with Eq. (4-82) and μ_{eff} is the

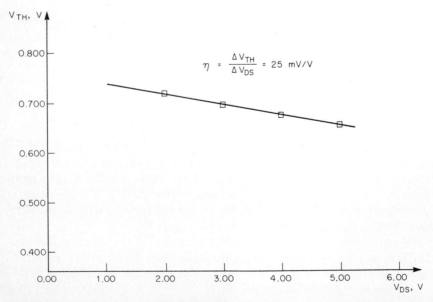

Figure 6-18 Evaluation of η with Eq. (6-29).

mobility calculated with Eq. (4-89):

$$\frac{1}{\mu_{\text{eff}}} - \frac{1}{\mu_s} = \frac{V_{DS}}{L_{\text{eff}}v_{\text{max}}} \tag{6-30}$$

The method explained is easily applicable, using a program that obtains μ_{eff} from the measurements; otherwise, it is better to proceed by trial and error.

In the example, the value of v_{max} has been calculated by trial and error in the way shown in Fig. 6-19. Then the value $v_{\text{max}} = 1.2 \times 10^5$ cm/s has been found, which is different from that found for the LEVEL 2 model. The calculation can be repeated again for the other values of V_{GS}.

6.4.4 Parameters for the model of the conductance in the saturation region

The parameter used for this model (see Sec. 4.4.4) is K.

Again a graphical method is proposed in Ref. [1]; it is advisable, however, to carry out some control simulations and, if necessary, correct the K value by trial and error.

The relationship between the value of K and the shortening of the channel is obtained from Eq. (4-57); from this equation we can get an approximate equation that gives the value of K starting from the short-

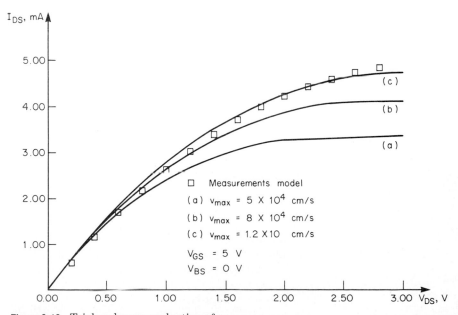

Figure 6-19 Trial-and-error evaluation of v_{max}.

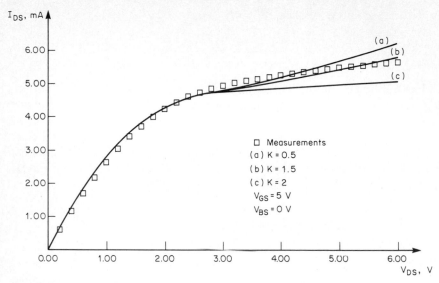

Figure 6-20   Trial-and-error evaluation of K.

ening of the channel in the saturation region:

$$KX_D^2 = \frac{\Delta(L_{\text{eff}} - L')^2}{\Delta V_{DS}} \tag{6-31}$$

where

$$L_{\text{eff}} - L' = L_{\text{eff}}\left(1 - \frac{I_{DS}}{I_{D,\text{sat}}}\right)$$

In Fig. 6-20 the measurements of an output characteristic and simulations with different values for K are shown. The simulation shown in Fig. 6-21, in which the value of K is 1.5, represents the best result obtained with the LEVEL 3 model for this example.

6.5   Measurements of Capacitance

6.5.1   Gate capacitance

The value of t_{ox} can be calculated from the measurements of the gate capacitance in the strong inversion region of a MOS with large dimensions (to avoid problems caused by parasitic capacitance) by using the equation

$$t_{\text{ox}} = \frac{\epsilon_s(\text{area})}{C_G} \tag{6-32}$$

The overlap capacitance between the gate and the diffusions is difficult to measure with precision, unless there are adequate test chips available. Knowing the lateral diffusion X_{jl} in the processes with a self-aligned gate, or the length of the overlap zone for the processes that are not self-aligned, we can calculate the parameters C_{GDO} and C_{GSO} according to the theoretical formula

$$C_{GDO} = C_{GSO} = C'_{ox}X_{jl} \tag{6-33}$$

Other, more precise formulas have been proposed [5, 6], to take into account the distortion of the electric field at the edge of the gate electrode:

$$C_{GSO} = C'_{ox}X_{jl}\left[1 + \frac{t_{ox}}{X_{jl}}\ln\left(1 + \frac{t_g}{t_{ox}}\right)\right.$$

$$\left. + \frac{t_g}{X_{jl}}\ln\left(1 + \frac{X_{jl}/2}{t_{ox} + t_g}\right)\right] \tag{6-34}$$

where t_g is the thickness of the gate electrode.

The parameter C_{GBO} can be calculated from the thickness of the thick oxide and from the length of the overlap between the gate and the substrate:

$$C_{GBO} = \frac{\epsilon_{ox}}{t_{oxf}}W_g \tag{6-35}$$

In most cases, however, the capacitance C_{GBO} can be neglected.

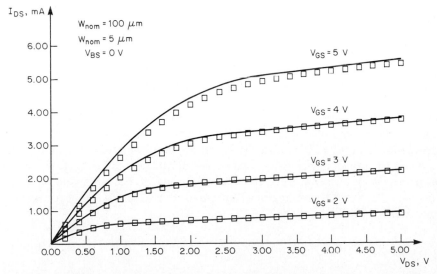

Figure 6-21 Measurements and simulation of the output characteristics of a short-channel MOS transistor with the LEVEL 3 model.

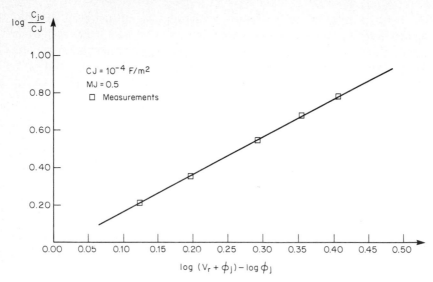

Figure 6-22   Graphic evaluation of the parameters *MJ* and *MJSW*.

6.5.2 Capacitance of the diffused regions

To evaluate the values per unit area and perimeter of the junction capacitances, we have to measure the capacitance vs. inverse voltage over two diffused regions, designed with different ratios between area and perimeter; these measurements make it possible to solve the following equations, for every voltage:

$$A_1 C_{ja} + P_1 C_{jp} = C_1$$
$$A_2 C_{ja} + P_2 C_{jp} = C_2$$

(6-36)

where the areas and perimeters of the diffused regions (A_1, A_2, P_1, P_2) are known, and C_1 and C_2 are the values of capacitance measured at the reverse voltage V_r.

Once we have determined C_{ja} and C_{jp} as a function of the inverse voltage V_r, the values of *CJ* and *CJSW* are those for $V_r = 0$ V; thus the exponents *MJ* and *MJSW* and the junction potential ϕ_j can be calculated by using the graph in Fig. 6-22. The experimental points in Fig. 6-23 have been obtained by separating the contributions of the area and of the perimeter of the diffused regions expressed in Eq. (4-80). On a *log* scale they are as follows:

$$\log \frac{C_{ja}}{CJ} = MJ \left[\log \phi_J - \log (V_r + \phi_j) \right]$$

(6-37)

$$\log \frac{C_{jp}}{CJSW} = MJSW[\log \phi_j - \log (V_r + \phi_j)]$$

(6-38)

The values calculated from the measurements using Eq. (6-35) and the results of the simulation are shown in Fig. 6-23.

Finally, the parameters I_S, the saturation current of the diode, and FC, the capacitance in forward bias, need not be indicated on the SPICE .MODEL card. If it is also necessary to simulate correctly the behavior of the junctions in forward bias, these parameters can be calculated in the way shown in Chap. 5.

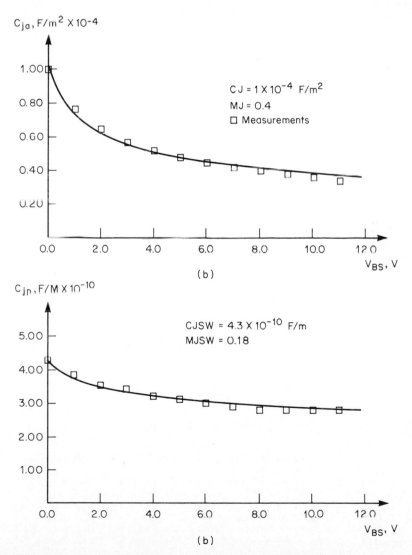

Figure 6-23   Measurements and simulations of (a) C_{ja} and (b) C_{jp}.

6.6   Final Considerations

Figures 6-4, 6-7, 6-14, and 6-21 are already sufficient to give an idea of the precision which can be obtained from the three models. As far as the measurements are concerned, we must note the high number of parameters to be calculated for the LEVEL 2 and LEVEL 3 models, with respect to those necessary for the LEVEL 1 model. The difficulty in calculating some parameters of the more complex models is the price to be paid if greater precision is needed.

An alternative solution is using an optimization program. In such a case we can use the methods in this chapter to supply a "starting point," which is already very near to the optimal solution.

One example is the SUXES program [7], which can be used to optimize the value of all parameters of the SPICE models or of any other model.

The optimization programs should be used with great care because the result of the fitting can be strongly influenced by the range of voltages. For example, by supplying the program measurements only with gate voltages that are too high with respect to the threshold voltage, the value of V_{T0} determined by the program will, in this case, be noticeably different from the real value. In the same way, calculating the optimal values of the parameters that have little influence on the values of the current calculated from the model can lead to solutions which are physically absurd; for this purpose it is possible to assign weight and ranges of variation to the single parameter. Another method calculates groups of parameters, following procedures similar to those seen in this chapter. In order to calculate the value of the parameters for the LEVEL 3 model, best suitable to this kind of approach, it is convenient to first find V_{T0}, KP, θ, γ, X_j, and $2\phi_p$, using only measurements with a low V_{DS}; then the other parameters v_{max}, η, and K can be found by using measurements of the output characteristics.

In conclusion, it is clear that using programs of this type is a great help, but the methods of calculation shown in this chapter are essential, both as experience for those who face these problems and as verification of the results obtained from the fitting program.

REFERENCES

1. A. Vladimirescu, A. R. Newton, and D. O. Pederson, *SPICE2 Version 2G User's Guide*, University of California, Berkeley, 1980.
2. P. I. Suciu and R. L. Johnston, Experimental Derivation of the Source and Drain Resistance of MOS Transistors, *IEEE Trans. Electron Devices*, **27**, 1980.
3. K. Terada and H. Muta, A New Method to Determine the Effective Channel Length, *Jpn. J. Appl. Phys.*, **18**(5), May 1979.
4. F. M. Klaassen, *Characterization and Measurements of MOS Transistors*, NATO Advanced Study Institute on Process and Device Modelling, Nijhoff, the Hague, 1978.

5. M. I. Elmasry, Capacitance Calculation in MOSFET VLSI, *IEEE Electron Device Lett.*, **EDL-3**(1), January 1982.
6. C. P. Yuan and T. N. Trik, A Simple Formula for the Estimation of the Capacitance of Two-Dimensional Interconnects in VLSI Circuits, *IEEE Electron Device Lett.*, **EDL-3**(12), December 1982.
7. D. Ward and K. Doganis, Optimized Extraction of MOS Model Parameters, *IEEE Trans. Comput. Aided Design Integrated Circuits Syst.* **CAD-1**(4), October 1982.

Noise and Distortion

Ermete Meda

Department of Electronics (DIBE), University of Genoa, Genoa, Italy

7.1 Noise

Noise phenomena are caused by the small current and voltage fluctuations generated within the devices themselves. This section deals with the physical causes that give rise to noise in semiconductor devices and reviews the noise models implemented in SPICE.

7.1.1 Noise spectral density

The noise voltage or current $x(t)$ is a quantity with an average value that is often zero. Thus the most significant mathematical characterization is obtained using the mean-square value of $x(t)$:

$$\overline{X^2} = \lim_{T \to \infty} \frac{1}{T} \int_0^T x^2(t) \, dt \qquad (7\text{-}1)$$

or its root-mean-square (rms) value:

$$X = \sqrt{\overline{X^2}} \qquad (7\text{-}2)$$

Moreover, as the variable $x(t)$ might contain frequency components distributed on a large spectrum, it is necessary to correlate $\overline{X^2}$ to its power spectral density $S_x(f)$.

It can be shown that $\overline{X^2}$ and $S_x(f)$ are related through the following equation:

$$\overline{X^2} = \int_0^\infty S_x(f)\ df \tag{7-3}$$

The physical meaning of this expression is clear: The mean-square value of the noise, $\overline{X^2}$, is equal to the sum, extended to the whole spectrum, of contributions at different frequencies.

7.1.2 Noise sources†

a. Shot noise.   *Shot noise* is always associated with a direct-current flow and is present in diodes and bipolar transistors. The origin of shot noise can be understood by considering the *pn* junction and the carrier concentrations in the device in the forward-bias condition (see Chap. 1 and Appendix A).

The flow of each carrier across the junction is a random event and is dependent on the carrier having sufficient energy and velocity directed toward the junction. Thus external current I, which appears to be a steady current, is, in fact, composed of a large number of random, independent current pulses [1].

The fluctuation in I is termed *shot noise* and is generally specified in terms of its mean-square variation with an average value of I_D. This is written as $\overline{i^2}$, where

$$\overline{i^2} = \overline{(I - I_D)^2} = \lim_{T \to \infty} \frac{1}{T} \int_0^T (I - I_D)^2\ dt \tag{7-4}$$

It can be shown that if a current I is composed of a series of random, independent pulses with average value I_D, then the resulting noise current has a mean-square value

$$\overline{i^2} = S_i(f)\ \Delta f = 2qI_D\ \Delta f \tag{7-5}$$

where Δf is the bandwidth and $S_i(f)$ is the power spectral density. Equation (7-5) is valid until the frequency becomes comparable to $1/\tau$, where τ is the carrier transit time through the depletion layer.

---

† The material in this section is derived primarily from P. R. Gray and R. G. Meyer, *Analysis and Design of Analog Integrated Circuits,* copyright © 1977 by John Wiley & Sons, Inc. Reprinted by permission.

The effect of shot noise can be represented in the low-frequency, small-signal equivalent circuit of the diode, as shown in Fig. 7-1.

b. Thermal noise. *Thermal noise* is generated by a mechanism different from that responsible for shot noise. In conventional resistors it is due to the random thermal motion of the electrons and is unaffected by the presence or absence of direct current, since typical electron drift velocities in a conductor are much less than electron thermal velocities. Since this source of noise is due to the thermal motion of electrons, it is related to absolute temperature T. In fact, thermal noise is *directly proportional* to T (unlike shot noise, which is *independent* of T), and as T approaches zero, thermal noise also approaches zero [1].

In a resistor R, thermal noise can be shown to be represented by a shunt current source $\overline{i^2}$, as in Fig. 7-2. This is equivalent to

$$\overline{i^2} = S_i(f)\,\Delta f = \frac{4kT}{R}\,\Delta f \qquad (7\text{-}6)$$

where k is Boltzmann's constant. From Eq. (7-6) it can be seen that thermal noise is frequency-independent.

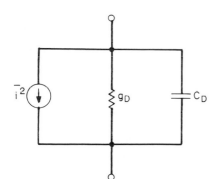

Figure 7-1 Junction diode small-signal equivalent circuit with noise.

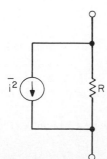

Figure   7-2 Representation   of thermal noise.

c. Flicker noise, or $1/f$ **noise.** *Flicker noise* is a type of noise found in all active devices (as well as in some discrete passive elements, such as carbon resistors). The origins of flicker noise are varied, but in BJTs it is caused mainly by traps associated with contamination and crystal defects in the emitter-base depletion layer. These traps capture and release carriers in a random fashion, and the time constants associated with the process give rise to a noise signal with energy concentrated at low frequencies [1].

Flicker noise is always associated with a flow of direct current and displays a spectral density of the form

$$\overline{i^2} = S_i(f)\,\Delta f = k_f \frac{I^{a_f}}{f^b}\,\Delta f \tag{7-7}$$

where Δf is a small bandwidth at frequency f, I is a direct current, k_f is a constant for a particular device (it depends on contamination and crystal imperfections), a_f is a constant in the range 0.5 to 2, and b is a constant approximately unity. If $b = 1$ in Eq. (7-7), the noise spectral density has $1/f$ frequency dependence, thus the alternative name $1/f$ noise.

d. Burst, or popcorn, noise. *Burst noise* is another type of low-frequency noise found in some integrated circuits and discrete transistors. The source of this noise is not fully understood, although it has been shown to be related to the presence of heavy-metal ion contamination. Gold-doped devices show very high levels of burst noise [1]. The spectral density of burst noise can be shown to be

$$\overline{i^2} = S_i(f)\,\Delta f = k_b \frac{I^c}{1 + (f/f_c)^2}\,\Delta f \tag{7-8}$$

where k_b is a constant for a particular device, I is a direct current, c is a constant in the range 0.5 to 2, and f_c is a particular frequency for a given noise process [2].

7.1.3   Noise models and their implementation in SPICE

The ac analysis portion of SPICE contains the capability of evaluating the noise characteristics of an electronic circuit. Every resistor in the circuit generates thermal noise, and every semiconductor device in the circuit generates both shot and flicker noise, in addition to the thermal noise generated by the ohmic resistances of the device [3].

The noise generated by a circuit element can be modeled as an electrical excitation to the small-signal circuit. Each noise source in the circuit is statistically uncorrelated to the other noise sources in the circuit. The contribution of each noise source in the circuit to the total noise, at a specified

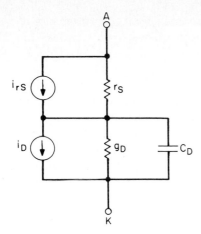

Figure   7-3 Complete   diode small-signal equivalent circuit with noise sources.

output, is determined separately. The total noise is then the rms sum of the individual noise contributions.

Noise is specified with respect to a certain noise bandwidth and has the units $V/Hz^{1/2}$ or $A/Hz^{1/2}$. The output noise of a circuit is the rms level of noise at the output divided by the square root of the bandwidth. The equivalent input noise is the output noise divided by the transfer function of the output with respect to the specified input.

In this section, the noise models for diodes, BJTs, and MOSFETs implemented in SPICE are presented.

It is important to note that the equations developed in the following pages for each device model are implemented in SPICE by their rms values [see (Eq. 7-2)], as shown in the figures.

a. Diode-noise model.†   The equivalent circuit for a junction diode was considered briefly in the description of shot noise. The basic equivalent circuit of Fig. 7-1 can be made complete by adding series resistance r_S as shown in Fig. 7-3. Since r_S is a physical resistor due to the resistivity of the silicon, it exhibits thermal noise. Experimentally it has been found that any flicker noise present can be represented by a current source in shunt with $\overline{i_D^2}$; this is conveniently combined with the shot-noise source as indicated by

$$\overline{i_{rS}^2} = \frac{4kT}{r_S} \Delta f$$

$$\overline{i_D^2} = 2qI_D \, \Delta f + k_f \frac{I_D^{af}}{f} \Delta f$$

(7-9)

---

† The material in Secs. 7.1.3a to c is taken from P. R. Gray and R. G. Meyer, *Analysis and Design of Analog Integrated Circuits,* copyright © 1977 by John Wiley & Sons, Inc. Reprinted by permission.

The model parameters k_f and a_f are estimated from measurements of the diode noise at low frequency. For silicon diodes, typical values for these parameters are $k_f = 10^{-16}$ and $a_f = 1$.

b. BJT noise model.   In a bipolar transistor in the forward active region, minority carriers diffuse and drift across the base region to be collected at the collector-base junction. Minority carriers entering the collector-base depletion region are accelerated by the field existing there and swept across this region to the collector. The time of arrival at the collector-base junction of the diffusion (or drifting) carriers is a random process, and thus the transistor collector current consists of a series of random current pulses. Consequently, collector current I_C shows shot noise, and this is represented by a shot-noise current source $\overline{i_C^2}$.

Base current I_B in a transistor is due to recombination in the base and base-emitter depletion regions and also to carrier injection from the base into the emitter. All of these are independent, random processes, and thus I_B also shows shot noise. This is represented by shot-noise current source $\overline{i_B^2}$ in Fig. 7-4.

Transistor base resistor r_B is a physical resistor and thus has thermal noise. Collector series resistor r_C also shows thermal noise, but since this is in series with the high-impedance collector node, this noise is negligible and is usually not included in the model. Note that resistors r_π and r_o in the model are *fictitious* resistors that are used for modeling purposes only, and they do *not* exhibit thermal noise [1].

Flicker noise and burst noise in a bipolar transistor have been found experimentally to be represented by current sources across the internal

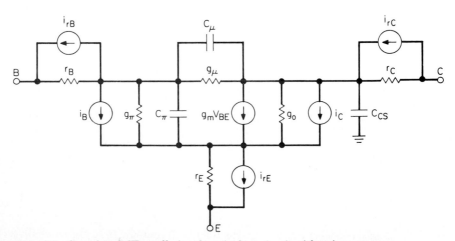

Figure 7-4   Complete BJT small-signal equivalent circuit with noise sources.

base-emitter junction. These are conveniently combined with the shot noise sources in $\overline{i_B^2}$.

The full small-signal equivalent circuit including noise for the BJT is shown in Fig. 7-4. Since they arise from separate, independent physical mechanisms, all the noise sources are independent of each other and have mean-square values.

$$\overline{i_{rB}^2} = \frac{4kT}{r_B}\,\Delta f$$

$$\overline{i_{rC}^2} = \frac{4kT}{r_C}\,\Delta f$$

$$\overline{i_{rE}^2} = \frac{4kT}{r_E}\,\Delta f \tag{7-10}$$

$$\overline{i_C^2} = 2qI_C\,\Delta f$$

$$\overline{i_B^2} = 2qI_B\,\Delta f + k_f\frac{I_B^{a_f}}{f}\,\Delta f + k_b\frac{I_B}{1 + (f/f_c)^2}\,\Delta f$$

This equivalent circuit is valid for both *npn* and *pnp* transistors. For *pnp* devices, the magnitudes of I_B and I_C are used in the above equations. In SPICE, however, the burst-noise effect is not modeled.

c. JFET and MOS noise models.   It was shown in Chap. 3 that the resistive channel joining source and drain is modulated by the gate-source voltage so that the drain current is controlled by the gate-source voltage. Since

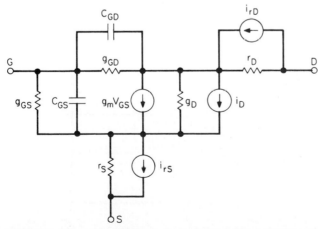

Figure 7-5 Complete JFET small-signal equivalent circuit with noise sources.

the channel material is resistive, it exhibits *thermal noise,* and this is the major source of noise in JFETs and MOS transistors. It can be shown that this noise source can be represented by a noise current generator $\overline{i_D^2}$ from drain to source in the JFET small-signal equivalent circuit of Fig. 7-5. Flicker noise in the JFET is also found experimentally to be represented by a drain-source current source, and these two can be lumped into one noise source $\overline{i_D^2}$. The other source of noise in JFETs is shot noise generated by the gate leakage current, but it is not modeled in SPICE.

In Fig. 7-6, the complete MOS small-signal equivalent circuit with noise source is shown. These sources in Figs. 7-5 and 7-6 have the following values:

$$\overline{i_{rD}^2} = \frac{4kT}{r_D} \Delta f$$

$$\overline{i_{rS}^2} = \frac{4kT}{r_S} \Delta f \qquad (7\text{-}11)$$

$$\overline{i_D^2} = \frac{8kTg_m}{3} \Delta f + k_f \frac{I_D^{a_f}}{f} \Delta f$$

where I_D is the drain bias current, k_f is a constant for a given device, a_f is a constant in the range 0.5 to 2, and g_m is the transconductance of the device at the operating point.

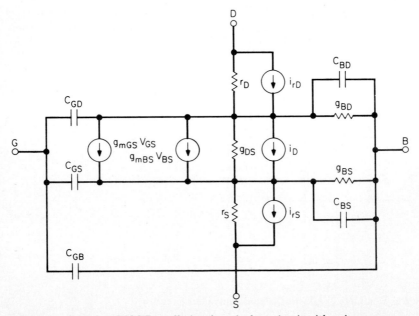

Figure 7-6 Complete MOST small-signal equivalent circuit with noise sources.

7.2 Distortion

This section deals first with the results necessary for a better understanding of harmonic distortion, at low and high frequency. Subsequently, the methodology used in SPICE for the calculation of distortion coefficients is outlined [4, 5].

7.2.1 Harmonic distortion coefficients

The methodology for determining the coefficients of harmonic distortion, at low and high frequency, is now briefly reviewed.

a. Low frequencies. At low frequencies, where reactive effects can be neglected, the output signal $y(t)$ (voltage or current) of a nonlinear circuit can be expressed in terms of its input $x(t)$ by a *Taylor* series [6]:

$$y(t) = a_1 x(t) + a_2 x^2(t) + a_3 x^3(t) + \cdots \qquad (7\text{-}12)$$

Coefficient a_1 represents the linear gain of the circuit, whereas coefficients a_2, a_3, ..., represent its distortion. It is important to note that, at low frequencies, a_1, a_2, a_3, ..., a_n are constants.

Applying a cosine wave of frequency ω and amplitude V_A at the input, the nth harmonic distortion (HD_n) is defined as the ratio of the component at frequency $n\omega$ to the one at the fundamental ω. This is obtained by a trigonometric manipulation [7] so that Eq. (7-12) becomes

$$y(t) = (a_1 + \tfrac{3}{4}a_3 V_A^2) V_A \cos \omega t + \tfrac{1}{2}a_2 V_A^2 \cos 2\omega t$$

$$+ \tfrac{1}{4}a_3 V_A^3 \cos 3\omega t + \cdots \qquad (7\text{-}13)$$

Under low-distortion conditions, only the second- and third-order distortion components are considered, so that harmonic distortion is then specified by

$$HD_2 = \frac{\text{amplitude 2d harmonic}}{\text{amplitude fundamental at the output}} = \frac{a_2 V_A}{2a_1} \qquad (7\text{-}14a)$$

$$HD_3 = \frac{\text{amplitude 3d harmonic}}{\text{amplitude fundamental at the output}} = \frac{a_3 V_A^2}{4a_1} \qquad (7\text{-}14b)$$

Applying now the sum of two cosine waves of frequencies ω_1 and ω_2, both of amplitude V_A at the input, gives rise to output signal components at all combinations of ω_1, ω_2, and their multiples. Under low-distortion conditions, the number of terms can be reduced to the ones caused by coefficients a_2 and a_3 only [7].

Second-order intermodulation distortion IM_2 is defined by the ratio of the component at frequency $\omega_1 + \omega_2$ to the one at ω_1 or ω_2, so that

$$IM_2 = \frac{a_2}{a_1} V_A \qquad (7\text{-}15)$$

and by comparison with Eq. (7-14a)

$$IM_2 = 2HD_2 \qquad (7\text{-}16)$$

Third-order intermodulation distortion IM_3 can be detected at the frequencies $2\omega_1 + \omega_2$ and $2\omega_2 + \omega_1$ and is given by

$$IM_3 = \frac{3}{4} \frac{a_3}{a_1} V_A^2 \qquad (7\text{-}17)$$

such that
$$IM_3 = 3HD_3 \qquad (7\text{-}18)$$

These components can be mapped vs. frequency, as shown in Fig. 7-7, by considering their Fourier transforms [6]

$$F(n\omega) = \int_{-\infty}^{\infty} V_A^n \cos n\omega t \, e^{-j\omega t} \, dt \qquad (7\text{-}19)$$

b. High frequencies.† At high frequencies the simple power-series approach cannot be used because reactive elements (linear and nonlinear) are important—for example, C_μ and C_π in the BJT small-signal model. Thus coefficients a_1, a_2, and a_3 are not constant but are functions of frequency [8].

Then, in this case, the linear term in Eq. (7-12), $a_1 x(t)$, is replaced by the convolution integral [$x(t) = 0$, for $t < 0$]:

$$y_1(t) = \int_0^t a_1(t - \tau) x(\tau) \, d\tau \qquad (7\text{-}20)$$

In the frequency domain, Eq. (7-20) may be written as

$$Y_1(\omega) = A_1(\omega) X(\omega) \qquad (7\text{-}21)$$

A generalization of the second-degree term $a_2 x^2(t)$ is the double convolution integral

$$y_2(t) = \int_0^t \int_0^t a_2(t - \tau_1, t - \tau_2) x(\tau_1) x(\tau_2) \, d\tau_1 \, d\tau_2 \qquad (7\text{-}22)$$

---

† The material in this section is taken from Narayanan [8]. Reprinted with permission from *The Bell System Technical Journal.* Copyright © 1967 by AT&T.

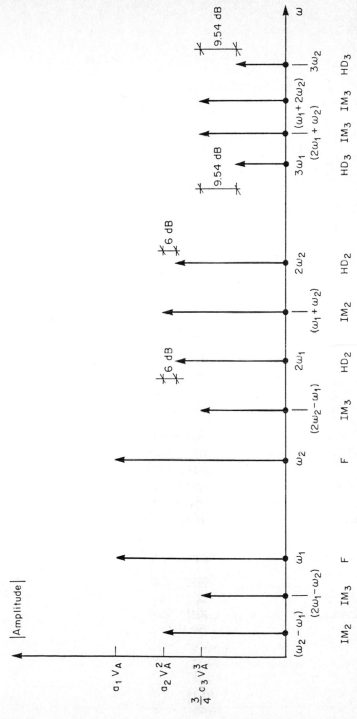

Figure 7-7 Intermodulation components in the output signal caused by coefficients a_1, a_2, and a_3 in the power series expansion. The input signals have frequencies ω_1 and ω_2 and amplitude V_A [6].

In the frequency domain, Eq. (7-22) can be written

$$Y_2(\omega_1, \omega_2) = A_2(\omega_1, \omega_2)X(\omega_1)X(\omega_2) \tag{7-23}$$

The expression in Eq. (7-23) involves a product of the input with itself, thus representing a quadratic system. $A_2(\omega_1, \omega_2)$ is known as the second-degree *Volterra kernel.*

It is obvious that when two sinusoidal signals at frequencies ω_a and ω_b are applied, the output at the harmonic frequencies $\omega_a \pm \omega_b$ is given by

$$|A_2(\omega_a \pm \omega_b)| \cos [(\omega_a \pm \omega_b)t + \varphi_{a \pm b}] \tag{7-24}$$

Since in general $A_2(\omega_a, \omega_b)$ will not be equal to $A_2(\omega_a, -\omega_b)$, different values of distortion at different harmonic frequencies are directly reflected in the kernel. Moreover, as in the low-frequency case of Eq. (7-16), the 2ω product is less by a factor of 2.

Likewise, the third-degree term $a_3x^3(t)$ can be generalized to a triple convolution integral

$$y_3(t) = \int_0^t \int_0^t \int_0^t a_3(t - \tau_1, t - \tau_2, t$$
$$- \tau_3)x(\tau_1)x(\tau_2)x(\tau_3) \, d\tau_1 \, d\tau_2 \, d\tau_3 \tag{7-25}$$

In the transform domain, Eq. (7-25) can be written as

$$Y_3(\omega_1, \omega_2, \omega_3) = A_3(\omega_1, \omega_2, \omega_3)X(\omega_1)X(\omega_2)X(\omega_3) \tag{7-26}$$

$A_3(\omega_1, \omega_2, \omega_3)$ is known as the third-degree Volterra kernel.

Thus, in this analysis, the output voltage

$$y(t) = y_1(t) + y_2(t) + y_3(t) \tag{7-27}$$

is expressed in terms of a Volterra series of the input signal $x(t)$, so that the kernels $A_1(\omega)$, $A_2(\omega_1, \omega_2)$, and $A_3(\omega_1, \omega_2, \omega_3)$ are the transfer ratios.

For example, the second and third harmonic distortions are given by

$$HD_2 = \frac{1}{2}\frac{A_2(2\omega)}{A_1(\omega)} V_A$$

$$HD_3 = \frac{1}{4}\frac{A_3(3\omega)}{A_1(\omega)} V_A^2 \tag{7-28}$$

7.2.2 Distortion models implemented in SPICE

The application of the distortion theory to circuit analysis is mathematically difficult and tedious, and it is beyond the aim of this book. This section provides only a brief overview of the methodology, followed by three examples to point out how SPICE analyzes a network. Finally, the models implemented in SPICE for the distortion analysis and computation are presented.

a. Methodology description. The distortion algorithm implemented in SPICE requires two independent steps: evaluation of the Taylor series coefficients for each nonlinearity and evaluation of the distortion current excitation vector for each distortion component frequency of analysis [9].

Then the Volterra series approach consists of expanding the equations describing the nonlinearities as Taylor series, with the BJT junction voltages as independent variables, and expressing each small-signal node voltage as a Volterra series expansion of the source voltage. In particular, the small-signal nonlinear equations of the circuit are formulated and separated into the usual linear part and a nonlinear part that is a function of the junction voltages expanded in a power series and that can be represented by equivalent current sources.

Thus the analysis of the circuit results in different sets of linear equations, one for each order, that are solved separately and successively beginning with the first order. The first-order solution is then used to calculate the equivalent current sources in order to determine the second-order solution. Finally, both first- and second-order solutions are used to find the third-order solution. The method shows that the equivalent current sources associated with each nonlinearity can be interpreted as the sources that produce distortion at the circuit output. This procedure amounts to evaluating, at each distortion component frequency of the analysis, the admittance matrix Y and finding the equivalent excitation vector in the node equations[10]

$$Y \cdot V = I \tag{7-29}$$

b. Methodology examples

Example 1: First, the simple case of a resistive nonlinearity is considered, as shown in Fig. 7-8. At node j, it can be written

$$I_{Dj} = I_D e^{qV_j/kT} \tag{7-30}$$

where I_{Dj} is the small-signal current, V_j is the small-signal voltage, and I_D is the bias current.

A useful way to handle this circuit analytically is to represent the nonlinearity as a voltage-controlled current source. Equation (7-30) can then be rewritten as an expansion of the exponential term in a Taylor series

$$I_{Dj} = a_1 V_j + a_2 V_j^2 + a_3 V_j^3 \tag{7-31}$$

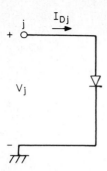

Figure 7-8   Resistive nonlinearity.

where the coefficients of the series are

$$a_1 = \frac{q}{kT} I_D$$

$$a_2 = \frac{1}{2!} \left(\frac{q}{kT} \right)^2 I_D \tag{7-32}$$

$$a_3 = \frac{1}{3!} \left(\frac{q}{kT} \right)^3 I_D$$

Equation (7-31) can be represented by the circuit shown in Fig. 7-9.

The Volterra kernels can now be evaluated by the following procedure.

1. Calculate the linear transfer functions from the sinusoidal input voltage V_{in} to the output voltage V_{out} and to every node voltage V_j of the circuit generating distortion, i.e., every node having nonlinear branches. Thus it can be written as

$$V_{out} = T_1(\omega) V_{in} \tag{7-33}$$

$$V_j = H_1(\omega) V_{in} \tag{7-34}$$

2. Consider two sinusoidal input signals $V_{in} = V_1(\sin \omega_1 t + \sin \omega_2 t)$ at frequencies ω_1 and ω_2. Since the two linear terms $H_1(\omega_1)$ and $H_1(\omega_2)$ are

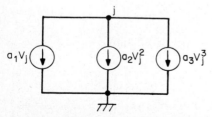

Figure 7-9   Equivalent circuit of resistive nonlinearity.

known from Eq. (7-34), the second-order (and third-order) distortion values can be calculated at the node j as

$$a_2 V_j^2 = a_2 V_{j1}^2 [\cos(\omega_1 t + \varphi_1) + \cos(\omega_2 t + \varphi_2)]^2 \qquad (7\text{-}35)$$

where
$$V_j = H_1(\omega) V_{in} \qquad (7\text{-}36)$$

so that
$$a_2 V_j^2 = a_2 H_1(\omega_1) H_1(\omega_2) V_{in}^2(\omega_1, \omega_2) \qquad (7\text{-}37)$$

3. Analyze the circuit with these second-order source values as input signals and calculate the second-order distortion HD_2 at the point of interest. Obviously, the linear transfer function $H_1(\omega)$ from the input voltage V_{in} to the voltage V_j is equivalent to the term $A_1(\omega)$ in Eq. (7-21). Similarly, this calculation can be repeated for the third-order distortion.

Example 2: Now the case of a capacitance nonlinearity, for instance, C_μ in the BJT small-signal model, is considered.

In Chap. 2 it was shown that this capacitance is defined by a relation of the form

$$C_\mu = \frac{K}{(\phi + V_{BC})^{1/n}} \qquad (7\text{-}38)$$

where
$$C_\mu = \frac{dQ_{BC}}{dV_{BC}}\bigg|_0 \qquad (7\text{-}39)$$

If $V_{BC} - V_0 + V_j$, with V_0 the base-collector bias voltage and V_j the small-signal voltage at node j, then

$$
\begin{aligned}
C_\mu &= \frac{K}{(\phi + V_0 + V_j)^{1/n}} \\
&= \frac{K}{(\phi + V_0)^{1/n}} \left(1 + \frac{V_j}{\phi + V_0}\right)^{-1/n}
\end{aligned}
\qquad (7\text{-}40)
$$

Equation (7-40) can now be expressed in terms of the power series of V_j, so that

$$C_\mu(V_j) = C_{\mu 0} + C_{\mu 1} V_j + C_{\mu 2} V_j^2 + \cdots \qquad (7\text{-}41)$$

and the total current flowing in the capacitance C_μ is

$$
I = \begin{cases}
\dfrac{dQ_{BC}}{dt} = \dfrac{dQ_{BC}}{dV_j}\dfrac{dV_j}{dt} = C_\mu(V_j)\dfrac{dV_j}{dt} & (7\text{-}42a) \\[3ex]
C_{\mu 0}\dfrac{dV_j}{dt} + \tfrac{1}{2} C_{\mu 1}\dfrac{dV_j^2}{dt} + \tfrac{1}{3} C_{\mu 2}\dfrac{dV_j^3}{dt} & (7\text{-}42b)
\end{cases}
$$

Equation (7-42) can be represented by the circuit shown in Fig. 7-10. Thus, for circuits containing nonlinear capacitances, the analysis procedure is as follows.

1. As above, solve the circuit to obtain the linear transfer function in the form

$$V_{\text{out}} = T_1(\omega) V_{\text{in}} \tag{7-43}$$

$$V_j = H_1(\omega) V_{\text{in}} \tag{7-44}$$

2. Insert the second-order distortion sources in the circuit. The preceding capacitive nonlinearity [see Eq. (7-41)] gives

$$\frac{C_{\mu 1}}{2} \frac{dV_j^2}{dt} = \frac{C_{\mu 1}}{2} \frac{d}{dt} [H_1(\omega_1) H_1(\omega_2) V_{\text{in}}^2]$$

$$= \frac{C_{\mu 1}}{2} j(\omega_1 + \omega_2) H_1(\omega_1) H_1(\omega_2) V_{\text{in}}^2(\omega_1, \omega_2) \tag{7-45}$$

Note that this case is different from simple resistive nonlinearity; the major difference consists of the presence of the factor $j(\omega_1 + \omega_2)$.

Example 3:   In the previous examples, we dealt with factors contributing to distortion for one nonlinear element at a time. Here, we analyze a general network containing two nonlinear elements (a capacitor C_X and a diode D) at the same time in order to evaluate distortion. Look at Fig. 7-11, taking into account that the diode bias current I_Q is not represented and that V_{in} and V_2 are the input and output voltages, respectively.

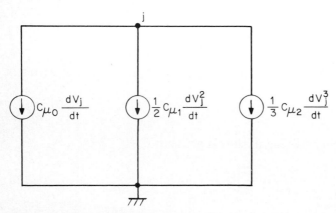

Figure 7-10   Equivalent circuit of capacitive nonlinearity.

As explained earlier, the first step is to express each small-signal node voltage as a Volterra series expansion, so that

$$V_2 = A_1(j\omega)V_{\text{in}} + A_2(j\omega_1, j\omega_2)V_{\text{in}}^2 + A_3(j\omega_1, j\omega_2, j\omega_3)V_{\text{in}}^3 \qquad (7\text{-}46)$$

$$V_1 = B_1(j\omega)V_{\text{in}} + B_2(j\omega_1, j\omega_2)V_{\text{in}}^2 + B_3(j\omega_1, j\omega_2, j\omega_3)V_{\text{in}}^3$$

Then, the equations describing the nonlinearities are expanded in a Taylor series, as follows.

For diode D (see Example 1):

$$I_D = a_1 V_2 + a_2 V_2^2 + a_3 V_2^3 \qquad (7\text{-}47a)$$

For capacitor C_X (see Example 2):

$$I_{CX} = C_0 \frac{dV_1}{dt} + \frac{C_1}{2}\frac{dV_1^2}{dt} + \frac{C_2}{3}\frac{dV_1^3}{dt} \qquad (7\text{-}47b)$$

Finally, the Kirchhoff nodal equations can be written as follows.

For node 1:
$$\frac{V_1 - V_{\text{in}}}{R} + C\frac{d}{dt}(V_1 - V_2) + I_{CX} = 0$$
$$\qquad (7\text{-}48)$$

For node 2:
$$-C\frac{d}{dt}(V_1 - V_2) + I_D = 0$$

where the terms I_{CX} and I_D are obtained by Eqs. (7-47) and V_1 and V_2 are obtained by Eqs. (7-46).

These small-signal nonlinear node equations must now be separated into their linear and nonlinear terms, so that the method of analysis involves three steps (one for each order), which are solved separately and successively, beginning with the first order.

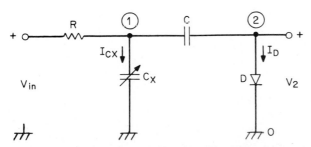

Figure 7-11   General network for evaluating distortion.

Linear (first-order) terms.   Considering only linear terms, Eqs. (7-48) become

$$\frac{B_1 V_{in} - V_{in}}{R} + j\omega C(B_1 V_{in} - A_1 V_{in}) + j\omega C_0 B_1 V_{in} = 0$$

$$-j\omega C(B_1 V_{in} - A_1 V_{in}) + a_1 A_1 V_{in} = 0$$

(7-49)

Dividing Eqs. (7-49) by V_{in} yields

$$\frac{B_1 - 1}{R} + j\omega C(B_1 - A_1) + j\omega C_0 B_1 = 0$$

$$-j\omega C(B_1 - A_1) + a_1 A_1 = 0$$

(7-50)

Solving Eqs. (7-50) with respect to A_1 and B_1,

$$A_1 = \frac{1}{R} \frac{j\omega C}{(1/R + j\omega C + j\omega C_0)(a_1 + j\omega C) + \omega^2 C^2}$$

$$B_1 = \frac{1}{R} \frac{a_1 + j\omega C}{(1/R + j\omega C + j\omega C_0)(a_1 + j\omega C) + \omega^2 C^2}$$

(7-51)

Second-order terms.   Considering only second-order terms, Eqs. (7-48) become

$$\frac{B_2 V_{in}^2}{R} + j(\omega_1 + \omega_2)C(B_2 V_{in}^2 - A_2 V_{in}^2) + j(\omega_1 + \omega_2)C_0 B_2 V_{in}^2$$
$$+ \tfrac{1}{2} C_1 j(\omega_1 + \omega_2)B_1(j\omega_1)B_1(j\omega_2) = 0 \quad \text{(7-52)}$$

$$-j(\omega_1 + \omega_2)C(B_2 V_{in}^2 - A_2 V_{in}^2) + a_1 A_2 V_{in}^2 + a_2 A_1(j\omega_1)A_1(j\omega_2) V_{in}^2 = 0$$

and solving as in the previous case

$$A_2 = \frac{\left[\dfrac{a_2 A_1(j\omega_1)A_1(j\omega_2)}{j(\omega_1 + \omega_2)C}\right]\left[\dfrac{1}{R} + j(\omega_1 + \omega_2)(C + C_0)\right] + \dfrac{C_1}{2} j(\omega_1 + \omega_2)B_1(j\omega_1)B_1(j\omega_2)}{\left[1 + \dfrac{a_1}{j(\omega_1 + \omega_2)C}\right]\left[\dfrac{1}{R} + j(\omega_1 + \omega_2)(C + C_0)\right] - j(\omega_1 + \omega_2)C}$$

(7-53)

$$B_2 = \left[1 + \frac{a_1}{j(\omega_1 + \omega_2)C}\right] A_2 + \frac{a_2 A_1(j\omega_1)A_1(j\omega_2)}{j(\omega_1 + \omega_2)C}$$

where the values of A_1 and B_1 are given by Eqs. (7-51).

Third-order terms. If we consider only third-order terms and second-order interactions, Eqs. (7-48) become

$$\frac{B_3}{R} V_{in}^3 + j(\omega_1 + \omega_2 + \omega_3)C(B_3 V_{in}^3 - A_3 V_{in}^3) + j(\omega_1 + \omega_2$$
$$+ \omega_3)C_0 B_3 V_{in}^3 + j(\omega_1 + \omega_2 + \omega_3)C_1 B_1 B_2 V_{in}^3 + j(\omega_1 + \omega_2$$
$$+ \omega_3)\frac{C_2}{3} B_1(j\omega_1)B_1(j\omega_2)B_1(j\omega_3) V_{in}^3 = 0$$

$$(7\text{-}54)$$

$$-j(\omega_1 + \omega_2 + \omega_3)C(B_3 V_{in}^3 - A_3 V_{in}^3) + a_1 A_3 V_{in}^3 + 2a_2 A_1 A_2 V_{in}^3$$
$$+ a_3 A_1(j\omega_1)A_1(j\omega_2)A_1(j\omega_3) V_{in}^3 = 0$$

Equations (7-54) can now be solved with respect to A_3 and B_3, taking into account the values of A_1, B_1 and A_2, B_2 given by Eqs. (7-51) and (7-53), respectively.

When the coefficients A_1, B_1, A_2, B_2, A_3, B_3 of the Volterra series have been calculated, the amount of distortion can be easily evaluated. For example, suppose we wish to calculate HD_2 when an input signal $V_{in} = V_A \cos \omega_1 t$ is applied to the circuit of Fig. 7-11. From Eq. (7-28) it follows that

$$HD_2 = \frac{1}{2} \frac{A_2(2\omega_1)}{A_1(\omega_1)} V_A \qquad (7\text{-}55)$$

where Eqs. (7-51) and (7-53) must be used.

c. Diode distortion model. The diode distortion model is shown in Fig. 7-12. Current source I_{DI} has been added to the small-signal diode model to represent the distortion introduced by the diode. The value of the current source is given by the following equations.

For second-order harmonic distortion (HD_2):

$$I_{DI}(2\omega_1) = \left(\frac{1}{2}\frac{d^2 I_D}{d V_D^2}\Big|_0 + j2\omega_1 \frac{1}{2}\frac{dC_D}{dV_D}\Big|_0\right) V_D^2(2\omega_1) \qquad (7\text{-}56)$$

For second-order intermodulation distortion (IM_2):

$$I_{DI}(\omega_1 \pm \omega_2) = \left(\frac{1}{2}\frac{d^2 I_D}{d V_D^2}\Big|_0\right.$$
$$\left. + j(\omega_1 \pm \omega_2)\frac{1}{2}\frac{dC_D}{dV_D}\Big|_0\right) V_D(\omega_1)V_D(\omega_2) \qquad (7\text{-}57)$$

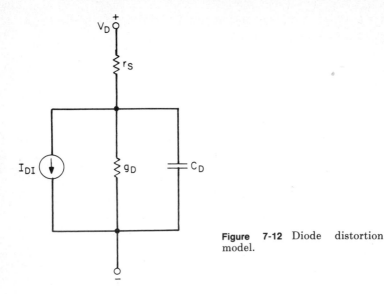

Figure 7-12 Diode distortion model.

where $V_D(\omega_n)$ is the voltage across the diode due to a response at frequency ω_n. All currents and voltages in these equations are expressed as phasors. The value of second-order harmonic distortion current source $I_{DI}(2\omega_1)$ is determined by the response at fundamental frequency ω_1.

It is important to note that the first term in Eq. (7-56) corresponds to a_2 in Eq. (7-31) and the second term in Eq. (7-56) corresponds to the second term in Eq. (7-42b).

Similarly, the value of second-order intermodulation distortion current source $I_{DI}(\omega_1 + \omega_2)$ is determined solely by the response at fundamental frequencies ω_1 and ω_2. The equations used to determine the third-order

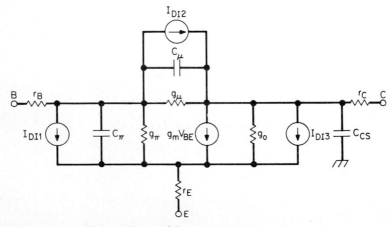

Figure 7-13 BJT distortion model.

distortion (HD_3 and IM_3) for a diode are similar to those given previously for second-order effects, but they are more complex.

The third-order harmonic distortion current source $I_{DI}(3\omega_1)$ is determined by the response at fundamental frequency ω_1 plus the response at the second harmonic frequency $2\omega_1$. Similarly, the third-order intermodulation distortion current source $I_{DI}(2\omega_1 - \omega_2)$ is determined by the response at fundamental frequencies ω_1 and ω_2, plus the response at the second-order harmonic frequency $2\omega_1$.

d. BJT distortion model.   The BJT distortion model is shown in Fig. 7-13. Current sources I_{DI1}, I_{DI2}, and I_{DI3} have been added to the small-signal BJT model to represent the distortion introduced by the transistor. The values of these sources are calculated from the following equations.

For second-order harmonic distortion (HD_2):

$$I_{DI1}(2\omega_1) = (g_{\pi 2} + j2\omega_1 C_{\pi 2}) V_{BE}^2(\omega_1)$$

$$I_{DI2}(2\omega_1) = (g_{\mu 2} + j2\omega_1 C_{\mu 2}) V_{BC}^2(\omega_1) \tag{7-58}$$

$$I_{DI3}(2\omega_1) = g_{m2} V_{BE}^2(\omega_1) + g_{o2} V_{CE}^2(\omega_1) + g_{mo2} V_{BE}(\omega_1) V_{CE}(\omega_1)$$

with
$$g_{\pi 2} = \frac{1}{2} \frac{d^2 I_B}{dV_{BE}^2} \bigg|_0 \qquad\qquad g_{\mu 2} = \frac{1}{2} \frac{d^2 I_B}{dV_{BC}^2} \bigg|_0$$

$$C_{\pi 2} = \frac{1}{2} \frac{d^2 C_{BE}}{dV_{BE}^2} \bigg|_0 \qquad\qquad C_{\mu 2} = \frac{1}{2} \frac{d^2 C_{BC}}{dV_{BC}^2} \bigg|_0 \tag{7-59}$$

$$g_{m2} = \frac{1}{2} \frac{d^2 I_C}{dV_{BE}^2} \bigg|_0 - g_{o2} \qquad\qquad g_{o2} = \frac{1}{2} \frac{d^2 I_C}{dV_{BC}^2} \bigg|_0 - g_{\mu 2}$$

$$g_{mo2} = \frac{d^2 I_C}{dV_{BE} dV_{BC}} \bigg|_0 - 2g_{o2}$$

For second-order intermodulation distortion (IM_2):

$$I_{DI1}(\omega_1 \pm \omega_2) = [g_{\pi 2} + j(\omega_1 \pm \omega_2)C_{\pi 2}] V_{BE}(\omega_1) V_{BE}(\omega_2)$$

$$I_{DI2}(\omega_1 \pm \omega_2) = [g_{\mu 2} + j(\omega_1 \pm \omega_2)C_{\mu 2}] V_{BC}(\omega_1) V_{BC}(\omega_2) \tag{7-60}$$

$$I_{DI3}(\omega_1 \pm \omega_2) = g_{m2} V_{BE}(\omega_1) V_{BE}(\omega_2) + g_{o2} V_{CE}(\omega_1) V_{CE}(\omega_2)$$
$$+ g_{mo2}[V_{BE}(\omega_1) V_{CE}(\omega_2) + V_{BE}(\omega_2) V_{CE}(\omega_1)]$$

where $V_{BC}(\omega_n)$, $V_{BE}(\omega_n)$, and $V_{CE}(\omega_n)$ are, respectively, the base-collector, base-emitter, and collector-emitter voltages due to a response at frequency ω_n. The admittances and capacitances are evaluated at the dc operating point of the network. All voltages and currents in these equations are expressed as phasors.

The values of second-order harmonic distortion current sources are determined by the response at the fundamental frequency ω_1. Similarly, the values of the second-order intermodulation distortion current sources are determined solely by the response at the fundamental frequencies ω_1 and ω_2.

The equations used to determine third-order distortion (HD_3 and IM_3) for a BJT are similar to those given previously for second-order effects, but they are more complex. Third-order harmonic distortion current sources, $I_{DI1}(3\omega_1)$, $I_{DI2}(3\omega_1)$, and $I_{DI3}(3\omega_1)$ are determined by the response at fundamental frequency ω_1 plus the response at the second-order harmonic frequency $2\omega_1$. Similarly, third-order intermodulation distortion current sources $I_{DI1}(2\omega_1 - \omega_2)$, $I_{DI2}(2\omega_1 - \omega_2)$, and $I_{DI3}(2\omega_1 - \omega_2)$ are determined by the response at fundamental frequencies ω_1 and ω_2, plus the response at the second-order harmonic frequency $2\omega_1$ [11].

7.2.3 .DISTO card

The general form of the .DISTO card is

.DISTO RLOAD   INTER SKW2 REFPWR SPW2

where $RLOAD$ is the name of the output load resistor into which all distortion power products are computed. Other coefficients are optional.

This card determines whether SPICE will compute the distortion characteristic of the circuit in a small-signal mode as part of the small-signal sinusoidal steady-state analysis [12–17].

The analysis is performed assuming that one or two signal frequencies be ω_1 (the nominal analysis frequency) and $\omega_2 = SKW_2 \cdot \omega_1$. The program then computes the following distortion measures: HD_2, HD_3, IM_2 as $SIM_2(\omega_1 + \omega_2)$ and $DIM_2(\omega_1 - \omega_2)$, and IM_3 as $DIM_3(2\omega_1 - \omega_2)$.

At the conclusion of the distortion analysis, any of these distortion components can be listed or plotted as a function of frequency. The contributions of every distortion source also can be printed at selected frequency points. The correct use of the $REFPWR$ parameter is very important in this card. It is the reference power level used in computing the distortion products; if the parameter is omitted, a value of 1 mW is used.

Consider, for example, an input signal $V_{\text{in}} = V_A \cos \omega_1 t$. The output signal applied on R_L (output load) will be $V_{\text{out}} = V_o \cos (\omega_1 t + \varphi)$, so that from Eq. (7-14a)

$$HD_2 = \frac{a_2 V_A}{2a_1} = \frac{1}{2}\frac{a_2}{a_1^2}\, V_o = \frac{1}{2}\frac{a_2}{a_1^2}\, \sqrt{2}\, V_{o,\text{eff}} \qquad (7\text{-}61)$$

Using the relation $P_{out} = V_{o,eff}^2/R_L$, it can be written

$$HD_2 = \frac{1}{2}\frac{a_2}{a_1^2}\sqrt{2R_L P_{out}} \qquad (7\text{-}62)$$

Thus, P_{out} represents the REFPWR parameter.

7.2.4 Fourier analysis

For relatively large levels of distortion, the method of fitting a Fourier series of a periodic waveform on the time-domain response yields a good approximation of the distortion components of the waveform.

For large-signal sinusoidal simulations, a Fourier analysis of the output waveform can be specified to obtain the frequency-domain Fourier coefficients: the first nine harmonic components of the specified output are computed by SPICE. The transient time interval and the Fourier analysis options are specified on the .TRAN and .FOUR control cards. It is very important to note that the sinusoidal waveform is the only periodic waveform allowed.

REFERENCES

1. P. R. Gray and R. G. Meyer, *Analysis and Design of Analog Integrated Circuits,* Wiley, New York, 1977.
2. S. M. Sze, *Physics of Semiconductor Devices,* Wiley, New York, 1969.
3. L. Nagel, SPICE2: A Computer Program to Simulate Semiconductor Circuits, Electronics Research Laboratory Rep. No. ERL-M520, University of California, Berkeley, 1975.
4. R. Roher, L. Nagel, R. Meyer, and L. Weber, Computationally Efficient Electronic-Circuit Noise Calculations, *IEEE J. Solid-State Circuits,* **SC-6,** 1971.
5. P. Antognetti, P. Antoniazzi, and E. Meda, Computer-Aided Evaluation of Transient Intermodulation Distortion (TIM) in Monolithic Integrated Amplifiers, *Proc. IEEEDA,* 1981.
6. W. Sansen, Optimum Design of Integrated Variable-Gain Amplifiers, Electronics Research Laboratory Rep. No. ERL-M367, University of California, Berkeley, 1972.
7. K. A. Simons, The Decibel Relationships between Amplifiers Distortion Products, *Proc. IEEE,* **58,** 1970.
8. S. Narayanan, Transistor Distortion Analysis Using Volterra Series Representations, *Bell Syst. Tech. J.,* **40,** 1967.
9. S. Chilsholm and L. Nagel, Efficient Computer Simulation of Distortion in Electronic Circuits, *IEEE Trans. Circuit Theory,* 1973.
10. L. Kuo, Distortion Analysis of Bipolar Transistor Circuits, *IEEE Trans. Circuit Theory,* 1972.
11. P. Antognetti, P. Antoniazzi, and E. Meda, TIM Distortion in Monolithic Integrated Circuits: Measurements and Simulation, *AES Convention,* Hamburg, 1981.
12. *SPICE Version 2G User's Guide.*
13. H. C. Poon, Modeling of Bipolar Transistor Using Integral Charge Control Model with Application to Third Order Distortion Studies, *IEEE Trans. Electron Devices,* **ED-19,** 1972.
14. S. Narayanan and H. C. Poon, An Analysis of Distortion in Bipolar Transistors Using

Integral Charge Control Model and Volterra Series, *IEEE Trans. Circuit Theory,* **CT-20,** 1973.

15. R. G. Meyer, M. Shensa, and R. Eschenbach, Cross Modulation and Intermodulation in Amplifiers at High Frequencies, *IEEE J. Solid-State Circuits,* **SC-7,** 1972.
16. H. C. Poon, Implication of Transistor Frequency Dependence on Intermodulation Distortion, *IEEE Trans. Electron Devices,* **ED-21,** 1974.
17. R. K. Brayton and R. Spence, *Sensitivity and Optimization,* Elsevier, Amsterdam, 1980.

The SPICE Program

Claudio Fasce

CAD Department, SGS Microelettronica, Milan, Italy

Giuseppe Massobrio

Department of Electronics (DIBE), University of
Genoa, Genoa, Italy

This chapter points out the rudiments needed to modify existing device models or to implement new device models in SPICE. A complete description of the development and design of SPICE can be found in the literature [1–6]; it is beyond the aim of this book. However, we will briefly describe the general capabilities of SPICE. We refer to SPICE version 2G.

8.1 SPICE Capabilities

SPICE, which stands for *simulation program* with *integrated circuit emphasis*, is a general-purpose circuit simulation program developed at the University of California at Berkeley for nonlinear dc, nonlinear transient, and linear ac analyses. Circuits can contain resistors, capacitors, inductors, mutual inductors, independent voltage and current sources, four types of dependent sources, transmission lines, and the four most common semiconductor devices: diodes, BJTs, JFETs, and MOSFETs.

SPICE has built-in models for the semiconductor devices, and the user need specify only the pertinent model parameter values. The model for the BJT is based on the integral charge model of Gummel-Poon; however, if the Gummel-Poon parameters are not specified, the model reduces to the simpler Ebers-Moll model. In either case, charge-storage effects, ohmic resistances, and a current-dependent output conductance can be included. The diode model can be used for either junction diodes or Schottky barrier diodes. The JFET model is based on the FET model of Shichman-Hodges. Three MOSFET models are implemented: MOS1 is described by a square-law IV characteristic, MOS2 is an analytical model, and MOS3 is a semiempirical model. Both MOS2 and MOS3 include second-order effects such as channel-length modulation, subthreshold conduction, scattering limited velocity saturation, small-size effects, and charge-controlled capacitances [1].

SPICE is node voltage–oriented, so any node voltage can be requested. Element currents flowing through independent voltage sources can also be requested. Tabular lists and printer plots are available.

8.1.1 Types of analysis

a. DC analysis. The *dc analysis* portion of SPICE determines the dc operating point of the circuit with inductors shorted and capacitors opened. A dc analysis is automatically performed prior to a transient analysis to determine the transient initial conditions, and prior to an ac small-signal analysis to determine the linearized, small-signal models for nonlinear devices. If requested, the dc small-signal value of a transfer function (ratio of output variable to input source), input resistance, and output resistance will also be computed as a part of the dc solution. The dc analysis can also be used to generate dc transfer curves: A specified independent voltage or current source is stepped over a user-specified range, and the dc output variables are stored for each sequential source value. If requested, SPICE also will determine the dc small-signal sensitivities of specified output variables with respect to circuit parameters. The dc analysis options are specified on the .DC, .TF, .OP, and .SENS control cards.

If one desires to see the small-signal models for nonlinear devices in conjunction with a transient analysis operating point, then the .OP card must be provided. The dc bias conditions will be identical for each case, but the more comprehensive operating-point information is not available for printing when transient initial conditions are computed [1].

b. AC small-signal analysis. The *ac small-signal* portion of SPICE computes the ac output variables as a function of frequency. The program first computes the dc operating point of the circuit and determines linearized, small-signal models for all the nonlinear devices in the circuit. The resultant linear circuit is then analyzed over a user-specified range of frequencies. The desired output of an ac small-signal analysis is usually a transfer function (voltage gain, transimpedance, and so forth). If the circuit has only one ac input, it is convenient to set that input to unity and zero phase so that output variables have the same value as the transfer function of the output variable with respect to the input.

The generation of *white noise* by resistors and semiconductor devices can also be simulated with the ac small-signal portion of SPICE. Equivalent noise-source values are determined automatically from the small-signal operating point of the circuit, and the contribution of each noise source is added at a given summing point. The total output noise level and the equivalent input noise level are determined at each frequency point. The output and input noise levels are normalized with respect to the square root of the noise bandwidth. The output noise and equivalent input noise can be printed or plotted in the same fashion as other output variables. No additional input data are necessary for this analysis.

Flicker-noise sources can be simulated during noise analysis by including values for the parameters KF and AF on the appropriate device model cards (see Chap. 7).

The *distortion* characteristics of a circuit in the small-signal mode can be simulated as a part of the ac small-signal analysis. The analysis is performed assuming that one or two signal frequencies are imposed at the input.

The frequency range and the noise and distortion analysis parameters are specified on the .AC, .NOISE, and .DISTO control cards [1].

c. Transient analysis. The *transient analysis* portion of SPICE computes the transient output variables as a function of time over a user-specified time interval. The initial conditions are automatically determined by a dc analysis. All sources that are not time-dependent (for example, power supplies) are set to their dc value. For large-signal sinusoidal simulations, a Fourier analysis of the output waveform can be specified to obtain the frequency-domain Fourier coefficients. The transient time interval and the Fourier analysis options are specified on the .TRAN and .FOUR control cards [1].

d. Analysis at different temperatures. All input data for SPICE are assumed to have been measured at 27°C (300 K). The simulation also assumes a nominal temperature of 27°C. The circuit can be simulated at other temperatures by using a .TEMP control card. Temperature appears explicitly in the exponential terms of the BJT and diode model equations. In addition, saturation currents have a built-in temperature dependence, as do the forward and reverse β and the junction potentials.

Temperature also appears explicitly in the value of surface mobility for the MOSFET model. The effects of temperature on resistors are also modeled. Temperature effects on the device model parameters are detailed in the chapters where each device is discussed.

8.1.2   Convergence

Both dc and transient solutions are obtained by an iterative process that is terminated when both the following conditions hold:

1. The nonlinear branch currents converge to within a tolerance of 0.1 percent or 10^{-12} A, whichever is larger.

2. The node voltages converge to within a tolerance of 0.1 percent or 10^{-6} V, whichever is larger.

Although the algorithm used in SPICE has been found to be very reliable, in some cases it will fail to converge to a solution. When this occurs, the program prints the node voltages at the last iteration and terminates the job. In such cases, the node voltages that are printed are not necessarily correct or even close to the correct solution.

Failure to converge in the dc analysis is usually due to an error in specifying circuit connections, element values, or model parameter values. Regenerative switching circuits or circuits with positive feedback probably will not converge in the dc analysis unless the OFF option is used for some of the devices in the feedback path or the .NODESET card is used to force the circuit to converge to the desired state (see Sec. 8.3) [1].

8.1.3   Input format

The input format for SPICE is of the free-format type. Fields on a card are separated by one or more blanks, a comma, an equal ($=$) sign, or a left or right parenthesis; extra spaces are ignored. A card can be continued by entering a plus sign ($+$) in column 1 of the subsequent card; SPICE continues reading beginning with column 2.

A name field must begin with a letter (A to Z) and cannot contain any delimiters. Only the first eight characters of the name are used.

A number field can be an integer field (12, -44), a floating-point field (3.14159), either an integer or a floating-point number followed by an integer exponent (1E-14, 2.65E3), or either an integer or a floating-point number followed by one of the following factors:

$$T = 1E12$$

$$G = 1E9$$

$$MEG = 1E6$$

$$K = 1E3$$
$$MIL = 25.4E\text{-}6$$
$$M = 1E\text{-}3$$
$$U = 1E\text{-}6$$
$$N = 1E\text{-}9$$
$$P = 1E\text{-}12$$
$$F = 1E\text{-}15$$

Letters immediately following a number that are not scale factors are ignored, and letters immediately following a scale factor are ignored. Hence, 10, 10V, 10VOLTS, and 10HZ all represent the same number, and M, MA, MSEC, and MMHOS all represent the same scale factor. Note that 1000, 1000.0, 1000HZ, 1E3, 1.0E3, 1KHZ, and 1K all represent the same number [1].

Control commands begin with a period (.); the other lines are assumed to be circuit elements.

8.1.4   Circuit description

The circuit to be analyzed is described to SPICE by a set of element cards, which define the circuit topology and element values, and a set of control cards, which define the model parameters and the run controls. The first card in the input deck must be a *title* card, and the last card must be an .END card. The order of the remaining cards is *arbitrary* (except, of course, that continuation cards must immediately follow the card being continued).

Each element in the circuit is specified by an element card that contains the element name, the circuit nodes to which the element is connected, and the values of the parameters that determine the electrical characteristics of the element. The first letter of the element name specifies the element type.

With respect to branch voltages and currents, SPICE uniformly uses the associated reference convention (current flows in the direction of voltage drop).

Nodes *must* be nonnegative integers but need not be numbered sequentially. The datum (ground) node *must* be numbered zero. The circuit *cannot* contain a loop of voltage sources and/or inductors and *cannot* contain a cutset of current sources and/or capacitors. Each node in the circuit *must* have a dc path to ground. Every node *must* have at least two connections, except for transmission-line nodes (to permit unterminated transmission lines), and MOSFET substrate nodes (which have two internal connections anyway) [1].

Prefixes used for element names are as follows:

R	Resistor
C	Capacitor
L	Inductor
K	Mutual inductor
T	Transmission lines (lossless)
V	Independent voltage source
I	Independent current source
G	(Non)linear voltage-controlled current source
E	(Non)linear voltage-controlled voltage source
F	(Non)linear current-controlled current source
H	(Non)linear current-controlled voltage source
Q	BJT
D	DIODE
J	JFET
M	MOSFET
X	Subcircuit

The R and T elements must be constants, while the C, L, G, E, F, H elements may be expressed as nonlinear polynomials. The elements Q, D, J, M are defined by model parameters. The subcircuit is a user-defined subcircuit.

The independent sources can be constants or any of the following functions. Any of these functions can be used as values for independent sources:

1. *Constants.*   AC or dc constant values. Constants can be suffixed with scaling factors (see Sec. 8.1.3).

2. *Pulse.*   Defined with high and low values, rise and fall times, and pulse width and repetition rate.

3. *Sinusoidal.*   Exponentially decaying sinusoid.

4. *Exp.*   Sum of two exponential waveforms.

5. *Piecewise linear.*   Tabular function of time.

6. *SFFM.*   Single-frequency, frequency-modulated waveform.

Figure 8-1 shows a schematic diagram for a single-stage BF preamplifier and the SPICE input deck that defines the topology of the circuit.

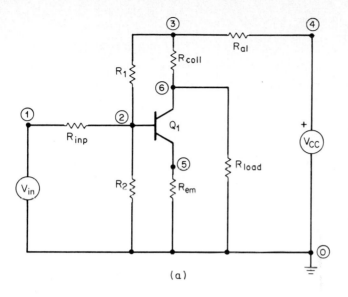

(a)

SPICE 2G.5

BF PREAMPLIFIER

•••• INPUT LISTING   TEMPERATURE = 27.000 DEG C

•••

```
VIN 1 0 SIN (0 0.1 5K 100NS) AC 0.1
VCC 4 0 12
Q1 6 2 5 TRN
RINP 1 2 20K
R1 2 3 10K
R2 2 0 3.3K
RCOLL 3 6 2.2K
REM 5 0 1K
RAL 3 4 100
RLOAD 6 0 100
•MODEL TRN NPN (BF = 80  RB = 100 CCS = 2PF TF = 0.3NS
+TR = 6NS CJE = 3PF CJC = 2PF VA = 50)
•OPT LIST ACCT NODE
•TF V (6) VIN
•DC VIN -0.25 0.25 0.01
•AC DEC 5 1 10GHZ
•TRAN 400US 20MS
•PLOT DC V(6)
•PLOT AC VM(6) VP(6)
•WIDTH OUT=80
•PLOT TRAN V(6)
•END
```

(b)

Figure 8-1 (a) Schematic diagram for a single-stage *BF* preamplifier and (b) SPICE input deck.

8.1.5 Semiconductor devices

The models for the four semiconductor devices included in SPICE require many parameter values. Moreover, many devices in a circuit often are defined by the same set of device model parameters. For these reasons, a set of device model parameters is defined on a separate .MODEL card and assigned a unique model name. The device element cards in SPICE then reference the model name. This scheme alleviates the need to specify all the model parameters on each device element card.

Each device element card contains the device name, the nodes to which the device is connected, and the device model name. In addition, other optional parameters can be specified for each device: geometric factors and an initial condition.

The AREA factor used on the DIODE, BJT, and JFET device cards determines the number of equivalent parallel devices of a specified model. Several geometric factors associated with the channel and the drain and source diffusions can be specified on the MOSFET device card.

Two different forms of initial conditions can be specified for devices. The first form is included to improve the dc convergence for circuits that contain more than one stable state. If a device is specified OFF, the dc operating point is determined with the terminal voltages for that device set to zero. After convergence is obtained, the program continues to iterate to obtain the exact value for the terminal voltages.

If a circuit has more than one dc stable state, the OFF option can be used to force the solution to correspond to a desired state. If a device is specified OFF when in reality the device is conducting, the program will still obtain the correct solution (assuming the solutions converge), but more iterations will be required since the program must independently converge to two separate solutions. The .NODESET card serves a purpose similar to that of the OFF option. The .NODESET card is easier to apply and is the preferred means to aid convergence.

The second form of initial conditions is specified for use with the transient analysis. These are true initial conditions as opposed to the convergence aids specified previously. See the description of the .IC card and the .TRAN card for a detailed explanation of initial conditions [1].

8.1.6 Subcircuits

A *subcircuit* that consists of SPICE elements can be defined and referenced in a fashion similar to that used for device models. The subcircuit is defined in the input deck by a group of element cards; the program then automatically inserts the elements wherever the subcircuit is referenced. There is no limit on the size or complexity of subcircuits, and subcircuits can contain other subcircuits [1].

8.2 SPICE Structure

The SPICE2 program consists of more than 15,000 FORTRAN statements divided into seven major independent modules together with a
main program (see Fig. 8-2). The names of the modules are READIN,
ERRCHK, SETUP, DCOP, DCTRAN, ACAN, and OVTPVT.

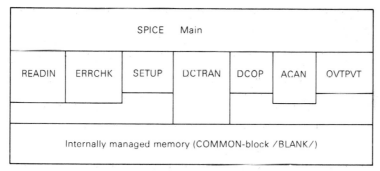

Figure 8-2   Memory map for SPICE2 [4].

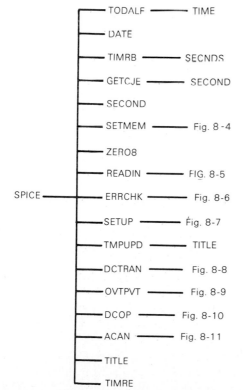

Figure 8-3   Map of SPICE2.

The complete SPICE2 map is outlined in Fig. 8-3. Individual modules are shown in Figs. 8-4 to 8-11. The ellipses (. . .) in these figures mean that the calling map for the present routine has already been shown elsewhere (i.e., previously). For the reader's convenience, the maps of individual modules of SPICE2 may be found as follows: SETMEM (Fig. 8-4, below), READIN (Fig. 8-5, pp. 298–300), ERRCHK (Fig. 8-6, p. 301), SETUP

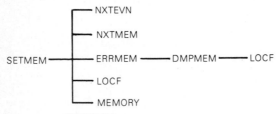

Figure 8-4 SETMEM map.

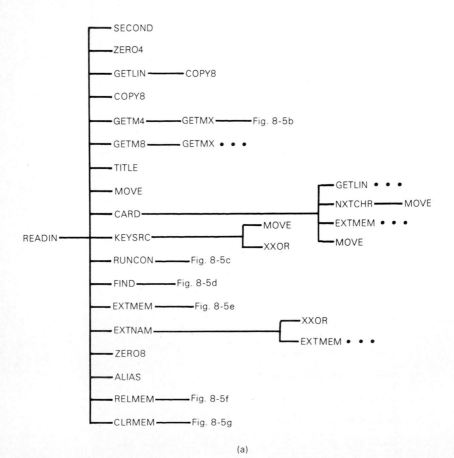

(a)

Figure 8-5 READIN map.

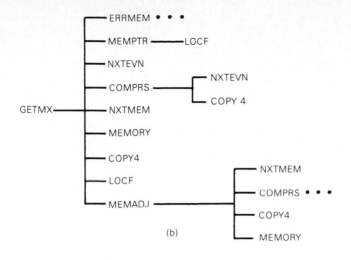

(b)

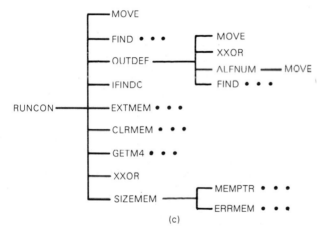

(c)

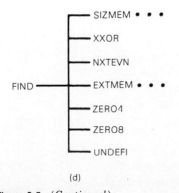

(d)

Figure 8-5 *(Continued)*

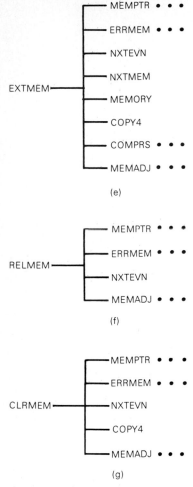

(e)

(f)

(g)

Figure 8-5 (*Continued*)

(Fig. 8-7, p. 302), DCTRAN (Fig. 8-8, pp. 303–304), OVTPVT (Fig. 8-9, p. 305), DCOP (Fig. 8-10, p. 306), and ACAN (Fig. 8-11, p. 307).

In the following sections we will describe the seven modules in addition to the only subroutines relevant to the aim of this book.

8.2.1 The SPICE main program

The SPICE main program contains the main control loop of the program. This control loop is illustrated by the flowchart in Fig. 8-12.

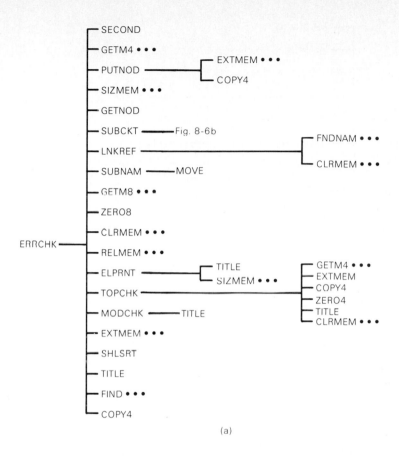

(a)

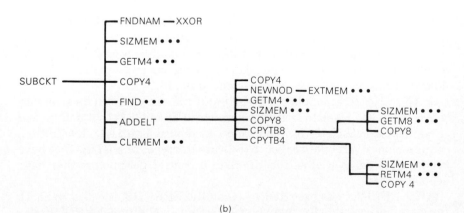

(b)

Figure 8-6 ERRCHK map.

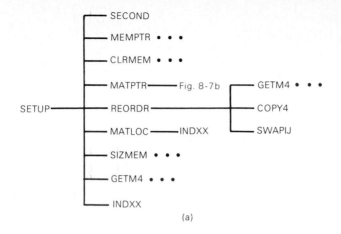

(a)

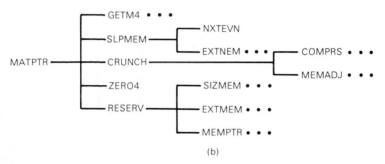

(b)

Figure 8-7 SETUP map.

The program begins by initializing some program constants and reading the job title card. If an end-of-file is encountered on the input file, the program terminates. Otherwise, the READIN module is called to read the remainder of the input file. The READIN program stops reading after an .END card or an input end-of-file is encountered.

As the READIN module reads the input file, the circuit data structure is constructed. This data structure consists of linked lists that define each circuit element, each device model, and each output variable. In addition, common block variables are set to indicate the types of analysis that have been requested as well as the simulation control parameters that have been specified.

After READIN has executed successfully, ERRCHK is called to check the circuit description for common user errors. In addition, ERRCHK also prints the circuit listing, the device model parameter listing, and the node table.

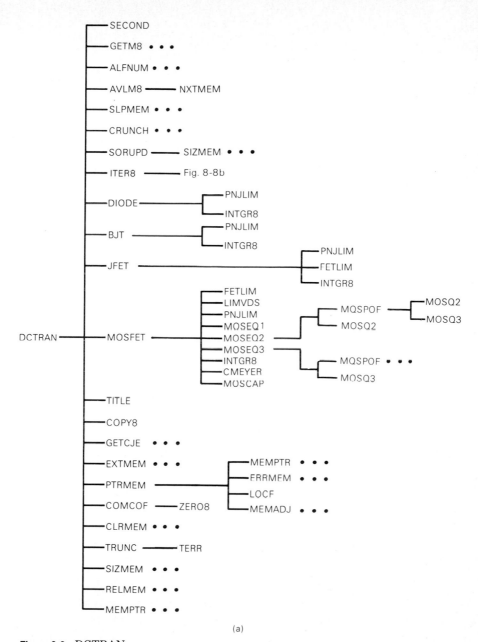

(a)

Figure 8-8 DCTRAN map.

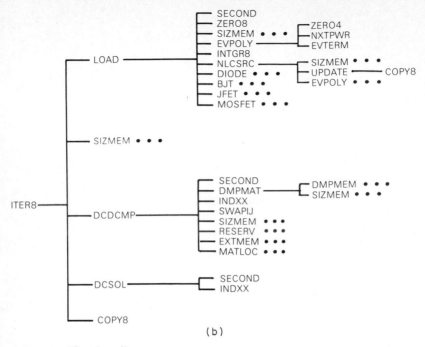

(b)

Figure 8-8   (*Continued*)

Once READIN and ERRCHK have executed, the preparation for circuit analysis can begin. SETUP is called to construct the integer pointer structure used by both DCTRAN and ACAN.

After SETUP has executed successfully, the circuit analysis can proceed. As shown in Fig. 8-12, the main analysis loop is repeated for each of the user-specified temperatures. After the main analysis loop is finished for each temperature, the program prints the job statistics and reads the next input deck [3].

8.2.2   The main analysis loop

The main analysis loop consists of three parts: the *dc transfer-curve loop* (shown in Fig. 8-13), the *dc operating point* and *ac analysis loop* (shown in Fig. 8-14), and the *transient analysis loop* (shown in Fig. 8-15). These loops are executed in succession.

The dc transfer-curve loop (see Fig. 8-13) first checks to ensure that a dc transfer curve is requested. The flags MODE and MODEDC are set to 1 and 3, respectively, and DCTRAN is called to perform the dc transfer curve analysis. OVTPVT is then called to construct the tabular listings and line-printer plots requested for the dc transfer-curve analysis.

The dc operating point and ac analysis are illustrated in Fig. 8-14. First, the loop checks if a dc operating point has been requested. If not, the dc operating point analysis will be performed only if an ac analysis of a non-linear circuit is requested, since, for this case, the linearized device model parameters are a necessary prerequisite for the ac analysis.

Once it is established that a dc operating point is necessary, the flags MODE and MODEDC are set to 1 and DCTRAN is called to compute the dc operating point. DCOP is then called to print the linearized device model parameters [3].

If requested, the ac analysis is performed next by setting the flag MODE to 3 and calling ACAN. After the ac analysis has been obtained, OVTPVT is called to generate the tabular listings and line-printer plots that are requested in conjunction with the ac analysis.

The final analysis loop, transient analysis, is shown in Fig. 8-15. First, the loop checks that a transient analysis is requested. Then, the flags MODE and MODEDC are set to 1 and 2, respectively, and DCTRAN is called to determine the transient initial conditions. Next, DCOP is

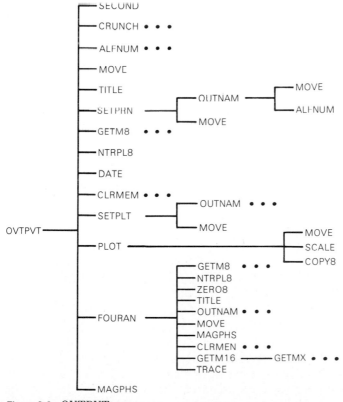

Figure 8-9 OVTPVT map.

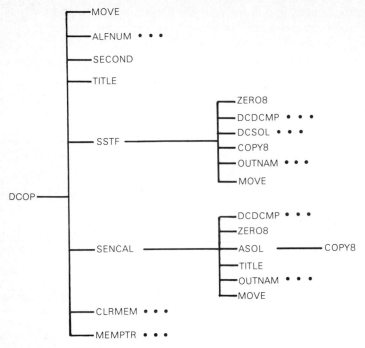

Figure 8-10   DCOP map.

invoked to print the nonlinear device operating points. Finally, the flag MODE is set to 2 and DCTRAN is called again to determine the transient analysis. OVTPVT then is called to generate the tabular listings and line-printer plots that are requested. In addition, OVTPVT will perform, if requested, the Fourier analysis of time-domain waveforms after the transient analysis has been determined [3].

8.2.3   The READIN module

The READIN module reads the input file, checks each line of input for syntax, and constructs the data structure for the circuit. The data structure that defines the circuit is a set of linked lists.

Each call to subroutine CARD returns the contents of the next line of input, together with the contents of any continuation lines. The input line is broken into fields, with each field containing a decimal number (a number field) or an alphanumeric name (a name field). The first field of the card determines whether the card is an element card or a control card; every control card must begin with a period. If the card is an element card or a .MODEL card, then it is processed by the READIN subroutine. If the card is a control card, then it is processed by the RUNCON subroutine.

The FIND subroutine handles all matters of the linked-list structure. Whenever a specific element of a linked list (that is, a bead) is desired, the FIND subroutine is called to locate the bead. If a bead is to be added to the data structure, then the FIND subroutine handles the linking and memory extension [3].

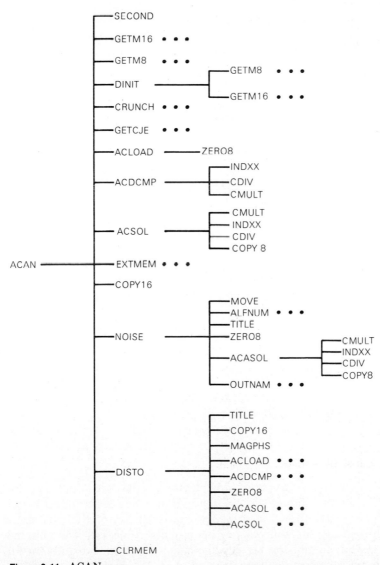

Figure 8-11 ACAN map.

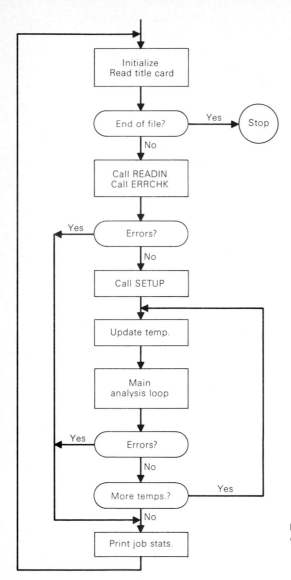

Figure 8-12 Main control flow-chart of SPICE2 [3].

8.2.4   The ERRCHK module

First, ERRCHK checks for undefined elements (as might happen, for example, if a coupled inductor element references an inductor that is not defined). Then, ERRCHK constructs an ordered list of the user node numbers, processes the sources in the circuits, processes the coupled inductors in the circuit, and establishes a list of source breakpoints for later use in the transient analysis.

The ELPRNT subroutine prints the circuit summary. The MODCHK subroutine assigns default values for model parameters that have not been specified, prints the summary of device model parameters, performs some additional model parameter processing (such as inverting series resistances), and, finally, reserves any internal device nodes that are generated by nonzero device ohmic resistances.

The TOPCHK subroutine constructs the node table, prints the node table, and checks that every node in the circuit has at least two elements connected to it, that every node in the circuit has a dc path to ground, and that there are no inductor and/or voltage source meshes and no capacitor and/or current source cuts.

8.2.5   The SETUP module

The SETUP routine is the control routine for this module. First, the subroutine MATPTR is called to determine the nodal admittances and matrix coefficients that are nonzero. MATPTR cycles through the circuit elements and, with the aid of the RESERV subroutine, establishes the initial integer pointer structure of nonzero Y-matrix terms. Next, the pointer structure is reordered to minimize fill-in, and the fill-in that does occur is added to the pointers. The reordering is accomplished by the REORDR subroutine. Finally, the MATLOC subroutine computes and stores the pointer locations that are used to construct the Y matrix. The INDXX routine is used by MATLOC to determine the location of a specific Y-matrix coefficient [3].

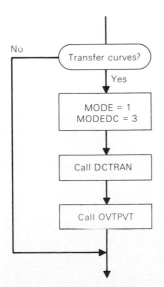

Figure 8-13 DC transfer-curve flowchart [3].

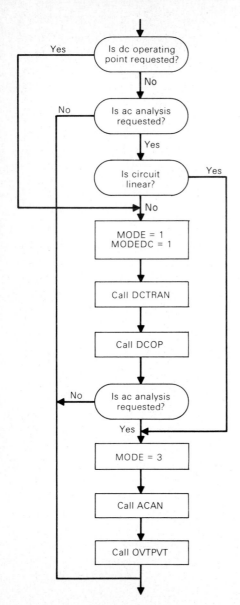

Figure 8-14 DC operating-point and ac analysis flowchart [3].

8.2.6   The DCTRAN module

DCTRAN is the largest and most complicated module in the SPICE2 program. This subroutine performs the dc operating-point analysis, the transient initial-condition analysis, the dc transfer-curve analysis, and the transient analysis.

The flowchart for the dc operating-point analysis is given in Fig. 8-16, for the transient initial time-point solution in Fig. 8-17, for the dc transfer-curve analysis in Fig. 8-18, for the transient analysis in Fig. 8-19 [3].

a. DC operating-point analysis. The logic for a dc operating point is uncomplicated. As shown in Fig. 8-16, the program first evaluates the source values at time zero. Next, the INITF flag is set to 2, and the subroutine ITER8 is called to determine the dc solution iteratively. If the solution converges, the INITF flag is reset to 4, and the device model routines DIODE, BJT, JFET, and MOSFET are called to compute the linearized, small-signal value of the nonlinear capacitors in the device models. The node voltages for the solution are then printed, and DCTRAN returns to the main program.

b. Transient initial time-point solution. The solution of the initial transient time point is simpler than that of the dc operating point. As shown in Fig.

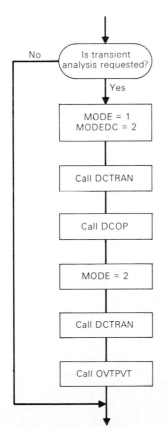

Figure 8-15 Transient analysis flowchart [3].

8-17, the difference between these two single-point dc analyses is that, for the case of the transient initial time point, the linearized capacitance values are not computed, since they are of no use in the transient analysis [3].

c. DC transfer-curve analysis. The dc transfer-curve analysis requires multiple-point analysis and the subsequent storage of output variables. As shown in Fig. 8-18, the dc transfer-curve analysis begins the same way as the dc operating-point analysis. Specifically, the variable source value

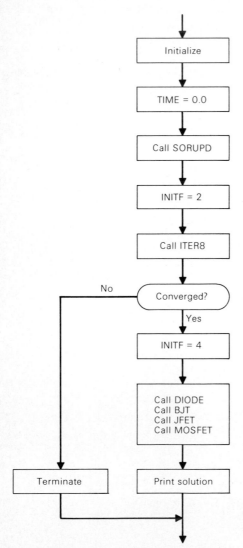

Figure 8-16 DC operating-point flowchart [3].

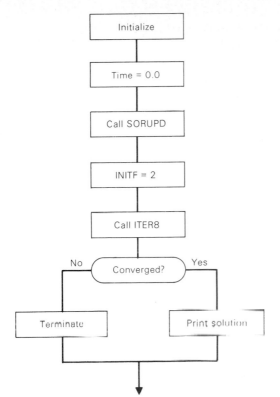

Figure 8-17 Transient initial condition flowchart [3].

is set to the first value, all other source values are set to their time-zero values, and ITER8 is called with INITF equal to 2 to obtain the first-point solution. The values of the specified output variables are stored in central memory after this transfer point has been computed. Then the variable source value is incremented, and the INITF flag is set to 6. This process then continues until the required number of dc transfer-curve points has been computed [3].

d. Transient analysis.   The transient analysis loop is shown in Fig. 8-19. As this figure shows, the transient analysis and the dc transfer-curve analysis loops are very similar. The time-zero solution has already been computed before this loop is entered. Therefore, the first step in the analysis is the storage of the time-zero output variables. The first time point is then computed by calling ITER8 with the flag INITF set to 5. Each subsequent time point is computed with the flag INITF set to 6. If the time point does not converge, the time step DELTA is reduced by a factor of 8 and the time point is reattempted. If the time point does converge, then the subroutine TRUNC is called to determine the new time step that is

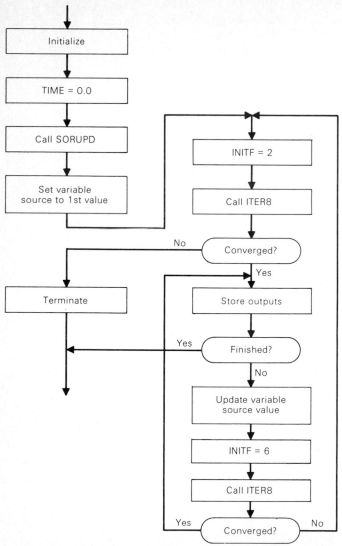

Figure 8-18   DC transfer-curve flowchart [3].

consistent with the truncation error tolerance that has been specified. As
long as the time step is larger than the time step presently in effect, the
time point is accepted and the output variables are stored. If the truncation error is too large, as indicated by a time step smaller than the present
time step, then the time point is rejected and reattempted with the
smaller time step. This process continues until the time interval specified
by the user has been covered [3].

e. Automatic time step.   There are two methods available to the user to control the computation of the appropriate transient analysis time step: *iteration count* and *truncation error estimation*.

The iteration count method uses the number of Newton-Raphson iterations required to converge at a time point as a measure of the rate of

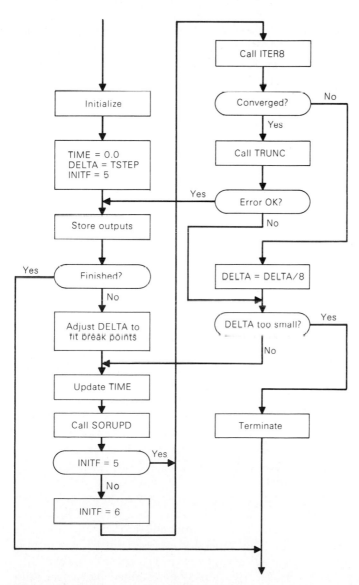

Figure 8-19   Transient analysis flowchart [3].

change of the circuit. If the number of iterations is less than ITL3, the time step is doubled; if the number is greater than ITL4, the time step is divided by 8. This method has the advantage of minimal computation overhead; however, it has the disadvantage that it does not take into account the true rate of change of circuit variables.

The second method of time-step determination is based on the estimation of local truncation error. The appropriate function derivative is approximated by a divided difference, hence the *estimation*. Although requiring somewhat more CPU time than iteration count, the truncation error estimation method allows a meaningful error bound to be set on the computed output values [3, 4].

f. Breakpoint table.   In transient analysis, SPICE2 always uses a program-calculated time step, regardless of the user-specified print interval. However, the independent source waveforms frequently have sharp transitions that could cause an unnecessary reduction in the time step in order to find the exact transition time. To overcome this problem, ERRCHK generates a *breakpoint table,* LSBKPT, which contains a sorted list of all the transition points of the independent sources. During transient analysis, whenever the next time point is sufficiently close to one of the breakpoints, the time step is adjusted so that the program lands exactly on the breakpoint [4].

g. ITER8 subroutine.   All four of the analysis procedures implemented in DCTRAN use the subroutine ITER8 to determine the solution of the circuit equations. The flowchart for the ITER8 subroutine is shown in Fig. 8-20. The solution commences as follows. First, the subroutine LOAD is called. This subroutine constructs the linearized system of nodal equations for the specific iteration. The contributions of the Y matrix from nonlinear devices are computed separately in the subroutines DIODE, BJT, JFET, and MOSFET. The NONCON flag is the number of nonconvergent nonlinearities in the circuit. This flag is incremented during the LOAD procedure, since the convergence is checked as an integral part of the linearization procedure. Immediately following the construction of the Y matrix and the forcing vector, a check on NONCON is performed to determine if the solution has converged. If convergence has been obtained, then the ITER8 subroutine is exited. Otherwise, the system of linearized equations is solved for the next iterate value, and the system of circuit equations is linearized again. This process continues until the solutions converge or until the number of iterations exceeds a preset number [3].

The solution of the linearized equations is determined by the routines DCDCMP and DCSOL.

8.2.7 The DCOP module

DCOP performs three functions: it prints out information about the device operating point, it computes the small-signal transfer function (subroutine SSTF), and it computes dc sensitivities of output variables with respect to circuit parameters (subroutine SENCAL).

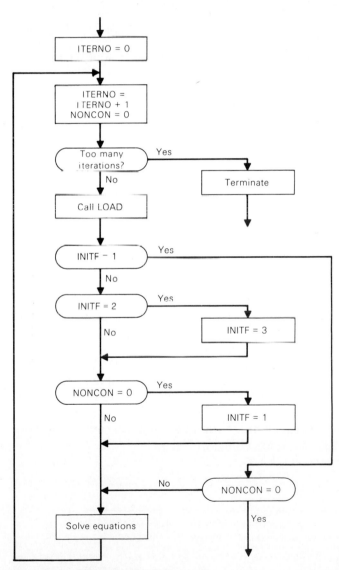

Figure 8-20 ITER8 flowchart [3].

8.2.8   The ACAN module

The ACAN module in SPICE2 determines the small-signal frequency-domain analysis. The ACLOAD subprogram constructs the complex system of linear equations for each frequency point. This system of equations is then solved with the routines ACDCMP and ACSOL. Subroutines DISTO and NOISE perform distortion and noise analysis, respectively.

8.2.9   The OVTPVT module

The last module in SPICE2 is OVTPVT. The OVTPVT subroutine interpolates the output variables to match the user-specified printing increment and generates the tabular listings specified by the user. The PLOT subroutine is called to generate the line-printer plots of the simulated results. If requested, the FOURAN subroutine performs the Fourier analysis for transient analysis.

8.3   Convergence Problems

It has been seen in Sec. 8.1.2 that problems with convergence can arise during simulation. In the following sections, some suggestions are given to improve and force convergence.

a. DC or transient analysis.   First, it is very important to point out that most convergence problems are due to errors in the description of the circuit, both in the value of some parameter (electrical, geometrical, or physical) or in the circuit description itself (wrong connections, wrong model references, floating subcircuit nodes). Consequently, it is very important to examine the input data for possible errors.

To facilitate this error check, use the LIST and NODE options in the .OPTIONS card. Once this check has been completed, you can decide if you want to modify some of the convergence parameters in the .OPTIONS card.

Those parameters that may have an influence on convergence in both dc and transient analysis are described in Table 8-1.

TABLE 8-1

Parameter	Default value	Suggested to help or to speed up convergence	Suggested to force convergence
RELTOL	0.001	0.01	0.1
ABSTOL	1 pA	1 nA	1 mA
VNTOL	1 μV	1 mV	10 mV

The three parameters in Table 8-1 give an estimate of the convergence conditions that must be satisfied. Their modification obtains two main results:

1. Speeding up the convergence in any case, if the suggested values are used, still assuring good solution values (this depends also on the type of circuit analyzed).
2. Forcing the convergence when nothing else seems to work. This can help to isolate the part of the circuit that is creating the problems.

To see the default values used in the actual simulation, the user can use option OPTS in the .OPTIONS card.

b. DC analysis. Something can be done specifically for dc convergence if this is the main problem.

1. Use the .NODESET card, giving reasonable initial values to the most critical nodes in the circuit. In some cases it may help to step the dc voltage generators from 0 to their dc value using the intermediate solution as initial conditions (with .NODESET) of the following solution (source stepping).
2. Use the OFF option for those devices in a positive feedback path.
3. Increase the dc iteration limit, ITL1, in the .OPTIONS card.
4. Step the dc voltage generators from 0 to their dc values via a transient analysis. In this way the sources are increased "slowly" and the capacitors are included in the solution. This is actually more similar to the real world and can be very appropriate for those circuits that need a finite setup time. When the dc solution is needed anyway (for an ac analysis), put in the .NODESET card the solution obtained in this way.

c. Transient analysis. The typical error message when convergence is not achieved in transient analysis is

INTERNAL TIMESTEP TOO SMALL IN TRANSIENT ANALYSIS

This is due to the fact that SPICE starts internally to use the time step specified in the .TRAN card, but if convergence is not achieved within a certain criteria (see below), it keeps decreasing the internal time step, dividing it by 8 each time until a certain lower limit (not accessible to the user) is exceeded.

Several other things can be done to help convergence:

1. Use the .IC card (or the IC = *option* in the transistor description) and/or the option UIC in the .TRAN card. A good method for avoiding

initial convergence problems (in the simulated time sense) is to use the option UIC without specifying any IC (which is to say they are all set to zero) and applying the power supply only. In this case, the circuit will reach the steady state (i.e., initial conditions for transient analysis) gradually. After the simulated time necessary for the circuit to reach the steady state (depending on the circuit), the time-dependent input sources can be applied.

2. Act on the parameters LVLTIM, TRTOL, and ITL4 of the .OPTIONS card. In the latest version of SPICE, the truncation error time-step control is the default method for deciding when to double the internal time step if convergence is easy and when to divide the internal time step by 8 if convergence is difficult. If this method is used (LVLTIM = 2, default), then TRTOL is the parameter that must be modified to have some influence on convergence; if TRTOL is increased, less importance is given to the truncation error, and then it is less likely that the internal time step will be divided by 8, which means that the condition described at the beginning is less likely to occur. If LVLTIM = 1, the doubling or the division by 8 of the internal time step depends only on the number of iterations needed to reach convergence, and then TRTOL has no more influence. In this case, ITL4 is the parameter to modify, and, as is evident, increasing ITL4 helps SPICE to converge.

3. Change the method of integration from TRAPEZOIDAL (default) to GEAR. The parameter is METHOD in the .OPTIONS card. If it is changed to GEAR, MAXORD must also be changed. The suggested value to give MAXORD is 4. If GEAR is chosen, the truncation error control is also automatically inserted.

8.4 Linked-List Specification

After reading the input data, SPICE builds a list of vectors that contain the basic information for all the elements and for all the analysis and output calls.

This section details an analysis of the vector structure limited to the most important ones, the semiconductor devices; for all the other elements, which are beyond the aim of this book, read Refs. [4, 5].

The vector LOCATE (50) contains the pointers to each list head element; element types are referenced through the ID identifier. Inside this vector all integer data are referenced using the LOC pointer, while all real data are accessed using the LOCV pointer. In the following section, the expression enclosed in parentheses specifies in which subroutine the data are defined.

8.4.1 Semiconductor devices

a. DIODE (identifier ID = 11)

	− 1	Subcircuit information	(FIND)
LOC	+ 0	Next pointer	(FIND)
	+ 1	LOCV pointer	(FIND)
	+ 2	Positive node np	(READIN)
	+ 3	Negative node nn	(READIN)
	+ 4	Model added node np'	(MODCHK)
	+ 5	Pointer to the device model	(FNDNAM)
	+ 6	OFF condition indicator	(READIN)
LOCV	+ 0	Element name	(FIND)
	+ 1	Area factor (default = 1)	(READIN)
	+ 2	Initial condition specification V_D	(READIN)

b. BJT (ID = 12)

	− 1	Subcircuit information	(FIND)
LOC	+ 0	Next pointer	(FIND)
	+ 1	LOCV pointer	(FIND)
	+ 2	Collector node nc	(READIN)
	+ 3	Base node nb	(READIN)
	+ 4	Emitter node ne	(READIN)
	+ 5	Model added node nc'	(MODCHK)
	+ 6	Model added node nb'	(MODCHK)
	+ 7	Model added node ne'	(MODCHK)
	+ 8	Pointer to the device model	(FNDNAM)
	+ 9	OFF condition indicator	(READIN)
LOCV	+ 0	Element name	(FIND)
	+ 1	Area factor (default = 1)	(READIN)
	+ 2	Initial condition specification V_{BE}	(READIN)
	+ 3	Initial condition specification V_{CE}	(READIN)

c. JFET (ID = 13)

	− 1	Subcircuit information	(FIND)
LOC	+ 0	Next pointer	(FIND)
	+ 1	LOCV pointer	(FIND)
	+ 2	Drain node nd	(READIN)

+	3	Gate node ng	(READIN)
+	4	Source node ns	(READIN)
+	5	Model added node nd'	(MODCHK)
+	6	Model added node ns'	(MODCHK)
+	7	Pointer to the device model	(FNDNAM)
+	8	OFF condition indicator	(READIN)
LOCV +	0	Element name	(FIND)
+	1	Area factor (default = 1)	(READIN)
+	2	Initial condition specification V_{DS}	(READIN)
+	3	Initial condition specification V_{GS}	(READIN)

d. MOS (ID = 14)

−	1	Subcircuit information	(FIND)
LOC +	0	Next pointer	(FIND)
+	1	LOCV pointer	(FIND)
+	2	Drain node nd	(READIN)
+	3	Gate node ng	(READIN)
+	4	Source node ns	(READIN)
+	5	Bulk node nb	(READIN)
+	6	Model added node nd'	(MODCHK)
+	7	Model added node ns'	(MODCHK)
+	8	Pointer to the device mode	(FNDNAM)
+	9	OFF condition indicator	(READIN)
LOCV +	0	Element name	(FIND)
+	1	Channel length L	(READIN)
+	2	Channel width W	(READIN)
+	3	Drain diffusion area A_D	(READIN)
+	4	Source diffusion area A_S	(READIN)
+	5	Initial condition specification V_{DS}	(READIN)
+	6	Initial condition specification V_{GS}	(READIN)
+	7	Initial condition specification V_{BS}	(READIN)
+	8	Device mode (normal = +1; inverse = −1)	(MOSFET)
+	9	Threshold voltage	(MOSFET)
+	10	Saturation voltage	(MOSFET)
+	11	Drain diffusion perimeter P_D	(READIN)
+	12	Source diffusion perimeter P_S	(READIN)
+	13	Drain diffusion equivalent squares NR_D	(READIN)

+	14	Source diffusion equivalent squares NR_S	(READIN)
+	15	Channel change parameter XQC	(READIN)

8.4.2 Semiconductor device models

a. DIODE model (ID = 21)

−	1	Subcircuit information	(FIND)
LOC +	0	Next pointer	(FIND)
+	1	LOCV pointer	(FIND)
+	2	Unused (zero)	(FIND)
+	3	Unused (zero)	(FIND)
+	4	Unused (zero)	(FIND)
LOCV +	0	Model name	(FIND)
+	1	IS	(READIN)
+	2	RS (1/RS in MODCHK)	(READIN)
+	3	N	(READIN)
+	4	TT	(READIN)
+	5	CJO	(READIN)
+	6	VJ	(READIN)
+	7	M	(READIN)
+	8	EG	(READIN)
+	9	XTI	(READIN)
+	10	KF	(READIN)
+	11	AF	(READIN)
+	12	FC (FC $*$ VJ in MODCHK)	(READIN)
+	13	BV	(READIN)
+	14	IBV	(READIN)
+	15	VJ $*$ (1 − exp ((1 − M) $*$ ln (1 − FC)))/(1 − M)	(MODCHK)
+	16	exp ((1 + M) $*$ ln (1 − FC))	(MODCHK)
+	17	1 − FC $*$ (1 + M)	(MODCHK)
+	18	N $* kT/q *$ ln (N $*$ (kT/q)/($\sqrt{2} *$ IS))	(MODCHK)

b. BJT model (ID = 22)

−	1	Subcircuit information	(FIND)
LOC +	0	Next pointer	(FIND)
+	1	LOCV pointer	(FIND)

+	2	Model type (NPN = 1; PNP = -1)	(READIN)
+	3	Unused (zero)	(FIND)
+	4	Unused (zero)	(FIND)
LOCV +	0	Model name	(FIND)
+	1	IS	(READIN)
+	2	BF	(READIN)
+	3	NF	(READIN)
+	4	VAF (1/VAF in MODCHK)	(READIN)
+	5	IKF (1/IKF in MODCHK)	(READIN)
+	6	ISE	(READIN)
+	7	NE	(READIN)
+	8	BR	(READIN)
+	9	NR	(READIN)
+	10	VAR (1/VAR in MODCHK)	(READIN)
+	11	IKR (1/IKR in MODCHK)	(READIN)
+	12	ISC	(READIN)
+	13	NC	(READIN)
+	14	0	(READIN)
+	15	0	(READIN)
+	16	RB	(READIN)
+	17	IRB	(READIN)
+	18	RBM	(READIN)
+	19	RE (1/RE in MODCHK)	(READIN)
+	20	RC (1/RC in MODCHK)	(READIN)
+	21	CJE	(READIN)
+	22	VJE	(READIN)
+	23	MJE	(READIN)
+	24	TF	(READIN)
+	25	XTF	(READIN)
+	26	VTF (1.44/VTF in MODCHK)	(READIN)
+	27	ITF	(READIN)
+	28	PTF [PTF/($2\pi * $ TF) in MODCHK]	(READIN)
+	29	CJC	(READIN)
+	30	VJC	(READIN)
+	31	MJC	(READIN)
+	32	XCJC	(READIN)
+	33	TR	(READIN)

+ 34	0		(READIN)
+ 35	0		(READIN)
+ 36	0		(READIN)
+ 37	0		(READIN)
+ 38	CJS		(READIN)
+ 39	VJS		(READIN)
+ 40	MJS		(READIN)
+ 41	XTB		(READIN)
+ 42	EG		(READIN)
+ 43	XTI		(READIN)
+ 44	KF		(READIN)
+ 45	AF		(READIN)
+ 46	FC (FC $*$ VJE in MODCHK)		(READIN)
+ 47	$VJE * (1 - \exp((1 - MJE) * \ln(1 - FC)))/(1 - MJE)$		(MODCHK)
+ 48	$\exp((1 + MJE) * \ln(1 - FC))$		(MODCHK)
+ 49	$1 - FC * (1 + MJE)$		(MODCHK)
+ 50	$FC * VJC$		(MODCHK)
+ 51	$VJC * (1 - \exp((1 - MJC) * \ln(1 - FC)))/(1 - MJC)$		(MODCHK)
+ 52	$\exp((1 + MJC) * \ln(1 - FC))$		(MODCHK)
+ 53	$1 - FC * (1 + MJC)$		(MODCHK)
+ 54	$kT/q * \ln((kT/q)/(\sqrt{2} * IS))$		(MODCHK)

c. JFET model (ID = 23)

	−	1	Subcircuit information	(FIND)
LOC	+	0	Next pointer	(FIND)
	+	1	LOCV pointer	(FIND)
	+	2	Model type (NJF = 1; PJF = −1)	(READIN)
	+	3	Unused (zero)	(FIND)
	+	4	Unused (zero)	(FIND)
LOCV	+	0	Model name	(FIND)
	+	1	VTO	(READIN)
	+	2	BETA	(READIN)
	+	3	LAMBDA	(READIN)

+	4	RD (1/RD in MODCHK)	(READIN)
+	5	RS (1/RS in MODCHK)	(READIN)
+	6	CGS	(READIN)
+	7	CGD	(READIN)
+	8	PB	(READIN)
+	9	IS	(READIN)
+	10	KF	(READIN)
+	11	AF	(READIN)
+	12	FC (FC $*$ PB in MODCHK)	(READIN)
+	13	PB $*$ $(1 - \exp(0.5 * \ln(1 - FC)))/0.5$	(MODCHK)
+	14	$\exp(1.5 * \ln(1 - FC))$	(MODCHK)
+	15	$1 - FC * 1.5$	(MODCHK)
+	16	$kT/q * \ln(kT/q)/(\sqrt{2} * IS))$	(MODCHK)

d. MOS model (ID = 24)

	−	1	Subcircuit information	(FIND)
LOC	+	0	Next pointer	(FIND)
	+	1	LOCV pointer	(FIND)
	+	2	Model type (NMOS = 1; PMOS = −1)	(READIN)
	+	3	Unused (zero)	(FIND)
	+	4	Unused (zero)	(FIND)
LOCV	+	0	Model name	(FIND)
	+	1	LEVEL	(READIN)
	+	2	VTO	(READIN)
	+	3	KP	(READIN)
	+	4	GAMMA	(READIN)
	+	5	PHI	(READIN)
	+	6	LAMBDA [2 $* K_{Si} * \epsilon_0/q *$ NSUB (LEVEL = 3) in MODCHK]	(READIN)
	+	7	RD (1/RD in MODCHK)	(READIN)
	+	8	RS (1/RS in MODCHK)	(READIN)
	+	9	CBD	(READIN)
	+	10	CBS	(READIN)
	+	11	IS	(READIN)
	+	12	PB	(READIN)
	+	13	CGSO	(READIN)

+ 14	CGDO	(READIN)
+ 15	CGBO	(READIN)
+ 16	RSH (1/RSH in MODCHK)	(READIN)
+ 17	CJ	(READIN)
+ 18	MJ	(READIN)
+ 19	CJSW	(READIN)
+ 20	MJSW	(READIN)
+ 21	JS	(READIN)
+ 22	TOX (COX = $K_{SiO_2} * \epsilon_0$/TOX in MODCHK)	(READIN)
+ 23	NSUB (NSUB * 10^6 in MODCHK)	(READIN)
+ 24	NSS (NSS * 10^4 in MODCHK)	(READIN)
+ 25	NFS (NFS * 10^4 in MODCHK)	(READIN)
+ 26	TPG	(READIN)
+ 27	XJ	(READIN)
+ 28	LD	(READIN)
+ 29	UO (UO * 10^{-4} in MODCHK)	(READIN)
+ 30	UCRIT [UCRIT * 10^2 * K_{Si} * ϵ_0/COX (LEVEL = 2) in MODCHK] [THETA (LEVEL = 3) in MODCHK]	(READIN)
+ 31	UEXP [ETA * 8.15 * 10^{-22}/COX (LEVEL = 3) in MODCHK]	(READIN)
+ 32	UTRA [KAPPA (LEVEL = 3) in MODCHK]	(READIN)
+ 33	VMAX	(READIN)
+ 34	NEFF	(READIN)
+ 35	XQC [1 (LEVEL = 3) in MODCHK]	(READIN)
+ 36	KF	(READIN)
+ 37	AF	(READIN)
+ 38	FC (FC * PB in MODCHK)	(READIN)
+ 39	DELTA [DELTA * 0.25 * 2π * K_{Si} * ϵ_0/COX (LEVEL = 3) in MODCHK]	(READIN)
+ 40	THETA [PB * (1 − exp (0.5 * ln (1 − FC)))/ 0.5 in MODCHK]	(READIN)
+ 41	ETA [exp (1.5 * ln (1 − FC)) in MODCHK]	(READIN)
+ 42	KAPPA (1 − FC * 1.5 in MODCHK)	(READIN)
+ 43	−1	(MODCHK)
+ 44	VTO − GAMMA * $\sqrt{PHI}$ + PHI	(MODCHK)
+ 45	$\sqrt{2 * K_{Si} * \epsilon_0/q * NSUB}$	(MODCHK)

TABLE 8-2

Flag	Value	Meaning
MODE	1	DC analysis (subtype defined by MODEDC)
	2	Transient analysis
	3	AC analysis (small-signal)
MODEDC†	1	DC operating point
	2	Initial operating point for transient analysis
	3	DC transfer-curve computation
INITF	1	Converge with OFF devices allowed to float
	2	Initialize junction voltages
	3	Converge with OFF devices held "off"
	4	Store small-signal parameters
	5	First time point in transient analysis
	6	Prediction step
NONCON		Number of nonconvergent branches
ITERNO		Iteration number for current sweep point
ITEMNO		Temperature number
NOSOLV	1	Use initial conditions (UIC) specified for transient analysis

† MODEDC is significant only if MODE = 1.

8.4.3 Useful flags

In analyzing the device model subroutines, the user must know the variables used by SPICE to keep track of the state of the analysis. The values of these flags (and the corresponding meanings) are as shown in Table 8-2.

Other useful flags the user can specify in the input deck are included in the .OPTIONS card [1]. This card allows the user to reset program control and options for specific simulation purposes. Any combination of the following options can be included in any order, where x represents some positive number.

Flag	Meaning
ACCT	Causes accounting and run-time statistics to be printed.
LIST	Causes the summary listing of the input data to be printed.
NOMOD	Suppresses printout of the model parameters.
NOPAGE	Suppresses page ejects.
NODE	Causes printing of the node table.
OPTS	Causes option values to be printed.

GMIN = x	Resets the value of GMIN, the minimum conductance allowed by the program; the default value is 10^{-12} mho.
RELTOL = x	Resets the relative error tolerance of the program; the default value is 0.001 (0.1 percent).
ABSTOL = x	Resets the absolute current error tolerance of the program; the default value is 10^{-12} A.
VNTOL = x	Resets the absolute voltage error tolerance of the program; the default value is 10^{-6} V.
TRTOL = x	Resets the transient error tolerance; the default value is 7.0. This parameter is an estimate of the factor by which SPICE overestimates the actual truncation error.
CHGTOL = x	Resets the charge tolerance of the program; the default value is 10^{-14}.
NUMDGT = x	Resets the number of significant digits printed for output variable values; x must satisfy the relation $0 < x < 8$; the default value is 4. This option is independent of the error tolerance used by SPICE.
TNOM = x	Resets the nominal temperature; the default value is 27°C (300 K).
PIVTOL = x	Resets the absolute minimum value for a matrix entry to be accepted as a pivot; the default value is 10^{-13}.
PIVREL = x	Resets the relative ratio between the largest column entry and an acceptable pivot value; the default value is 10^{-3}. In the numerical pivoting algorithm, the allowed minimum pivot value is determined by EPSREL = AMAX1 (PIVREL * MAXVAL, PIVTOL), where MAXVAL is the maximum element in the column where a pivot is sought.
ITL1 = x	Resets the dc iteration limit; the default is 100.
ITL2 = x	Resets the dc transfer-curve iteration limit; the default is 50.
ITL3 = x	Resets the lower transient analysis iteration limit; the default value is 4.
ITL4 = x	Resets the transient analysis time-point iteration limit; the default is 10.
ITL5 = x	Resets the transient analysis total iteration limit; the default is 5000. Set ITL5 = 0 to omit this test.
LIMTIM = x	Resets the amount of CPU time reserved by SPICE for generating plots should a limit on CPU time cause job termination; the default value is 2 s.
LIMPTS = x	Resets the total number of points that can be printed or plotted in dc, ac, or transient analysis; the default value is 201.

LVLTIM = x If x is 1, the iteration time-step control is used. If x is 2, the truncation error time step is used. The default value is 2. If METHOD = GEAR and MAXORD > 2, SPICE sets LVLTIM to 2.

METHOD = name   Sets the numerical integration method used by SPICE. Possible names are GEAR or TRAPEZOIDAL. The default is TRAPEZOIDAL.

MAXORD = x Sets the maximum order for the integration method if GEAR's variable-order method is used; x must be between 2 and 6. The default value is 2.

DEFL = x Resets the value for MOS channel length; the default is 10^{-4} m.

DEFW = x Resets the value for MOS channel width; the default is 10^{-4} m.

DEFAD = x Resets the value for MOS drain diffusion area; the default is 0.

DEFAS = x Resets the value for MOS source diffusion area; the default is 0.

REFERENCES

1. *SPICE Version 2G User's Guide.*
2. L. W. Nagel and D. O. Pederson, Simulation Program with Integrated Circuit Emphasis (SPICE), Electronics Research Laboratory Rep. No. ERL-M382, University of California, Berkeley, 1973.
3. L. W. Nagel, SPICE2: A Computer Program to Simulate Semiconductor Circuits, Electronics Research Laboratory Rep. No. ERL-M520, University of California, Berkeley, 1975.
4. E. Cohen, Program Reference for SPICE2, Electronics Research Laboratory Rep. No. ERL-M592, University of California, Berkeley, 1976.
5. M. Santomauro, L. Schnickel, and M. Tomaini, SPICE2: Un Programma per la Simulazione Circuitale, Thesis, Politecnico di Milano, Istituto di Elettrotecnica ed Elettronica, Milan, Italy, 1982.
6. P. Antognetti, D. O. Pederson, and H. De Man, *Computer Design Aids for VLSI Circuits*, Nijhoff, the Hague, 1981.
7. W. J. McCalla and D. O. Pederson, Elements of Computer-Aided Circuit Analysis, *IEEE Trans. Circuit Theory*, **CT-18**, 1971.
8. R. D. Berry, An Optimal Ordering of Electronic Circuit Equations for a Sparse Matrix Solution, *IEEE Trans. Circuit Theory*, **CT-18**, 1971.
9. L. O. Chua and P. M. Lin, *Computer-Aided Analysis of Electronic Circuits: Algorithms and Computational Techniques*, Prentice-Hall, Englewood Cliffs, N.J., 1975.
10. D. A. Calahan, *Computer-Aided Network Design*, McGraw-Hill, New York, 1968.
11. H. Shichman, Integration System of a Nonlinear Network Analysis Program, *IEEE Trans. Circuit Theory*, **CT-17**, 1970.
12. H. Shichman, Computation of DC Solutions for Bipolar Transistor Networks, *IEEE Trans. Circuit Theory*, **CT-16**, 1969.
13. G. D. Hachtel, R. K. Brayton, and F. G. Gustavson, The Sparse Tableau Approach to Network Analysis and Design, *IEEE Trans. Circuit Theory*, **CT-18**, 1971.
14. C. W. Ho, A. E. Ruehli, and P. A. Brennan, The Modified Nodal Approach to Network Analysis, *Proc. 1974 International Symposium on Circuits and Systems*, San Francisco, 1974.

15. E. Isaacson and H. B. Keller, *Analysis of Numerical Methods,* Wiley, New York, 1966.
16. M. Nakhla, K. Singhal, and J. Vlach, An Optimal Pivoting Order for the Solution of Sparse Systems of Equations, *IEEE Trans. Circuit Systems,* **CAS-21,** 1974.
17. R. Fletcher and M. J. D. Powell, A Rapidly Convergent Descent Method for Minimization, *Comput. J.,* **6,** 1963.
18. D. Agnew, Iterative Improvement of Network Solutions, *Proc. Sixteenth Midwest Symposium on Circuit Theory,* Waterloo, Ontario, 1973.
19. C. G. Broyden, A New Method of Solving Nonlinear Simultaneous Equations, *Comput. J.,* **12,** 1969.
20. C. W. Gear, *Numerical Initial Value Problems in Ordinary Differential Equations,* Prentice-Hall, Englewood Cliffs, N.J., 1971.
21. R. K. Brayton, F. G. Gustavson, and G. D. Hachtel, A New Efficient Algorithm for Solving Differential-Algebraic Systems Using Implicit Backward Differentiation Formulas, *Proc. IEEE,* **60,** 1972.
22. H. H. Rosenbrock, Some General Implicit Processes for the Numerical Solution of Differential Equations, *Comput. J.,* **5,** 1963.
23. R. A. Rohrer, Successive Secants in the Solution of Nonlinear Equations, *AMS,* 1970.
24. H. Y. Hsieh and M. S. Ghausi, A Probabilistic Approach to Optimal Pivoting and Prediction of Fill-in for Random Sparse Matrices, *IEEE Trans. Circuit Theory,* **CT-19,** 1972.
25. C. W. Gear, Numerical Integration of Stiff Ordinary Equations, Rep. No. 221, University of Illinois at Urbana-Champaign, 1967.
26. A. Nordsieck, On Numerical Integration of Ordinary Differential Equations, *Math. Comp.,* **16,** 1962.
27. I. S. Duff, A Survey of Sparse Matrix Research, *Proc. IEEE,* 1977.
28. W. Liniger, Multistep and One-Leg Methods for Implicit Mixed Differential Algebraic Systems, *IEEE Trans. Circuit Syst.,* **CAS-26,** 1979.
29. G. Dahlquist et al., *Numerical Methods,* Prentice-Hall, Englewood Cliffs, N.J., 1974.

The Two-Terminal
PN Structure

Giuseppe Massobrio

*Department of Electronics (DIBE), University of
Genoa, Genoa, Italy*

A.1   Elements of Semiconductor Physics

A brief review of semiconductor physics is presented. Generally, only those results that will be needed in the subsequent treatment and concerning the aim of the book will be given. Not all the results, however, will be proved here; such proofs can be found in most classic books on semiconductor devices [1–7].

A.1.1   Carrier concentrations for semiconductors in equilibrium†

Four classes of charged particles can be observed in a semiconductor:

Particles that have positive charge $\left\{ \begin{array}{l} \text{1. Mobile holes } p \\ \text{2. Immobile donor ions } N_D \end{array} \right.$

Particles that have negative charge $\left\{ \begin{array}{l} \text{1. Mobile electrons } n \\ \text{2. Immobile acceptor ions, } N_A \end{array} \right.$

---

† The material preceding Sec. A.1.1a is taken from P. E. Gray and C. L. Searle, *Electronic Principles: Physics, Models, and Circuits*, copyright © 1969 by John Wiley & Sons, Inc. Reprinted by permission.

In each case the symbol used above represents the *volume concentration* of the corresponding charged particle. Each of these types of charged particles carries a charge q ($q = 1.60 \times 10^{-19}$ C), the electronic charge. Therefore, the local charge density ρ can be written as

$$\rho = q(p + N_D - n - N_A) \qquad (A\text{-}1)$$

In a homogeneous (uniformly doped) semiconductor, the space-charge density must vanish at every point; local neutrality is required as a consequence of Gauss's law. Therefore, the concentrations of the several charged particles must satisfy this condition, which is obtained by setting $\rho = 0$ in Eq. (A-1). Then

$$n - p = N_D - N_A \qquad (A\text{-}2)$$

The impurity concentrations N_D and N_A are determined solely by the fabrication and subsequent processing of the semiconductor. Therefore, Eq. (A-2) can be regarded as one constraint on the concentrations p and n of the mobile carriers.

A second constraint can be obtained from considerations, on a quantum-statistical basis, of the hole and electron concentrations in a semiconductor. The result, which is introduced here as a postulate, is: The *equilibrium* hole and electron concentrations are coupled in such a way that for any particular semiconductor in equilibrium, the *product* of the hole and electron concentrations is a function of temperature alone and is independent of the concentrations of donor and acceptor impurities. That is,

$$np = f(T) \qquad (A\text{-}3)$$

where T denotes the absolute temperature and $f(T)$ is a function independent of the concentrations of the impurities. By common convention, the equilibrium np product is denoted by $n_i^2(T)$. That is,

$$np = n_i^2(T) \qquad (A\text{-}4)$$

The equilibrium carrier concentrations in a semiconductor are then governed by Eqs. (A-2) and (A-4). These equations can be solved for n and p explicitly. However, in many practical circumstances the impurity concentrations are large enough compared with the intrinsic concentration n_i to make simple approximate solutions quite accurate [1].

a. Intrinsic material. No dopant (impurity) atoms are present in *intrinsic* (pure) semiconductors, and the ionized crystal atoms contribute electron-hole *pairs;* from this fact and from Eq. (A-4), it follows that

$$n = p = n_i \qquad (A\text{-}5)$$

For this reason n_i is called the *intrinsic carrier concentration;* for Si it has the value of 1.5×10^{10} cm^{-3} at 300 K.

b. *n*-Type material.   Consider a material uniformly doped with an impurity that can provide free electrons to the crystal; such an impurity is called a *donor* impurity (it comes from group V of the periodic table).

The donor atoms donate electrons to the crystal and increase n at the expense of p (as compared to the case of intrinsic material), while the product np is still given by Eq. (A-4). These electrons become the *majority carriers,* and holes become the *minority carriers.* At room temperature practically all the donor atoms are ionized, each contributing one free electron; if $N_D \gg n_i$, practically all the free electrons originate from the donor atoms, and it can be written that

$$n \simeq N_D \qquad\qquad (A-6)$$

Combining Eq. (A-4) with Eq. (A-6) yields

$$p \simeq \frac{n_i^2}{N_D} \qquad\qquad (A-7)$$

If the doping concentration is very high ($> 10^{18}$ cm^{-3}), the preceding two relations will not hold, since the assumptions used in deriving them are not valid; semiconductors with very high N_D are called *degenerate.* Also, the two relations will not hold at extremely low temperatures, where the dopant atoms will not all be ionized, or at very high temperatures, where n_i rises to the point that the assumption $N_D \gg n_i$ is not valid. Whenever the preceding relations are used, it is implied that none of these extreme situations is in effect.

c. *p*-Type material.   Consider now a material doped with an impurity that can capture free electrons from the crystal, which is equivalent to providing free holes to it; such an impurity is called an *acceptor* impurity (it comes from group III of the periodic table).

The acceptor atoms donate holes to the crystal, thus increasing p at the expense of n; the majority carriers are now holes, and the minority carriers are electrons. It is assumed again that practically all the dopant atoms are ionized; each contributes one hole to the material.

If $N_A \gg n_i$, practically all the holes originate from the acceptor atoms, and it can be written that

$$p \simeq N_A \qquad\qquad (A-8)$$

Combining Eq. (A-4) with Eq. (A-8) yields

$$n \simeq \frac{n_i^2}{N_A} \qquad\qquad (A-9)$$

The preceding two approximations will fail at extreme temperatures or if N_A is too high, as explained for the case of n-type semiconductors.

d. Electric field influence. The preceding discussion has been limited to the case where no electric fields exist within the semiconductor material. Consider now the case where such fields do exist.

First, note that the existence of electric fields does not necessarily imply current flow; this statement is clearly true for a capacitor, for example. It is, however, valid for semiconductor materials as well. In such materials, current flow is the result of two effects: the *diffusion* of carriers, which is caused by carrier concentration gradients within the material, and the *drift* of carriers, which is caused by electric fields in the material.

Knowledge of the electric field intensity by itself is *not* enough to predict current flow; one must take into account diffusion, too. In many cases, an electric field can exist *without* any associated current flow, since its effect is balanced by that of the carrier concentration gradient (this is sometimes described by saying *the diffusion current cancels the drift current*). Such a situation occurs, for example, in the depletion region of a *pn* junction in the absence of external bias [3].

A.1.2 Transport of electric current†

The mobile charge carriers in a semiconductor are in constant motion, even under thermal equilibrium conditions. This motion is a manifestation of the random thermal energy of atoms and electrons that comprise the semiconductor. The motion of an individual hole or electron is tortuous and erratic; the charge carrier has frequent collisions with the atoms, including impurities, of the semiconductor. The direction of motion of the charge carrier usually changes as a consequence of these collisions.

Under equilibrium conditions, no average electric current results from this random thermal motion. The equilibrium situation can be disturbed in two ways:

1. An electric field can be applied

2. The carrier distributions can be made nonuniform

In both of these cases, a motion of the charge carriers results, thus producing electric currents [1].

---

† The material that follows, including Sec. A.1.2a, is derived in part from P. E. Gray and C. L. Searle, *Electronic Principles: Physics, Models, and Circuits,* copyright © 1969 by John Wiley & Sons, Inc. Reprinted by permission.

a. Drift current. An electric field affects the random thermal motion of charge carriers in the intervals between collisions by giving a small but uniform acceleration to all the carriers exposed to the field. Although, for any carrier, the velocity increment produced by this acceleration is wiped out by the collisions that the carrier has with its environment, the electric field has a net effect simply because the velocity increments that it gives to carriers are all directed along the field. This net effect is called *drift*.

The net effect of the collisions and the intervals of acceleration between collisions can be described by assigning to a group of carriers a *drift velocity* v_p, proportional to the electric field [1].

Thus the drift velocity of the holes is

$$v_p = \mu_p E \tag{A-10a}$$

where the parameter μ_p, which is independent of the electric field, is called the *hole mobility*.

Similarly, the drift velocity of the electrons is

$$v_n = -\mu_n E \tag{A-10b}$$

where μ_n is the *electron mobility*. The minus sign appears here because the electrons, having negative charge, are accelerated in a direction opposite to the field.

This treatment assumes that the mobility μ is independent of the applied electric field. This is a reasonable assumption only so long as the drift velocity is small in comparison to the thermal velocity of carriers, which is about 10^7 cm/s for Si at room temperature. As the drift velocity becomes comparable to the thermal velocity, its dependence on the electric field will begin to depart from the simple relationships given above. This is illustrated by experimental measurements of the drift velocity of electrons and holes in Si as a function of the electric field, shown in Fig. A-1. Evidently an initial straight-line dependence is followed by a less rapid increase as the electric field is increased. At large enough fields, a maximum drift velocity seems to be approached [2, 4].†

Thus the *hole current density* can be written as

$$J_{p,\text{drift}} = qp\mu_p E \tag{A-11a}$$

Similarly, the *electron current density* can be written as

$$J_{n,\text{drift}} = qn\mu_n E \tag{A-11b}$$

---

† The material in this paragraph is taken from A. S. Grove, *Physics and Technology of Semiconductor Devices,* copyright © 1967 by John Wiley & Sons, Inc. Reprinted by permission.

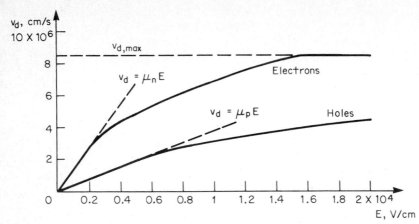

Figure A-1 Effect of an electric field on the magnitude of the drift velocity of carriers in Si. *(From A. S. Grove, Physics and Technology of Semiconductor Devices, copyright © 1967 by John Wiley & Sons, Inc. Used by permission.)*

The electron current density has the same direction as the electric field because, while the electrons move opposite to the field, they carry negative charge: the two minus signs cancel.

The *total electric current* produced by an electric field in a semiconductor is the sum of the currents carried by the holes and electrons. Thus

$$J_{\text{drift}} = q(\mu_p p + \mu_n n)E \qquad (A\text{-}12)$$

The coefficient of E in this result is simply the electrical *conductivity* σ of the semiconductor:

$$\sigma = q(\mu_p p + \mu_n n) \qquad (A\text{-}13)$$

b. Diffusion current.†   Diffusion is a manifestation of the random thermal motion of particles; it shows up as a particle current that appears whenever mobile particles are nonuniformly distributed in a system.

To discuss the diffusion process, consider it as two boxlike volume elements of a semiconductor with cross-sectional area A and width Δx, which lie on either side of an imaginary plane at x_0 normal to the x coordinate. Suppose that to the left there is a carrier concentration of holes greater than that to the right of x_0.

Thus there is a net flow of carriers from left to right because there are more carriers in the left volume element than in the right one. The net

---

† The material in this section is derived in part from P. E. Gray and C. L. Searle, *Electronic Principles: Physics, Models, and Circuits*, copyright © 1969 by John Wiley & Sons, Inc. Reprinted by permission.

rate of flow of holes is proportional to the concentration imbalance. More precisely, if p^- denotes the hole concentration in the left volume element and p^+ the concentration in the right volume element, the rate at which holes cross the boundary at x_0 is

$$\text{Hole flux} = (\text{constant})(p^- - p^+) \qquad \text{(A-14a)}$$

In the limit, as Δx approaches zero, this rate is proportional to the first derivative of the concentration at x_0. That is,

$$\text{Hole flux} = -D_p \left(\frac{dp}{dx} \right)_{x_0} \qquad \text{(A-14b)}$$

The constant of proportionality is called the *diffusion coefficient* for holes. The minus sign appears because carriers flow down the concentration slope from regions of high concentration to regions of lower concentration.

The *hole current density* associated with diffusion is then

$$J_{p,\text{diff}} = -qD_p \frac{dp}{dx} \qquad \text{(A-15a)}$$

The corresponding *electron current density,* which carries negative charge, is

$$J_{n,\text{diff}} = qD_n \frac{dn}{dx} \qquad \text{(A-15b)}$$

Diffusive flow is in no sense a cooperative process, and it has nothing to do with the fact that diffusing carriers are charged. It occurs simply because the *number of carriers* that (as a result of random thermal motion) have velocity components directed from the region of high concentration toward a region of lower concentration is greater than the number of carriers that have oppositely directed velocity components.

It is important to notice that the particle flux density that results from diffusion depends on the carrier concentration *gradient* and *not* on the concentration itself; it is the concentration imbalance that matters, not the value of the concentration [1].

c. Simultaneous presence of drift and diffusion currents. In many situations, an electric field and carrier concentration gradients are simultaneously present in a semiconductor. For small deviations from equilibrium, it is reasonable to regard the total hole or electron current density as a linear combination of the drift and diffusion current densities. Thus the *net hole current density* can be written, using Eqs. (A-11) and (A-15), as

$$J_p = q \left(p\mu_p E - D_p \frac{dp}{dx} \right) \tag{A-16a}$$

and the *net electron current density* can be written as

$$J_n = q \left(n\mu_n E + D_n \frac{dn}{dx} \right) \tag{A-16b}$$

This description of hole and electron motion in nonequilibrium situations in terms of drift and diffusion can be justified in detail in terms of statistical-mechanical concepts and techniques [1]. Here these descriptions will be adopted as postulates.

d. Einstein relations. *Drift* and *diffusion* are both manifestations of the random thermal motion of the carriers. Consequently, the mobility μ and the diffusion coefficient D are not independent. More precisely, they are related as follows:

$$\frac{D_p}{\mu_p} = \frac{kT}{q}$$

$$\frac{D_n}{\mu_n} = \frac{kT}{q} \tag{A-17}$$

These equations are known as *Einstein relations*. The proportionality constant kT/q, which has the dimensions of voltage, is called the *thermal voltage*. The factors that appear in the thermal voltage are Boltzmann's constant k ($k = 1.38 \times 10^{-23}$ J/K), electronic charge q ($q = 1.60 \times 10^{-19}$ C), and absolute temperature T. The thermal voltage is $kT/q = 25.86 \times 10^{-3}$ V at 300 K, while $\mu_n \simeq 1300$ cm^2/(V·s), $\mu_p \simeq 500$ cm^2/(V·s), $D_n \simeq 34$ cm^2·s, and $D_p \simeq 13$ cm^2·s.

The Einstein relations can be fully justified by considering the statistical-mechanical implications of the equilibrium situation in a semiconductor that is not uniformly doped. In accordance with our previous position, we will adopt these relations as postulates.

A.1.3   Carrier concentrations for semiconductors in nonequilibrium

Most semiconductor devices operate under nonequilibrium conditions, that is, under conditions in which the carrier concentration product np differs from its equilibrium value n_i^2. The performance of many semiconductor devices is determined by their tendency to return to equilibrium.

Consider, then, nonequilibrium situations in which Eq. (A-4) is violated. Accordingly, we can distinguish between two types of deviation from equilibrium:

$$np > n_i^2 \quad \textit{injection of excess carriers}$$

$$np < n_i^2 \quad \textit{extraction of carriers}$$

a. Injection-level concept.   Consider now the case when somehow excess carriers of both types are introduced into the semiconductor in equal concentrations in order to preserve space-charge neutrality. Two situations are relevant:

1. The excess minority-carrier concentration is much less than the equilibrium majority-carrier concentration. This condition is referred to as *low-level injection.*

2. The excess minority-carrier concentration approaches the majority-carrier concentration. This condition is referred to as *high-level injection.*

b. Return to equilibrium.   Whenever the carrier concentrations are disturbed from their equilibrium values, they attempt to return to equilibrium.

In the case of *injection* of excess carriers, return to equilibrium is through *recombination* of the injected minority carriers with the majority carriers. In the case of *extraction* of carriers, return to equilibrium is through the process of *generation* of electron-hole pairs.

The rate at which the recombination occurs is decided by the minority-carrier *lifetime* τ, which represents the reciprocal of the probability of an electron (hole) colliding with a hole (electron) in unit time. Since this collision ends up in recombination, this probability represents the probability of an electron (hole) recombination in unit time. Theory on recombination is widely discussed in the literature [2, 4–6], and it is beyond the scope of this chapter; only the expressions for the carrier lifetime and recombination rate (which will be used later) are given here.

Analysis of the process by which excess carriers recombine shows that the average time-space rate of recombination is in many practical cases approximately proportional to the excess minority-carrier concentration; this is true in the assumption of low-level injection. That is, in an n-type material, for example, the *net rate of recombination, U,* of excess holes with excess electrons can be approximated by

$$G_p - R_p = U_p = \frac{p_n - p_{n0}}{\tau_p} = \frac{p_n'}{\tau_p} \tag{A-18a}$$

where τ_p is the *lifetime* of the excess carriers and $p_n - p_{n0}$ denotes the *excess* hole concentration, i.e., the amount by which the actual minority-carrier hole concentration p_n exceeds the equilibrium hole concentration p_{n0}.

Similarly, in a p-type material, the recombination rate would be described by

$$G_n - R_n = U_n = \frac{n_p - n_{p0}}{\tau_n} = \frac{n_p'}{\tau_n} \tag{A-18b}$$

G is the rate of generation due to thermal mechanisms and R is the total rate of recombination (all per unit time and unit volume). A more accurate expression for U, following the Shockley-Hall-Read generation-recombination theory, can be found in Ref. [2].

A.2   Physical Operation of a *PN* Junction

To emphasize the salient features of the *pn*-junction behavior without introducing unessential and complicated details not within the aim of this book, we will consider a *step, or abrupt, pn junction*. This is a junction in which the transition from a p-type to an n-type semiconductor occurs over a region of negligible thickness, *abruptly*. It is assumed that the *metallurgical junction* or boundary between the n-type and p-type region is a plane, that all carrier distributions and currents are uniform on planes parallel to the boundary plane, and that all currents are directed perpendicular to the boundary plane. This *one-dimensional* model is reasonable for the representative structure shown in Fig. A-2 and is justifiable for most *pn* junctions.

Deviations and limitations of this idealized *pn*-junction model are not discussed here because they are developed step by step in Chap. 1, which related to *building* the real model. Here, only the basic physical operation of a *pn* junction is discussed.

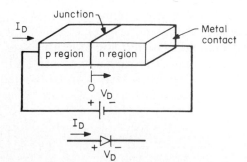

Figure A-2 Physical structure of a *pn* junction.

A.2.1   *PN* junction in equilibrium

A *pn* junction is said to be in *equilibrium* when it is at a uniform temperature and when no external disturbances, such as light or a bias voltage, are acting on it. Under equilibrium conditions the hole current and the electron current must *each* vanish at every point in the semiconductor.

It can be imagined, now, that a *barrier* exists at the boundary plane at $x = 0$ in the structure of Fig. A-2 so that no carrier (electrons or holes) flows between the p region and n region (see Fig. A-4a). The two regions, then, behave like two homogeneous and separated semiconductor materials whose electron and hole concentrations are uniform and are determined solely by the acceptor or donor concentration and the temperature.

Thus, from Eqs. (A-6) to (A-9), it can be written:

For $x > 0$ (n region):

$$n_{n0} \simeq N_D \tag{A-19}$$

$$p_{n0} \simeq \frac{n_i^2(T)}{N_D}$$

For $x < 0$ (p region):

$$n_{p0} \simeq \frac{n_i^2(T)}{N_A}$$

$$p_{p0} \simeq N_A \tag{A-20}$$

Clearly, concentration gradients of both holes and electrons must exist at the junction, as indicated in Fig. A-3.

Now, if the *barrier* (in reality it does not exist) at $x = 0$ is removed, the *pn* junction is formed and diffusion takes place in its vicinity. Electrons diffuse from the region where they are in high concentration (n side) to the region where their concentration is low (p side). Each such electron leaves behind a donor atom with a net positive charge, since the donor atoms are initially neutral; such ionized donor atoms are indicated by the $\oplus$ signs in Fig. A-4b, the circle suggesting a bound charge (i.e., one that cannot move). Similarly, holes diffuse from the p to the n region. They leave behind acceptor atoms with a net negative charge, since the acceptor atoms are initially neutral; such ionized acceptor atoms are indicated by the $\ominus$ signs in Fig. A-4b.

In the region near the junction, then, ionized atoms are left *uncovered*, as shown in Fig. A-4b. The electric field $E(x)$ created by these atoms has such a direction as to inhibit the diffusion of free carriers. The field lines extend from the donor ions on the n-type side of the junction to the acceptor ions on the p-type side. The presence of this field causes a potential barrier ϕ_0 between the two types of material; this potential is frequently referred to as the *contact potential* or the *built-in potential*.

As more carriers diffuse to the opposite side, more dopant atoms are uncovered and the field increases; eventually, the two effects balance each other and no further net movement of carriers takes place.

Therefore, a *space-charge region* where, by diffusion, the neutrality condition is violated would be established at the junction. Let the depths of this space-charge region at the two sides of the junction be $+x_n$ and $-x_p$.

a. Built-in potential. As stated before, when equilibrium is reached, the magnitude of the field is such that the tendency of electrons (holes) to *diffuse* from the n-type (p-type) region into the p-type (n-type) region is balanced by the tendency of the electrons (holes) to *drift* in the opposite direction under the influence of the built-in field.

This equilibrium condition results in zero electron and hole currents at any point of the structure. Thus, it can be written

$$J_p = J_{p,\text{drift}} + J_{p,\text{diff}} = 0$$

$$J_n = J_{n,\text{drift}} + J_{n,\text{diff}} = 0$$

(A-21)

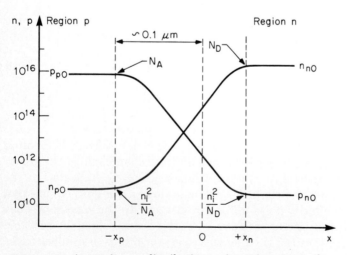

Figure A-3 Approximate distributions of carriers near the boundary plane of a *pn* junction in equilibrium. *(From Semiconductor Electronics Education Committee [5]. Copyright © 1964, Education Development Center, Inc., Newton, Mass. Used by permission.)*

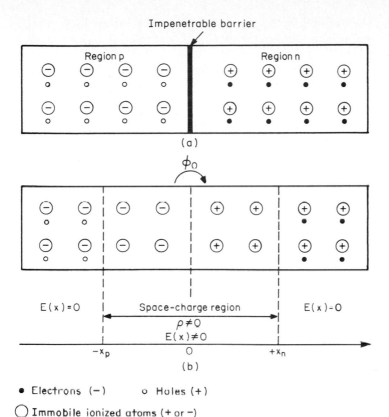

Figure A-4 What happens when a *pn* junction is formed, supposing (*a*) the presence of an imaginary barrier at $x = 0$ and (*b*) the removal of the imaginary barrier (real situation).

Taking into account Eqs. (A-16) yields

$$q\left[-D_p\frac{dp(x)}{dx} + \mu_p p(x)E(x)\right] = 0$$

$$q\left[D_n\frac{dn(x)}{dx} + \mu_n n(x)E(x)\right] = 0$$

(A-22)

From Eqs. (A-22),

$$E(x) = \frac{D_p}{\mu_p}\frac{1}{p(x)}\frac{dp(x)}{dx} = -\frac{D_n}{\mu_n}\frac{1}{n(x)}\frac{dn(x)}{dx}$$

(A-23)

Since the electric field is the negative of the gradient of the potential, the built-in potential ϕ_0 is found from Eq. (A-23) to be

$$\phi_0 = -\int_{-x_p}^{x_n} E(x)\, dx = -\frac{D_p}{\mu_p} \int_{-x_p}^{x_n} \frac{1}{p(x)} \frac{dp(x)}{dx}\, dx$$

$$= \frac{D_p}{\mu_p} [\ln p(-x_p) - \ln p(x_n)]$$

$$\phi_0 = -\int_{-x_p}^{x_n} E(x)\, dx = \frac{D_n}{\mu_n} \int_{-x_p}^{x_n} \frac{1}{n(x)} \frac{dn(x)}{dx}\, dx \qquad \text{(A-24)}$$

$$= \frac{D_n}{\mu_n} [\ln n(x_n) - \ln n(-x_p)]$$

As the carrier concentrations outside the space-charge region are known [see Eqs. (A-19) and (A-20)], the total built-in potential ϕ_0 can be written as

$$\phi_0 = \frac{D_p}{\mu_p} \ln \frac{p_{p0}}{p_{n0}} \simeq \frac{kT}{q} \ln \frac{N_A N_D}{n_i^2}$$

$$\phi_0 = \frac{D_n}{\mu_n} \ln \frac{n_{n0}}{n_{p0}} \simeq \frac{kT}{q} \ln \frac{N_A N_D}{n_i^2} \qquad \text{(A-25)}$$

where Einstein relations have been used. The built-in potential lies in the range 0.2 to 1 V for typical pn-junction diodes. Because of the logarithmic dependence, wide variations of the product $N_A N_D$ are needed to obtain appreciable variation of ϕ_0.

Inspection of Eqs. (A-25) shows that ϕ_0 can be resolved into two components:

$$\phi_n = \frac{kT}{q} \ln \frac{N_D}{n_i} \qquad \text{(A-26}a\text{)}$$

where ϕ_n is the potential at the neutral edge of the space-charge region in the n-type material, and

$$\phi_p = -\frac{kT}{q} \ln \frac{N_A}{n_i} \qquad \text{(A-26}b\text{)}$$

where $\phi_p < 0$ is the potential at the neutral edge of the space-charge region in the p-type material.

Thus the total potential change ϕ_0 from the neutral p-type region to the neutral n-type region is

$$\phi_0 = \phi_n - \phi_p = \frac{kT}{q} \ln \frac{N_D}{n_i} + \frac{kT}{q} \ln \frac{N_A}{n_i} = \frac{kT}{q} \ln \frac{N_A N_D}{n_i^2} \quad \text{(A-27)}$$

just as obtained in Eqs. (A-25).

The built-in potential ϕ_0 depends on the dopant concentration in each region.

The major portion of the potential change occurs in the region with the lower dopant concentration, and the space-charge region is wider in the same region (as shown later).

Besides, it must be noted that the potential at the junction plane ($x = 0$) is not exactly zero unless the junction is symmetrical (that is, $N_A = N_D$).

Of course, no *terminal voltage* results from the contact potential, because such contact potentials exist at *every* junction of dissimilar materials. If any conductor is connected to the two sides of the pn junction, contact potentials will exist at the two new junctions thereby created, and the sum of the contact potentials around the closed loop will be zero. If this were not so, a current would exist and self-heating would occur in violation of the second law of thermodynamics. It follows that the contact potential of a pn junction in thermal equilibrium *cannot* be measured by any voltmeter requiring a steady current, however small.

From Eqs. (A-25) we can also determine the carrier concentrations at the edges of the space-charge region (i.e., at $-x_p$ and at x_n). Then

$$n_{p0}(-x_p) = n_{n0}(x_n)e^{-q\phi_0/kT}$$
$$p_{n0}(x_n) = p_{p0}(-x_p)e^{-q\phi_0/kT} \quad \text{(A-28)}$$

These equations are known as *Boltzmann's relations*.

b. Space-charge region.†   The space-charge region is often referred to as a *depletion region* in which the space charge is overwhelmingly made up of dopant ions.

In the *depletion approximation* it is assumed that the semiconductor may be divided into distinct regions that are either neutral or completely depleted of mobile carriers. These regions join each other at the edges of

---

† The material in this section is derived in part from R. S. Muller and T. I. Kamins, *Device Electronics for Integrated Circuits*, copyright © 1977 by John Wiley & Sons, Inc. Reprinted by permission.

the depletion or space-charge region, where the majority-carrier density is assumed to change abruptly from the dopant concentration to zero. The depletion approximation will appreciably simplify the solution of *Poisson's equation*. (This equation is useful when it is necessary to determine the change in the carrier concentrations with respect to position.)

Since the carrier concentrations are assumed to be much less than the net ionized dopant density in the depletion region, Poisson's equation can be written as

$$\frac{d^2\phi}{dx^2} = -\frac{\rho}{\epsilon_s} = -\frac{q}{\epsilon_s}(p - n + N_D - N_A) \simeq -\frac{q}{\epsilon_s}(N_D - N_A) \quad \text{(A-29)}$$

In general, N_D and N_A may be functions of position, and thus Eq. (A-29) cannot be solved explicitly; ϵ_s is the permittivity of the semiconductor ($\epsilon_s = 1.04 \times 10^{-12}$ F/cm for Si). Equation (A-29), however, can be solved for the *step junction* shown in Fig. A-5, where the dopant concentration changes abruptly from N_A to N_D at $x = 0$. Using the depletion approximation, it is assumed that the region between $-x_p$ and x_n is totally depleted of mobile carriers, as shown in Fig. A-5b, and that the mobile majority-carrier densities abruptly become equal to the respective dopant concentrations at the edges of the depletion region. The charge density is, therefore, zero everywhere except in the depletion region, where it takes the value of the ionized dopant concentration (see Fig. A-5c). In the *n*-type material ($x > 0$) Eq. (A-29) becomes [6]

$$\frac{d^2\phi}{dx^2} = -\frac{dE(x)}{dx} = -\frac{qN_D}{\epsilon_s} \quad \text{(A-30)}$$

which may easily be integrated from an arbitrary point in the *n*-type depletion region to the edge of the depletion region at x_n, where the material becomes neutral and the field vanishes.

Carrying through this integration, we can find the *field* as

$$E(x) = -\frac{qN_D}{\epsilon_s}(x_n - x) \qquad \text{for } 0 < x < x_n \quad \text{(A-31a)}$$

The field is negative throughout the depletion region and varies linearly with x, having its maximum magnitude at $x = 0$ (see Fig. A-5d). The direction of the field toward the left is physically reasonable since the force it exerts must balance the tendency of the negatively charged electrons to diffuse toward the left out of the neutral *n*-type material.

The field in the *p*-type region is similarly found to be

$$E(x) = -\frac{qN_A}{\epsilon_s}(x + x_p) \qquad \text{for } -x_p < x < 0 \quad \text{(A-31b)}$$

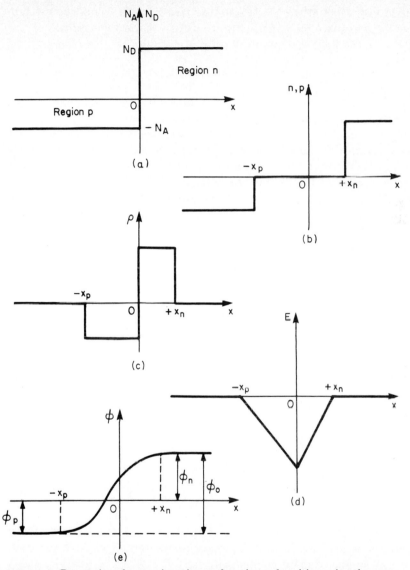

Figure A-5 Properties of a step junction as functions of position using the complete depletion approximation: (*a*) net dopant concentration, (*b*) carrier densities, (*c*) space charge used in Poisson's equation, (*d*) electric field found from first integration of Poisson's equation, and (*e*) potential obtained from second integration. *(From R. S. Muller and T. I. Kamins, Device Electronics for Integrated Circuits, copyright © 1977 by John Wiley & Sons, Inc. Used by permission.)*

The field in the p-type region is also negative in order to oppose the tendency of the positively charged holes to diffuse toward the right.

At $x = 0$, the field must be *continuous,* so that

$$N_A x_p = N_D x_n \qquad \text{(A-32)}$$

Thus, the width of the depleted region on each side of the junction varies inversely with the magnitude of the dopant concentration; the *higher* the dopant concentration, the *narrower* the space-charge region. In a highly asymmetrical junction where the dopant concentration on one side of the junction is much higher than that on the other side, the depletion region penetrates primarily into the lightly doped material, and the width of the depletion region in the heavily doped material may often be neglected. The expressions for the field may be integrated again to obtain the potential variation across the junction.

In the n-type material

$$\phi(x) = \phi_n - \frac{qN_D}{2\epsilon_s}(x_n - x)^2 \qquad \text{for } 0 < x < x_n \qquad \text{(A-33}a\text{)}$$

as shown in Fig. A-5e, where ϕ_n is given by Eq. (A-26a).

Similarly, in the p-type material

$$\phi(x) = \phi_p + \frac{qN_A}{2\epsilon_s}(x + x_p)^2 \qquad \text{for } -x_p < x < 0 \qquad \text{(A-33}b\text{)}$$

where ϕ_p is given by Eq. (A-26b).

Taking into account Eqs. (A-25), (A-26), (A-32), and (A-33), the total *depletion-region width* is found to be

$$W = x_n + x_p = \sqrt{\frac{2\epsilon_s}{q}\phi_0\left(\frac{1}{N_A} + \frac{1}{N_D}\right)} \qquad \text{(A-34)}$$

Thus, the depletion-region width depends most strongly on the material with the lighter doping and varies approximately as the inverse square root of the smaller dopant concentration.

The widths of the p-type and n-type portions of the depletion region are

$$x_p = W\frac{N_D}{N_A + N_D}$$

$$\qquad \text{(A-35)}$$

$$x_n = W\frac{N_A}{N_A + N_D}$$

The analysis of the *abrupt pn junction* has employed the depletion approximation to solve Poisson's equation and to specify the boundary condition. Usually, the step or abrupt junction is not an adequate representation for a *pn* junction made by diffusion. This is the case of *double-diffused junctions,* that is, junctions formed by two successive diffusions of opposite-type dopant atoms. A general analytical solution of Poisson's equation for double-diffused junctions, however, is not possible, and specific cases are usually considered only approximately. If more accurate results are needed, numerical techniques are used.

In addition to the abrupt junction, there is another doping profile that can be treated exactly and that gives useful results for approximating real *pn* junctions. It is the *linearly graded* junction where the net dopant concentration varies linearly from the *p*-type material to the *n*-type material. A description of this type of junction can be found in Refs. [2, 4, 6].

A.2.2 *PN* junction in nonequilibrium: Effect of a bias voltage

As for equilibrium, to analyze the junction under a bias voltage V_D, we will again make use of the depletion approximation together with the following assumptions about the applied bias.

1. *Ohmic contacts* connect the p and n regions to the external voltage source so that negligible voltage is dropped at the contacts.

2. *Small currents* flow through the neutral regions and cause only very low voltage drops.

3. The *n*-region is *grounded* and voltage V_D is applied to the p region.

With these assumptions, the entire applied voltage appears *across* the junction. Furthermore, under these assumptions, the solutions of Poisson's equation that were found for thermal equilibrium conditions in the previous section apply also to the junction under bias. Only the total potential across the junction changes from the built-in value ϕ_0 to $\phi_0 - V_D \equiv \phi_J$.

If V_D is *positive,* the barrier to majority carriers at the junction is reduced and the depletion region is narrowed. In this case, the junction is said to be *forward-biased.* Under forward bias, appreciable currents can flow even for small values of V_D.

If *negative* voltage is applied to the p region, the barrier to majority flow increases and the depletion region widens. In this case, the junction is said to be *reverse-biased.* Under reverse bias, majority carriers are pulled away from the edges of the depletion region, which therefore widens. There is very little current flow since the bias polarity aids the transfer of electrons from the p side to the n side and holes from the n side to the p side. Since these are minority carriers in each region, they are very low in density [6].

a. Space-charge region. For an abrupt pn junction, we can find the space-charge region width as a function of voltage by replacing the built-in potential ϕ_0 in Eq. (A-34) with $\phi_0 - V_D$ so that

$$W = x_n + x_p = \sqrt{\frac{2\epsilon_s}{q}(\phi_0 - V_D)\left(\frac{1}{N_A} + \frac{1}{N_D}\right)} \qquad \text{(A-36)}$$

As V_D becomes considerably larger than ϕ_0, the space-charge region begins to vary as the square root of the reverse-bias voltage.

It is important to know the maximum field, $E_{\max}$, at the junction and its relationship to applied voltage. For the step-junction case, taking into account that the field varies linearly with distance [see Eqs. (A-31)], it can be written that

$$\tfrac{1}{2}E_{\max}W = \phi_0 - V_D \qquad \text{(A-37)}$$

so that

$$E_{\max} = \frac{2(\phi_0 - V_D)}{W} \qquad \text{(A-38)}$$

b. Junction capacitance. If the voltage applied to the pn junction is changed by a small dV_D, then the depletion-region charge will change by an amount dQ'_J, where Q'_J is the charge per unit area. Hence a depletion layer or junction capacitance can be defined as

$$C'_J = \frac{dQ'_J}{dV_D} \qquad \text{(A-39)}$$

Since $Q'_J = qN_D x_n = qN_A x_p$ in the depletion approximation, then

$$C'_J = qN_D \frac{dx_n}{dV_D} = qN_A \frac{dx_p}{dV_D} \qquad \text{(A-40)}$$

From Eqs. (A-35) and (A-36), it follows that

$$\frac{dx_n}{dV_D} = \frac{1}{N_D}\sqrt{\frac{\epsilon_s}{2q(\phi_0 - V_D)(1/N_A + 1/N_D)}} \qquad \text{(A-41)}$$

and

$$C_J' = \sqrt{\frac{q\epsilon_s}{2(\phi_0 - V_D)(1/N_A + 1/N_D)}} = \frac{C_J'(0)}{\sqrt{1 - V_D/\phi_0}} \quad \text{(A-42)}$$

In this equation, the junction capacitance is defined in terms of the physical parameters of the device. For the circuit designer, it is more convenient to express this capacitance in terms of electrical parameters (as the second formulation) where $C_J'(0)$ is the capacitance at equilibrium, that is, at $V_D = 0$ [7].

Figure A-6 shows a plot of the junction capacitance as a function of junction voltage. Thus for $V_D > \phi_0$, the capacitance of a step *pn* junction [see Eq. (A-42)] will decrease approximately inversely with the square root of the reverse bias.

Using Eq. (A-36) in Eq. (A-42) yields

$$C_J' = \frac{\epsilon_s}{W} \quad \text{(A-43)}$$

which is the general relationship for small-signal capacitance valid for an arbitrarily doped junction.

A.2.3 Currents in a *pn* junction

The subject of this section is the current flow across a *pn* junction under both forward and reverse bias.

As a first step toward analyzing current flow, we derive the *continuity equation* for free carriers, that is, an equation that takes account of the various mechanisms affecting the population of carriers in an infinitesimal volume inside a semiconductor [2, 4–7].

Through the use of the continuity equation, together with the concept

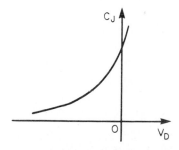

Figure A-6 Junction capacitance as a function of junction voltage.

of excess-carrier lifetime (previously developed), we can be in a position to find expressions for the current in a *pn* junction under bias.

Solutions of the continuity equations in the quasi-neutral regions give carrier densities in terms of position and time. Expressions for current are then obtained directly by using Eqs. (A-16), which define carrier flow in terms of carrier densities.

a. Continuity equation. The *continuity equation* can be written in terms of carrier densities as follows.

For *electrons:*

$$\frac{\partial n}{\partial t} = \mu_n n(x) \frac{\partial E(x)}{\partial x} + \mu_n E(x) \frac{\partial n(x)}{\partial x}$$

$$+ D_n \frac{\partial^2 n(x)}{\partial x^2} + (G_n - R_n) \quad \text{(A-44}a)$$

For *holes:*

$$\frac{\partial p}{\partial t} = -\mu_p p(x) \frac{\partial E(x)}{\partial x} - \mu_p E(x) \frac{\partial p(x)}{\partial x}$$

$$+ D_p \frac{\partial^2 p(x)}{\partial x^2} + (G_p - R_p) \quad \text{(A-44}b)$$

where G_n, R_n and G_p, R_p represent the generation and recombination rates per unit volume for electrons and holes, respectively.

In Eqs. (A-44) it has been assumed that mobility μ and diffusion constant D are not functions of x. Although this assumption is not valid in a number of important cases, the major physical effects are included in Eqs. (A-44) and the more exact formulations are seldom considered [6].

If the electric field is zero or negligible in the region under consideration, the first two terms on the right-hand side of Eqs. (A-44) can be neglected and the analysis is greatly simplified. Even if the field is not negligible, some of the terms in Eqs. (A-44) may be unimportant. For example, if the field is constant, the first term on the right-hand side in each equation drops out.

Rarely is it necessary to deal with the full complexity of Eqs. (A-44).

b. Current-voltage characteristics.† The *total current* consists, in general, of the *sum* of four components: hole and electron drift currents and hole and electron diffusion currents.

We now consider a *pn*-junction diode, connected to a voltage source

---

† The material in Secs. A.2.3*b* to *d* is derived primarily from R. S. Muller and T. I. Kamins, *Device Electronics for Integrated Circuits,* copyright © 1977 by John Wiley & Sons, Inc. Reprinted by permission.

with the n region grounded and the p region at V_D volts relative to ground. The diode structure has a constant cross-sectional area A_J and it is sketched in Fig. A-7. The junction is not illuminated, and carrier densities within the diode are influenced only by the applied voltage. The applied voltage V_D is dropped partially across the quasi-neutral regions and partially across the junction itself. Since the voltage drops across the quasi-neutral region are ohmic, they are very small at low currents [6].

If it is assumed, then, that V_D is sustained entirely at the junction, the total junction voltage will be $\phi_0 - V_D = \phi_J$ under bias, where ϕ_0 is the built-in voltage.

If V_D is *positive,* that is, for *forward bias,* the applied voltage reduces the barrier to the diffusion flow of majority carriers at the junction. The reduced barrier, in turn, permits a net transfer rate of holes from the p side into the n side and of electrons from the n side into the p side. When these transferred carriers enter the quasi-neutral regions, they are minority carriers and are very quickly neutralized by majority carriers that enter the quasi-neutral regions from the ohmic contacts at the ends. Once the minority carriers have been injected across the space-charge region, they tend to diffuse into the neutral interior region [6].

If V_D *is negative,* that is, for *reverse bias,* the barrier height to diffusing majority carriers is increased. Since equilibrium is disturbed, minority carriers near the junction space-charge region tend to be depleted. The majority-carrier concentration is likewise reduced.

From these few remarks, it can be seen that the minority-carrier densities deserve special attention because they really determine what currents flow in a pn junction. The majority carriers act only as suppliers of the injected minority-carrier current or as charge neutralizers in the quasi-neutral regions. It is helpful to consider the majority carriers as will-

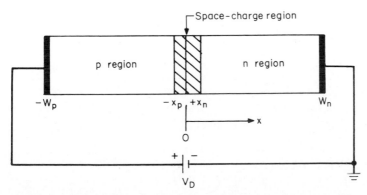

Figure A-7 Sketch of a pn-junction diode structure used in the discussion of currents. *(Adapted from Muller and Kamins [6]. Copyright © 1977 by John Wiley & Sons, Inc. Used by permission.)*

ing "slaves" of the minority carriers. Accordingly, solutions of the continuity equations for the minority-carrier densities in each of the quasi-neutral regions will be sought [6].

c. Boundary values of minority-carrier densities.

To write the solutions of the continuity equation in a useful fashion, however, it will be necessary to relate the boundary values of the minority-carrier densities to the applied voltage V_D. The most straightforward way of doing this is to make two additional assumptions: first, that the applied bias leads to low-level injection and, second, that the applied bias is small enough so that the detailed balance between majority and minority concentrations across the junction regions is not appreciably disturbed.

The first assumption implies that the majority-carrier concentrations are negligibly changed at the edge of the quasi-neutral regions by the applied bias. The assumption that detailed balance nearly applies allows the use of Eq. (A-28) across the junction where the potential difference is known to be $\phi_0 - V_D$.

Both of these assumptions are certainly valid when V_D is very small ($|V_D| \ll \phi_0$). Their validity at higher biases deserves more careful consideration, as given in Chap. 1.

By the assumption of low-level injection, the electron density at the quasi-neutral boundary in the n region next to the pn junction is equal to the dopant density whether at equilibrium or under bias. As previously stated, this position is indicated as x_n (see Fig. A-7), and the thermal equilibrium density is denoted with the subscript 0.

Likewise, the boundary of the quasi-neutral p region is at $-x_p$ and the hole density there is equal to the acceptor dopant density at equilibrium as well as under bias. The following equations summarize these statements [6]:

$$n_{p0}(-x_p) = n_{n0}(x_n)e^{-q\phi_0/kT} = N_D(x_n)e^{-q\phi_0/kT}$$
$$p_{n0}(x_n) = p_{p0}(-x_p)e^{-q\phi_0/kT} = N_A(-x_p)e^{-q\phi_0/kT}$$

(A-45)

and

$$n_p(-x_p) = N_D(x_n)e^{-q(\phi_0 - V_D)/kT}$$
$$p_n(x_n) = N_A(-x_p)e^{-q(\phi_0 - V_D)/kT}$$

(A-46)

These four equations can be combined to express the excess minority-carrier densities at the boundaries in terms of their thermal equilibrium values. If we define the excess densities by

$$n_p' \equiv n_p - n_{p0}$$
$$p_n' \equiv p_n - p_{n0}$$

(A-47)

then

$$n_p'(-x_p) = n_{p0}(-x_p)(e^{qV_D/kT} - 1)$$
$$p_n'(x_n) = p_{n0}(x_n)(e^{qV_D/kT} - 1)$$

(A-48)

Equations (A-48) are extremely important results that can be used to define specific solutions for the continuity equation for minority carriers in the quasi-neutral regions near a *pn* junction. The equations show that the minority-carrier density is an exponential function of applied bias, while the majority-carrier density has been assumed to be insensitive (to first order) to it. Since the minority-carrier density at thermal equilibrium is typically 11 or 12 orders of magnitude below the majority-carrier density, Eqs. (A-48) will not be in conflict with the low-level injection assumption until the exponential factor is typically of the order 10^{11} or 10^{12} [6].

d. Long-base diode analysis.   Taking into account what was described in the previous sections, we can now consider solutions of the continuity equations [see Eqs. (A-44)] in the quasi-neutral region.

First consider excess holes injected into the *n* region, where bulk recombination through generation-recombination centers is dominant. Thus, the term $G_p - R_p$ in Eq. (A-44b) can be expressed through Eqs. (A-18) because of the assumption of low-level injection. Therefore, the continuity equation in the quasi-neutral approximation becomes [6]

$$\frac{\partial p_n}{\partial t} - D_p \frac{\partial^2 p_n}{\partial x^2} = \frac{p_n - p_{n0}}{\tau_p} \tag{A-49}$$

where the subscript *n* emphasizes that the holes are in the *n* region.

Fortunately, the simplest case of impurity doping, that of a constant donor density along *x*, is frequently encountered in practice. Taking this case and considering the steady state ($\partial p / \partial t = 0$), Eq. (A-49) can be rewritten as a total differential equation in terms of the excess density p', which was defined in Eq. (A-47). Equation (A-49) becomes

$$0 = D_p \frac{d^2 p_n'}{dx^2} - \frac{p_n'}{\tau_p} \tag{A-50}$$

and it has the following exponential solution [6]:

$$p_n'(x) = A e^{-(x - x_n)/\sqrt{D_p \tau_p}} + B e^{(x - x_n)/\sqrt{D_p \tau_p}} \tag{A-51}$$

where *A* and *B* are constants determined by the boundary conditions of the problem.

The characteristic length $\sqrt{D_p \tau_p}$ in Eq. (A-51) is called the *diffusion length* and is denoted by L_p. (The diffusion length of an electron in a *p*-type region is denoted by L_n.) For specific applications of the solution to the continuity equation given in Eq. (A-51), two limiting cases based upon the length W_n of the *n* region from the junction to the ohmic contact (see Fig. A-7) can be considered: the *long-base diode* and the *short-base diode*.

Now if W_n is very long compared to the diffusion length L_p, essentially all the injected holes recombine before traveling completely across it. This case is known as the *long-base diode*.

For the long-base diode, L_p is the average distance traveled in the neutral region before an injected hole recombines. Since p_n' must decrease with increasing x, the constant B in Eq. (A-51) must be zero. The constant A in the solution is determined by applying Eq. (A-48), which specifies $p_n'(x_n)$ as a function of applied voltage. The complete solution is therefore [6]

$$p_n'(x) = p_{n0}(e^{qV_D/kT} - 1)e^{-(x-x_n)/L_p} \tag{A-52}$$

as shown in Fig. A-8. Thus, an expression for hole current can now be obtained.

The hole current flows only by diffusion, since it has been assumed that the field is negligible in the neutral regions; therefore, from Eqs. (A-16) and (A-52) [6],

$$J_p(x) = -qD_p \frac{dp_n}{dx} = qD_p \frac{p_{n0}}{L_p} (e^{qV_D/kT} - 1)e^{-(x-x_n)/L_p} \tag{A-53}$$

$$= qD_p \frac{n_i^2}{N_D L_p} (e^{qV_D/kT} - 1)e^{-(x-x_n)/L_p}$$

The hole current is therefore a maximum at $x = x_n$ and decreases away from the junction (see Fig. A-9) because the hole gradient decreases as carriers are lost by recombination. Since the total current must remain constant with distance from the junction in the steady-state case, the electron current must increase when moving away from the junction. This electron current supplies the electrons with which the holes are recombining [6].

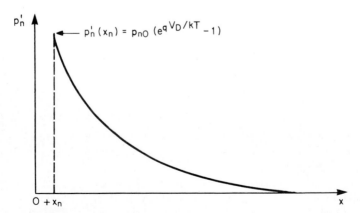

Figure A-8 Spatial variation of holes in the quasi-neutral n-region of a long-base diode under forward bias V_D. The excess density p_n' has been calculated from Eq. (A-52). *(Adapted from Muller and Kamins [6]. Copyright © 1977 by John Wiley & Sons, Inc. Used by permission.)*

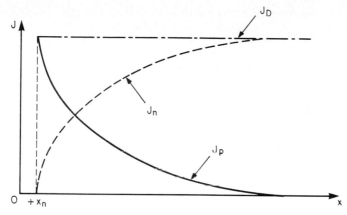

Figure A-9   Hole current (solid line) and recombining electron current (dotted line) in the quasi-neutral *n* region of the long-base diode. The sum of the two currents J_D (dot-dash line) is constant. The hole current is calculated from Eq. (A-53). *(Adapted from Muller and Kamins [6]. Copyright © 1977 by John Wiley & Sons, Inc. Used by permission.)*

The total current is carried entirely by electrons at the ohmic contact at W_n. Moving toward the junction, the electron current decreases as recombination occurs with the injected holes. At the junction, the only electron current flowing is the one injected into the *p* region. The electrons injected into the *p* region constitute the minority-carrier current there. Thus, the total current is obtained by summing the two minority-carrier injection currents: holes into the *n* side plus electrons into the *p* side [6].

The minority-carrier electron current that is injected into the *p* region can be found by a treatment analogous to the analysis used to obtain Eq. (A-53). If the ohmic contact is at $-W_p$, where $W_p \gg L_n \equiv \sqrt{D_n\tau_n}$, then

$$J_n = qD_n \frac{n_i^2}{N_A L_n} (e^{qV_D/kT} - 1)e^{(x+x_p)/L_n} \qquad \text{(A-54)}$$

Since the origin for x has been chosen at the physical junction (see Fig. A-7), x is a negative number throughout the *p* region. Hence, J_n decreases away from the junction as did J_p in the *n* region. To obtain an expression for the total current J_D, it is sufficient to sum the minority-carrier components at $-x_p$ and $+x_n$, respectively, as expressed in Eqs. (A-53) and (A-54):

$$J_D = J_p(x_n) + J_n(-x_p) = qn_i^2 \left(\frac{D_p}{N_D L_P} + \frac{D_n}{N_A L_n} \right) (e^{qV_D/kT} - 1)$$

$$= J_S(e^{qV_D/kT} - 1)$$

$$\text{(A-55)}$$

where J_S is called *reverse saturation current density* predicted by this theory when a negative bias of a few kT/q volts is applied.

The second limiting case (that of the *short-base diode*) occurs when the lengths W_n and W_p of the n- and p-type regions are much shorter than the diffusion lengths L_p and L_n. In this case, there is a little recombination in the bulk of the quasi-neutral regions. In the limit, all injected minority carriers recombine at the ohmic contacts at either end of the diode. In this case, we can obtain solutions most easily by approximating the exponentials in Eq. (A-51) by the first two terms of a Taylor series expansion.

The development and analysis of the short-base diode is beyond the scope of this appendix, but it is widely discussed in Chap. 1.

Comparing Eqs. (1-17) and (A-55) for the currents in the short- and long-base diodes, we can see that the solutions obtained are similar.

In the long-base diode, the characteristic length is that of the minority-carrier diffusion; in the short-base diode, it is the length of the quasi-neutral region. Of course, a given diode may be approximated by a combination of these two limiting cases; that is, it may be short-base in the p region and long-base in the n region or vice versa.

A.2.4   Transient behavior of the *pn* junction†

So far our study of the electrical and physical properties of the *pn* junction has been concerned mostly with the static, or steady-state, characteristics. Now the switching transients in the *pn* junction (the action of the diode turning on and off) are considered.

The ideal diode equation [see Eq. (A-55)] is a fundamental equation that relates the diode current to the diode voltage. The diode current may also be related to the excess minority-carrier charge. This is another fundamental equation, which introduces the concept of charge control. This concept is a most useful one for analysis of switching transients in both diodes and bipolar junction transistors.

With forward bias applied to the *pn* junction there is an excess minority-carrier concentration in the neutral regions near the depletion region. Each of these excess mobile carriers carries a charge q. Thus the excess hole charge Q_p in the neutral n region is

$$Q_p = qA_J \int_{x_n}^{W_n} p_n'(x) \, dx \qquad \text{(A-56a)}$$

where A_J is the cross-sectional area of the junction.

An equation similar to Eq. (A-56a) can be written for the excess elec-

---

† The material in this section is taken from Hodges and Jackson [7]. Used by permission.

tron charge Q_n in the neutral p region. However, for a practical diode, one side of the junction is much more heavily doped than the other. It is assumed that $N_A \gg N_D$; then $p_{n0} \gg n_{p0}$ and $Q_p \gg Q_n$. Henceforth, the minority-carrier holes in the neutral n region are considered.

In carrying out the integration in Eq. (A-56a), assume that the depletion-region width is much less than the hole diffusion length L_p. Taking into account Eqs. (A-52), Eq. (A-56a) becomes

$$Q_p = qA_J L_p p_{n0}(e^{qV_D/kT} - 1) \qquad \text{(A-56b)}$$

From this equation, it can be seen that the excess minority-carrier charge is related to the diode voltage. Furthermore, this charge is positive for forward bias and negative for reverse bias.

To relate the excess minority-carrier charge to the diode current, note that the forward current I_D is the product of the cross-sectional area of the junction, A_J, and the current density at the origin, $J_p(0)$. Thus from Eq. (A-55), and eliminating the exponential term by substitution from Eq. (A-56b), it follows that [7]

$$I_D - Q_p \frac{D_p}{L_p^2} = \frac{Q_p}{\tau_p} \qquad \text{(A-57)}$$

This equation relates, on a static basis, the diode current to the excess minority-carrier charge and the excess minority-carrier lifetime τ_p. In the steady state, the current I_D supplies holes to the neutral n region at the same rate as they are being lost by recombination.

To obtain a dynamic, or time-dependent, relation, we must include the time rate of change of the excess minority-carrier charge. Then an equation for the time-dependent current can be obtained as [7]

$$I_D(t) = \frac{Q_p}{\tau_p} + \frac{dQ_p}{dt} \qquad \text{(A-58)}$$

This is a fundamental *charge-control equation* for the *pn*-junction diode. It states that current $I_D(t)$ supplies holes to the neutral n region at the rate at which the stored charge increases plus the rate at which holes are being lost by recombination. Use of Eq. (A-58) can be made in the calculation of the switching times for a *pn*-junction diode.

A.2.5 Approximation considerations.† Though Eq. (A-55) has been obtained on two assumptions—first, that ohmic drops in the quasi-neu-

---

† The material in this section is due in part to R. S. Muller and T. I. Kamins, *Device Electronics for Integrated Circuits,* copyright © 1977 by John Wiley & Sons, Inc. Reprinted by permission.

tral regions are small so that V_D is sustained entirely across the junction space-charge region and, second, that applied bias does not greatly alter the detailed balance between drift and diffusion tendencies that exist at thermal equilibrium—it can be shown [6] that these assumptions are reasonable for many practical cases.

With regard to the first assumption, however, it must be noted that if the applied forward voltage is nearly as large as the built-in potential, the barrier to majority-carrier flow is substantially reduced and large currents may flow. In that case a significant portion of the applied voltage is dropped across the neutral regions, and the series resistance of the neutral regions must be considered. The effect of series resistance can easily be included using circuit analysis (see Chap. 1).

With regard to the second assumption, it proves it is reasonable to treat the case of small and moderate biases by considering only slight deviations from thermal equilibrium. This means that we can relate the carrier concentrations on either side of the junction space-charge region by considering the effective barrier height to be $\phi - V_D$ and by using Eqs. (A-48). These equations depend also on the validity of the low-level injection assumption. The current tendencies at pn junctions that are in detailed balance at thermal equilibrium are so large relative to currents that flow in practice that a more general relationship than Eqs. (A-48) is valid. This relationship, which will not be proved here, is that the carrier densities at either side of a biased pn junction (and also within the space-charge region) are dependent on applied voltage according to the equation [6]

$$pn = n_i^2 e^{qV_D/kT} \tag{A-59}$$

In the low-level injection case, Eq. (A-59) reduces to the boundary conditions for minority carriers, which were obtained previously. Equation (A-59) has special utility in cases where injected carrier densities approach the thermal-equilibrium densities of majority carriers.

Besides these two assumptions, another one must be taken into account. The analysis of the pn junction that led to Eq. (A-55) was based on events in the quasi-neutral regions. The space-charge region was treated solely as a barrier to the diffusion of majority carriers, and it played a role only in the establishment of minority-carrier densities at its boundaries [see Eqs. (A-48)]. This is a reasonable first-order description of events, and the equations derived from it [see Eq. (A-55)] are referred to as the *ideal diode* equations [6].

Over a significant range of useful biases, however, the ideal diode equations are quite inaccurate. It is necessary to consider corrections to these equations that arise for a more complete treatment of events in the space-charge region at the junction. Like the quasi-neutral regions of the diode, the space-charge region contains generation-recombination centers. Unlike the quasi-neutral regions, it is a region of steep impurity gradients and very rapidly changing populations of carriers. Since injected carriers, under forward bias, must cross through this region, some carriers may be lost by recombination. Under reverse bias, generation of carriers in the region leads to excess current above the saturation value predicted by the ideal diode equations [6].

To find expressions for generation and recombination in the space-charge region, we must use the *Shockley-Hall-Read (SHR)* theory in conjunction with Eq. (A-59). The recombination-generation rate can be positive or negative; that is,

$$V_D > 0 \rightarrow pn > n_i^2 \rightarrow U > 0 \rightarrow \text{recombination}$$

$$V_D < 0 \rightarrow pn < n_i^2 \rightarrow U < 0 \rightarrow \text{generation}$$

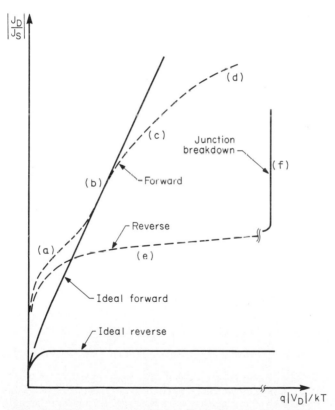

Figure A-10 Current-voltage characteristic of a real silicon diode: (*a*) generation-recombination current region, (*b*) diffusion current region, (*c*) high-level injection region, (*d*) series resistance effect, (*e*) reverse leakage current due to generation-recombination and surface effects, and (*f*) breakdown region. (*From S. M. Sze, Physics of Semiconductor Devices, copyright © 1969 by John Wiley & Sons, Inc. Used by permission.*)

The total current arising from generation and recombination in the space-charge region is given by the integral of the generation-recombination rate across the space-charge region; that is,

$$J_{r,g} = q \int_{-x_p}^{x_n} U \, dx = \begin{cases} I_r = f\left(e^{qV_D/2kT}\right) & \text{for } V_D > 0 \\[2em] I_g = f\left(\sqrt{\dfrac{qV_D}{kT}}\right) & \text{for } V_D < 0 \end{cases} \tag{A-60}$$

Thus, unlike current arising from carrier diffusion and recombination in the quasi-neutral regions, the current resulting from recombination in the space-charge region varies with applied voltage as $e^{qV_D/2kT}$; this different exponential behavior can be observed in real diodes, especially at low currents (see Fig. A-10).

The generation current in the space-charge region, on the contrary, is only a slight function of the reverse bias, varying roughly as the square root of applied voltage [6]. It is possible to estimate the relative importance of the contributions to the current from the quasi-neutral regions and from the space-charge region for an ideal diode by taking the ratio of Eq. (A-55) to Eq. (A-60) under forward and reverse bias. Hence, the current in the reverse-biased diode is generated predominantly in the space-charge region; the current in the forward-biased diode is dominated by the ideal diode current, especially as bias increases. More detailed considerations can be found in Refs. [2, 4, 6].

In Fig. A-10 a current-voltage characteristic of a real silicon diode is shown in order to point out the influence of the described effects on the total diode current.

REFERENCES

1. P. E. Gray and C. L. Searle, *Electronic Principles: Physics, Models, and Circuits,* Wiley, New York, 1969.
2. A. S. Grove, *Physics and Technology of Semiconductor Devices,* Wiley, New York, 1967.
3. Y. P. Tsividis, *Operation and Modeling of the MOS Transistor,* McGraw-Hill, New York, 1987.
4. S. M. Sze, *Physics of Semiconductor Devices,* Wiley, New York, 1969.
5. P. E. Gray, D. DeWitt, A. R. Boothroyd, and J. F. Gibbons, *Physical Electronics and Circuit Models of Transistors,* vol. 2, Semiconductor Electronics Education Committee, Wiley, New York, 1964.
6. R. S. Muller and T. I. Kamins, *Device Electronics for Integrated Circuits,* Wiley, New York, 1977.
7. D. A. Hodges and H. G. Jackson, *Analysis and Design of Digital Integrated Circuits,* McGraw-Hill, New York, 1983.
8. P. R. Gray and R. G. Meyer, *Analysis and Design of Analog Integrated Circuits,* Wiley, New York, 1977.

The Two-Terminal
MOS Structure

Giuseppe Massobrio

*Department of Electronics (DIBE), University of
Genoa, Genoa, Italy*

B.1   MOS Junction

B.1.1   Contact potentials

As in the discussion of the *pn* junction, in our treatment of MOS theory, we will use the concept of *contact potentials* instead of the traditional *energy bands.* Thus some basic results of the contact potential theory will be described first.

Consider the junction of two different neutral materials M_1 and M_2; either material can be a semiconductor or a metal. When the two materials are brought together, at first carriers diffuse from one to the other because the concentration of these carriers is, in general, different in M_1 and M_2, and no opposing field exists in the initially neutral materials.

Note: The material in this appendix is taken from the book by Y. P. Tsividis, *Operation and Modeling of the MOS Transistor*, McGraw-Hill, New York, 1987.

However, as each charged carrier crosses the junction, it leaves behind a net charge of opposite polarity, and an electric field is thus established in the vicinity of the junction, which tends to inhibit the movement of carriers. Eventually, the field intensity increases to the point that it counteracts the tendency of carriers to diffuse, and a balance is achieved at which there is no more net carrier movement; an electrostatic potential change is then encountered when going from one material, through the junction, to the other material. The total potential drop in going from M_1 to M_2 is called the *contact potential* of material M_1 to material M_2, and will be denoted by $\phi_{M1,M2}$.

It can be demonstrated [1] that if a material is chosen as a reference material, then the contact potential of material M_1 to material M_2 can be expressed by

$$\phi_{M1,M2} = \phi_{M1} - \phi_{M2} \tag{B-1}$$

where ϕ_{M_i} is the contact potential of a generic material M_i to the reference material; ϕ_{M_i} can be indirectly evaluated and tabulated.

If n materials in series are considered, the potential ϕ_{M1,M_n} between the first and the last material is expressed in terms of contact potentials in the loop. Using Eq. (B-1) yields

$$\phi_{M1,M_n} = (\phi_{M1} - \phi_{M2}) + (\phi_{M2} - \phi_{M3}) + \cdots + (\phi_{M_{n-1}} - \phi_{M_n}) \tag{B-2}$$

It is clear that, with the exception of ϕ_{M1} and ϕ_{M_n}, each contact potential appears twice in the sum: once with a plus sign and once with a minus sign. Therefore

$$\phi_{M1,M_n} = \phi_{M1} - \phi_{M_n} \tag{B-3}$$

Thus no matter how many materials are in the loop, the electrostatic potential difference between its two ends depends *only* on the first and the last material.

B.1.2   Flat-band voltage

Most of the phenomena that make possible the operation of the MOS transistor can be understood by considering only the part of the device between *source* and *drain*, i.e., a two-terminal structure (gate-insulator-substrate). This structure is sometimes called the *MOS junction.*

The simplest *two-terminal MOS structure* is shown in Fig. B-1a. Here it is assumed that the gate is made of the same material as the substrate (in this case, p-type silicon) with the same doping concentration. It is also assumed that somehow the same material is used to connect the gate to the substrate. No charges are shown in the silicon, since to each positively charged hole of the p-type material there corresponds a negatively charged acceptor atom from which the hole has originated; therefore the charges cancel and the material is shown as electrically neutral everywhere. Then no field can exist in the insulator, and there is no reason for carriers to be attracted toward the insulator-substrate interface.

We now consider a realistic case [1], as shown in Fig. B-1b. The gate is made of a certain material, not necessarily the same as the substrate, and a metal is used to contact the gate material and form the *gate terminal* G. The body (substrate) is contacted through a back metal plate; this plate is in turn contacted through some metal, which thus forms the *body terminal B*. The subscripts G and B will be used to denote the gate and body terminals rather than the regions themselves.

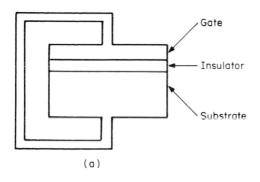

(a)

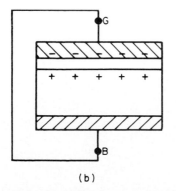

(b)

Figure B-1 (*a*) A two-terminal MOS structure with gate, substrate, and shorting external connection; (*b*), (*c*) effects of contact potentials; and (*d*), (*e*) effects of the interface charge [1].

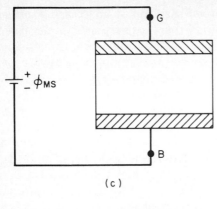

(c)

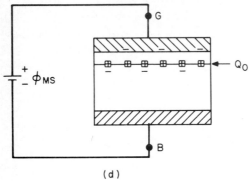

(d)

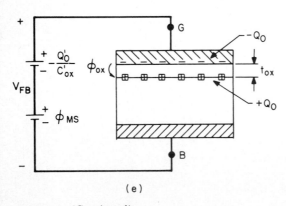

(e)

Figure B-1 *(Continued)*

a. Effect of contact potentials.   Now the gate terminal is shorted to the body terminal with a wire. Going from the gate, through the external connection, to the bulk and taking into account that the sum of the encountered *contact potentials* depends only on the first and the last material [see Eq. (B-3)], being independent of any material in between, it can be written [1]

$$\sum_{\text{gate}}^{\text{bulk}} (\text{contact potentials}) = \phi_{\text{gate material}} - \phi_{\text{bulk material}} \qquad \text{(B-4)}$$

The existence of a nonzero potential between the gate and the bulk causes net charges to appear on both sides of the oxide. If, for example, this potential is negative, the polarity of the charges will be as shown in Fig. B-1*b*. These net charges disappear if the total potential from the gate, through the external connection, to the bulk equals zero. This is possible if a voltage of an external source is applied to cancel the sum of the contact potentials (see Fig. B-1*c*).

From Eq. (B-4), the external voltage source must have a value given by

$$V_{GB} = \phi_{MS} \qquad \text{(B-5)}$$

where ϕ_{MS} is a widely used symbol defined as

$$\phi_{MS} = \phi_{\text{gate material}} - \phi_{\text{bulk material}} \qquad \text{(B-6)}$$

From Sec. B.1.1 it should be clear that the value of V_{GB} needed will be given by Eq. (B-6), independently of how many different materials are encountered in the external circuit.

b. Effect of fixed charges.   The effect of contact potentials is not the only one that can cause a net concentration of charges in the semiconductor in the absence of external bias. Another effect is that of the *fixed charges* which exist at the oxide-semiconductor interface or within the oxide itself. Part of these charges are there because they are trapped at broken bonds at the interface. Such broken bonds exist because the semiconductor crystal lattice is abruptly terminated there. Another part of the fixed charges is due to the fact that while the oxide is grown during fabrication, some ionic silicon is introduced into it. Contamination and ionizing radiation can cause additional charges to appear within the oxide.

Provided these charges are to a large extent immobile and voltage-independent, their total effect can be modeled by a single layer of charge at the Si-SiO$_2$ interface. These charges will be denoted by Q_0 and are shown inside little squares in Fig. B-1*d*. A battery of value ϕ_{MS} is used in that figure to cancel the effect of the contact potentials discussed previously, so that we can study the effect of Q_0 by itself. The charge Q_0 is called the *equivalent interface charge* and is always positive, for both p-type and n-type substrates. The equivalent interface charge Q_0 will cause a total charge $- Q_0$ to appear in the system, as required from charge neutrality. As shown in Fig. B-1d, part of that charge will appear at the gate and part in the semiconductor. If Q_0 is positive, the charge in the semiconductor will consist of both ionized acceptor atoms and free electrons. Therefore, a *channel* tends to form. If it is desired to eliminate this effect, if all the required balancing charge $- Q_0$ were provided on the gate, no charge would be induced in the semiconductor. To provide a charge $- Q_0$ on the gate, a battery can be connected in series with the external circuit, with the negative terminal toward the gate. Since at the gate and substrate ends of the oxide the charge must be $- Q_0$ and Q_0, respectively, the potential drop across the oxide, ϕ_{ox}, defined from the gate, through the oxide, to the substrate must be equal to $- Q_0/C_{ox}$, where C_{ox} is the total capacitance between the two ends of the oxide. This is exactly the voltage that must be provided by the battery, as shown in Fig. B-1e.

Rather than use the total charge Q_0, it is more convenient to use the equivalent interface charge per unit area, $Q_0' \equiv Q_0/A_J$, where A_J is the cross-sectional area. Similarly, instead of the total capacitance C_{ox}, use the oxide capacitance per unit area, $C_{ox}' \equiv C_{ox}/A_J$.

From basic physics [Eq. (A-43)], it can be written that

$$C_{ox}' = \frac{\epsilon_{ox}}{t_{ox}} \qquad (B\text{-}7)$$

where t_{ox} is the *oxide thickness* and ϵ_{ox} is the *permittivity of the oxide*, given by

$$\epsilon_{ox} = k_{ox}\epsilon_0 \qquad (B\text{-}8)$$

in which ϵ_0 is the *permittivity of free space* ($\epsilon_0 = 8.86 \times 10^{-14}$ F/cm), and k_{ox} is the *dielectric constant of the oxide* ($k_{ox} = 3.9$ for SiO_2).

The potential drop, ϕ_{ox}, across the oxide can now be written as

$$\phi_{ox} = - \frac{Q_0'}{C_{ox}'} \qquad (B\text{-}9)$$

Therefore it has been seen that an external voltage is required between the gate and substrate terminals, to keep the semiconductor neutral everywhere by canceling the effects of the contact potentials and Q_0'.

This voltage is called the *flat-band voltage* and is denoted by V_{FB}. From Fig. B-1e, the expression for the flat-band voltage becomes

$$V_{FB} = \phi_{MS} - \frac{Q_0'}{C_{ox}'} \qquad (B\text{-}10)$$

Of the two terms on the right-hand side of the above equation, the first takes care of the contact potentials and the second takes care of the equivalent interface charge.

B.1.3 Potential balance and charge balance for arbitrary V_{GB}

We will now discuss how the substrate is affected when the externally applied voltage V_{GB} assumes values different from the flat-band voltage V_{FB}. The general case is shown in Fig. B-2. An arbitrary value of V_{GB} in general causes charges to appear in the semiconductor; practically all these charges are contained within a region adjacent to the surface of the semiconductor, called the *space-charge region,* shaded in Fig. B-2.

Outside the space-charge region, the substrate is neutral, and since that region contains charges, a potential change is expected across it. The total potential drop across the space-charge region, defined from the surface to a point in the bulk outside that region, is called *surface potential* ϕ_s.

Four kinds of potential changes are encountered in the loop, as seen in Fig. B-2 [1]:

1. The voltage of the external source, V_{GB}
2. The potential drop across the oxide, ϕ_{ox}
3. The surface potential, ϕ_s
4. Several contact potentials; their sum, when going clockwise, is ϕ_{MS}, which is known from Eqs. (B-4) and (B-6)

Going around the loop, it can be written that

$$V_{GB} = \phi_{ox} + \phi_s + \phi_{MS} \qquad (B\text{-}11)$$

In Eq. (B-11), ϕ_{MS} is a known constant and any changes in V_{GB} must be balanced by changes in ϕ_{ox} and ϕ_s. The gate, the bulk contact, and the wires all are assumed to be made of the same material; then the only potential is between the bulk and its metal cap.

Three kinds of charge are encountered in the type of system depicted in Fig. B-2:

1. The charge on the gate, Q_G
2. The equivalent interface charge, Q_0
3. The charge in the space-charge region, Q_{SC}

These charges must balance one another for overall charge neutrality in the system.

Using charges per unit area yields

$$Q'_G + Q'_0 + Q'_{SC} = 0 \tag{B-12}$$

If Q'_G is changed, the balance required by Eq. (B-12) will be achieved through a change in Q'_{SC}, since the equivalent interface charge Q'_0 is practically fixed.

B.1.4   Effect of V_{GB} and V_{FB} on surface condition

Depending on whether V_{GB} is equal to, less than, or greater than the flat-band voltage V_{FB}, three cases can be distinguished, as indicated in the following sections.

a. Flat-band condition.   The flat-band case has already been discussed in detail in Sec. B.1.2 and is illustrated in Fig. B-1e. In the flat-band condition, therefore,

$$V_{GB} = V_{FB}$$

$$Q'_{SC} = 0 \tag{B-13}$$

$$\phi_s = 0$$

b. Accumulation.   The case in which $V_{GB} < V_{FB}$ is considered now. The negative change of V_{GB} (relative to the flat band) will cause a negative change in Q'_G which, according to Eq. (B-12), must be balanced by a positive change in Q'_{SC} above the value given by Eq. (B-13). Thus, holes will accumulate at the surface. This condition is called *accumulation* and is

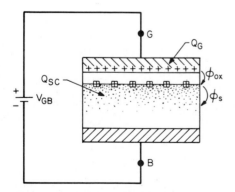

Figure B-2   A two-terminal MOS structure under general gate-substrate bias [1].

illustrated in Fig. B-3a. The negative change in V_{GB} will be shared by negative changes in ϕ_{ox} and ϕ_s, so that Eq. (B-11) will remain valid.

In accumulation, therefore,

$$V_{GB} < V_{FB}$$

$$Q'_{SC} > 0 \qquad\qquad (B\text{-}14)$$

$$\phi_s \quad < 0$$

c. Depletion and inversion. The case in which $V_{GB} > V_{FB}$ is considered now. The total charge on the gate Q'_G will become more positive than the value in the flat band (that value is $-Q_0$ in Fig. B-1e). An example is shown in Fig. B-3b, where it is assumed that the resulting Q'_G is positive. The positive change in Q'_G (relative to the flat band) must be balanced by a negative change in Q'_{SC} from the value of Eq. (B-13), so that Eq. (B-12) is still valid. Also, the positive change in V_{GB} will be shared among ϕ_{ox} and ϕ_s, while Eq. (B-11) remains valid. In *depletion* and *inversion*, therefore,

$$V_{GB} > V_{FB}$$

$$Q'_{SC} < 0 \qquad\qquad (B\text{-}15)$$

$$\phi_s \quad > 0$$

Depletion condition. Now the nature of the negative charge Q'_{SC} is considered. If V_{GB} is only slightly higher than V_{FB}, the positive potential at the surface with respect to the bulk will drive holes away from the surface, leaving it depleted; this condition is called *depletion* and is illustrated in Fig. B-3b. The space charge Q'_{SC} is then due to the *uncovered* acceptor atoms, each of which contributes a charge $-q$.

Inversion condition. As V_{GB} is increased further, more acceptor atoms are uncovered, and ϕ_s becomes sufficiently positive to attract a significant number of free electrons to the surface; each of these electrons will also contribute a charge $-q$ to Q'_{SC}. Eventually, with a sufficiently high V_{GB}, the density of electrons will exceed that of holes at the surface; this is a situation opposite from that normally expected in a p-type material. For this reason, this condition is called *inversion*. The situation is illustrated in Fig. B-3c.

B.1.5   Inversion

The inversion regime is investigated carefully in this section. Taking into account Eqs. (B-15), the discussion focuses on the relative contributions of uncovered acceptor atoms and of free electrons to Q'_{SC}, which are func-

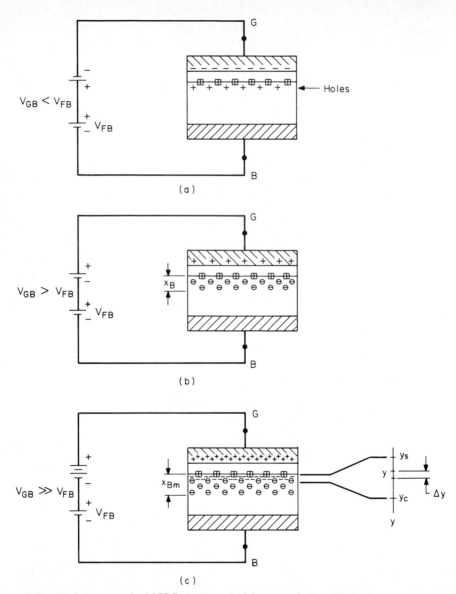

Figure B-3 A two-terminal MOS structure in (a) accumulation, (b) depletion, and (c) inversion [1].

tions of the surface potential ϕ_s. Expressions for this functional dependence will then be derived [1].

a. Charge due to the free electrons Q_I'. The electron concentration at the surface, denoted by n_s, can be related to the electron concentration in the bulk, n_b, through the Boltzmann factor [see Eqs. (A-28)]

$$n_s = n_b e^{q\phi_s/kT} \tag{B-16}$$

where ϕ_s is the potential at the surface with respect to the bulk. Since the substrate is neutral deep in the bulk, from Eqs. (A-8) and (A-9) it follows that

$$p_b \simeq N_A \tag{B-17}$$

$$n_b \simeq \frac{n_i^2}{N_A}$$

Using Eqs. (A-26b) and (B-17) and taking into account the convention on potentials used in this section, Eq. (B-16) can be written in the following two forms:

$$n_s = n_i e^{q(\phi_s - \phi_p)/kT} \tag{B-18}$$
$$= N_A e^{q(\phi_s - 2\phi_p)/kT}$$

The surface electron concentration has been plotted against the surface potential in Fig. B-4, using Eq. (B-18). At $\phi_s = \phi_p$, n_s becomes equal to the intrinsic concentration; then $n_s = p_s$. This is the limit point between the depletion and inversion regimes, as indicated in the figure.

Equation (B-18) gives the electron concentration per unit volume at the surface. At any other point of ordinate y, the electron concentration $n(y)$ will be given by equations like (B-18), when ϕ_s is replaced by $\phi(y)$, the potential at point y with respect to the bulk. Going away from the surface, $\phi(y)$ decreases toward zero and $n(y)$ vanishes rapidly due to its exponential dependence on $\phi(y)$. Hence a point $y = y_c$, below which the electron concentration will be negligible, can be chosen. Practically all the free electrons are then contained in a narrow layer (the inversion layer) between $y = y_s$ and $y = y_c$.

Therefore, the inversion-layer charge per unit area, Q_I', due to the free electrons, is given by

$$Q_I' = -q \int_{y_c}^{y_s} n(y)\, dy \tag{B-19}$$

b. Charge due to the uncovered acceptor atoms Q_B'. Evaluating Q_I' in this manner can be done only through numerical integration. Here we will

determine an accurate expression for Q_B' (the charge of the ionized acceptor atoms in the depletion region) and then evaluate Q_i'.

The depletion region is then considered. As in the case of the *pn* junction, we will regard this region as being defined by a sharp boundary at a
depth x_B below the surface. The inversion layer is at the top of this region;
numerical calculations have shown that most of this layer is concentrated
very close to the surface. Since the depth x_B of the depletion region is
usually much larger, it is assumed that the inversion layer is a sheet of
negligible thickness [1]. This has been called the *charge sheet approximation* and implies that practically all the depletion region is free of
minority carriers (it is the same as the *depletion approximation* used in
Sec. A.2.1 in the discussion of the *pn* junction).

For a negligible thickness, the potential drop across the inversion layer
will also be negligible, and it can be assumed that all the surface potential
ϕ_s is dropped across the depletion region in the *p*-type substrate. Thus
from Eqs. (A-35), with the assumption that $N_D \gg N_A$, it follows that

$$x_B \simeq \sqrt{\frac{2\epsilon_s}{qN_A}\phi_s} \qquad \text{(B-20)}$$

Let Q_B be the charge due to the uncovered acceptor atoms in the depletion region, and $Q_B' \equiv Q_B/A_J$ the corresponding charge per unit area. As

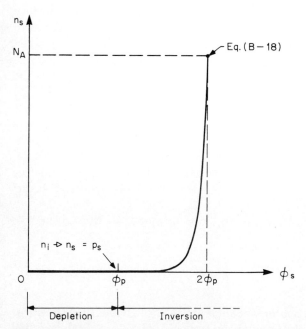

Figure B-4   Electron concentration at the surface vs. surface potential [1].

the total charge per unit area due to these uncovered acceptor atoms can be expressed as $-qN_Ax_B$, then

$$Q'_B = -\sqrt{2q\epsilon_s N_A \phi_s} \tag{B-21}$$

The total charge per unit area in the space-charge region, Q'_{SC}, is then the sum of the charge of electrons in the inversion layer, Q'_I, and the charge of the ionized acceptor atoms in the depletion region, Q'_B. Thus

$$Q'_{SC} = Q'_I + Q'_B \tag{B-22}$$

The evaluation of Q'_{SC} can be made directly (that is, without regarding Q'_I and Q'_B individually). Such an evaluation can be found by using the basic laws of electrostatics; the result is

$$Q'_{SC} = -\sqrt{2q\epsilon_s N_A} \sqrt{\phi_s + \frac{kT}{q} e^{q(\phi_s - 2\phi_p)/kT}} \tag{B-23}$$

It should be noted that the above expression is derived by assuming that no holes are present in the space-charge region; thus, it will be valid in inversion as well as in the upper part of the depletion region. However, in accumulation or in the bottom part of the depletion region, holes are present in significant numbers and the above expression will not hold; a more general expression is then needed.

The inversion layer charge Q'_I can then be evaluated by taking into account Eqs. (B-21) to (B-23). Thus

$$Q'_I = -\sqrt{2q\epsilon_s N_A} \left(\sqrt{\phi_s + \frac{kT}{q} e^{q(\phi_s - 2\phi_p)/kT}} - \sqrt{\phi_s} \right) \tag{B-24}$$

$|Q'_B|$ and $|Q'_I|$ vs. ϕ_s are plotted in Fig. B-5, using Eqs. (B-21) and (B-24); their sum, $|Q'_{SC}|$, is also shown.

It is convenient to divide the inversion region into three subregions: these are marked *weak, moderate,* and *strong inversion* in Fig. B-5. ϕ_{MO} and ϕ_{HO} indicate the onset of moderate and strong inversion, respectively. It is thus seen that for surface potentials less than about $2\phi_p$, practically all the surface charge is due to the charge in the depletion region. The corresponding inversion-layer charge is too small to be shown in the scale of the figure, but can nevertheless cause nonnegligible conduction when the MOS junction is part of a transistor. As ϕ_s is raised above $2\phi_p$, $|Q'_I|$ starts becoming significant due to the exponential in Eq. (B-24). For ϕ_s exceeding $2\phi_p$ by a few kT/q, Q'_I becomes a very strong function of ϕ_s. From Fig. B-4 it can be seen that at $\phi_s = 2\phi_p$ the surface electron concentration is already as high as the concentration of acceptor atoms; an increase of ϕ_s above $2\phi_p$ by several kT/q would be enough to provide very large n_s due to the exponential dependence of the latter on ϕ_s [see Eq. (B-

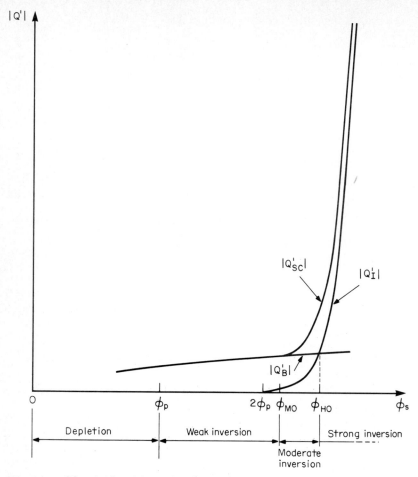

Figure B-5 Magnitude of inversion-layer charge, depletion-region charge, and their sum (all per unit area) vs. surface potential [1].

18)]. The concentration $n(y)$ at points very close to the surface would also increase drastically, and Q'_I would rise to the point that it would dominate Q'_B.

B.1.6   Relations of charges to their potentials

In the previous sections, it has been found that the equations that characterize MOS systems are

$$V_{GB} = \phi_{\text{ox}} + \phi_s + \phi_{MS} \qquad \text{(potential balance)} \qquad \text{(B-25)}$$

and
$$Q'_G + Q'_0 + Q'_I + Q'_B = 0 \qquad \text{(charge balance)} \qquad \text{(B-26)}$$

Here, three equations that relate each charge to the potential that causes its presence are analyzed:

1. The charge on the top *plate* of the *oxide capacitor* is related to the potential across the oxide by

$$Q_G' = C_{ox}' \phi_{ox} \qquad \text{(B-27)}$$

2. The *inversion-layer charge* is related to its cause, the surface potential, by Eq. (B-24), which is of the form

$$Q_I' = Q_I'(\phi_s) \qquad \text{(B-28)}$$

3. The *depletion-region charge* is related to the potential across that region by Eq. (B-21), which is of the form

$$Q_B' = Q_B'(\phi_s) \qquad \text{(B-29)}$$

Equations (B-25) to (B-29) contain six variables, three of which are potentials (V_{GB}, ϕ_{ox}, ϕ_s) and three of which are charges per unit area (Q_G', Q_I', Q_B'). Among the five equations, four of the variables can be eliminated and a relation between the other two can be obtained. For example, a relation between V_{GB} and ϕ_s can be developed. Eliminating the other variables among Eqs. (B-25) to (B-29), it follows that [1]

$$V_{GB} = V_{FB} + \phi_s + \gamma \sqrt{\phi_s + \frac{kT}{q} e^{q(\phi_s - 2\phi_p)/kT}} \qquad \text{(B-30)}$$

where

$$\gamma = \frac{\sqrt{2q\epsilon_s N_A}}{C_{ox}'} \qquad \text{(B-31)}$$

The parameter γ is called the *body-effect coefficient*, or *substrate-effect* coefficient. Equation (B-30) is plotted in Fig. B-6 as ϕ_s vs. V_{GB}.

The goal in this development is to find a relation of the form

$$Q_I' = Q_I'(V_{GB}) \qquad \text{(B-32)}$$

The reason is that the inversion-layer charge is the cause of the conduction when the MOS junction is part of a MOS transistor and the gate-to-substrate voltage is an external bias voltage at the disposal of the circuit designer. Equation (B-32) can be used to derive complete expressions that relate currents to voltages in MOS transistors. Unfortunately, if one attempts to derive a relation in the form of Eq. (B-32) from Eqs. (B-25) to (B-29), an implicit expression results; that is, Q_I' cannot be expressed in closed form as a function of V_{GB}. The solution of the resulting complicated equation has to be obtained numerically, if a value for V_{GB} is given and the value for Q'_I is desired. Instead of Eq. (B-32), we can consider its parametric representation, which consists of Eqs. (B-24) and (B-30). If values are assumed for ϕ_s, the corresponding Q'_I and V_{GB} can be found from these equations. Q'_I can then be plotted against V_{GB} as in Fig. B-7.

Before discussing the shape of the curve in Fig. B-7, an alternative expression for Q'_I is developed. If Q'_G and ϕ_{ox} are eliminated among Eqs. (B-25) to (B-27), and if Eqs. (B-21) and (B-10) are used, then [1]

$$Q'_I = -C'_{ox}(V_{GB} - V_{FB} - \phi_s - \gamma\sqrt{\phi_s}) \tag{B-33}$$

where γ has been defined in Eq. (B-31). If Eq. (B-33) is used in conjunction with Eq. (B-30), the plot of Fig. B-7 is again obtained.

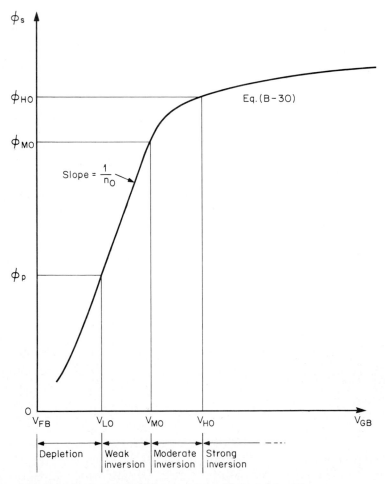

Figure B-6 Surface potential versus gate-substrate bias [1].

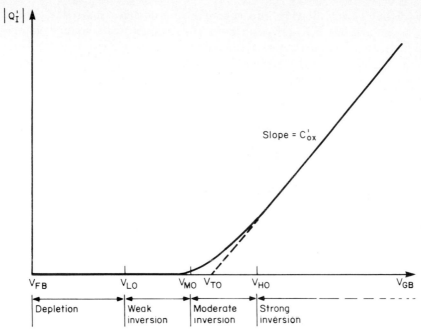

Figure B-7   Magnitude of inversion-layer charge per unit area vs. gate-substrate bias [1].

Equations that describe the behavior of Q_I' in terms of V_{GB} have been developed. They are general; i.e., they hold for any value of V_{GB}, irrespective of whether the channel is weakly, moderately, or strongly inverted. Equations (B-24) and (B-30), however, are too complex to be used for simple analysis by hand. If some generality can be sacrificed, then much simpler results can be obtained. Specifically, by considering one regime at a time, simplifications that will result in explicit expressions of the form of Eq. (B-32) can be made; these expressions will be simpler, but will be valid *only* in that regime [1].

B.1.7   Incremental capacitance

If V_{GB} is increased by a small amount dV_{GB}, a positive charge dQ_G' will flow into the gate terminal. For overall charge neutrality, a charge of equal value must flow out of the substrate terminal, or, equivalently, a charge of value $-dQ_G'$ must flow into the substrate terminal. An incremental *(small-signal)* capacitance per unit area, C_{GB}', can thus be defined to relate charge changes to voltage changes. Thus

$$C_{GB}' = \frac{dQ_G'}{dV_{GB}} \tag{B-34}$$

The charge $-dQ'_G$ flowing into the substrate changes the space charge Q'_{SC} by an amount dQ'_{SC}. Furthermore, the gate-substrate voltage change will be distributed partly across the oxide and partly across the semiconductor as a change in the surface potential. Inverting Eq. (B-34) and taking into account what was stated above, it follows that

$$\frac{1}{C'_{GB}} = \frac{d\phi_{\text{ox}}}{dQ'_G} + \frac{d\phi_s}{dQ'_G} \tag{B-35}$$

The preceding equation can be written as

$$\frac{1}{C'_{GB}} = \frac{1}{dQ'_G/d\phi_{\text{ox}}} + \frac{1}{-dQ'_{SC}/d\phi_s} \tag{B-36}$$

To interpret this equation, it must be noted that the denominator of the last fraction in Eq. (B-36) can be interpreted as a capacitance

$$C'_{SC} = -\frac{dQ'_{SC}}{d\phi_s} \tag{B-37}$$

corresponding to the space-charge region (it relates the changes of the potential across that region to the resulting changes in its charge), while the denominator of the first fraction can be interpreted as a capacitance, C'_{ox}, following Eq. (B-27). Thus Eq. (B-36) becomes

$$\frac{1}{C'_{GB}} = \frac{1}{C'_{\text{ox}}} + \frac{1}{C'_{SC}} \tag{B-38}$$

Therefore the incremental capacitance C'_{GB} is the same as that exhibited by two linear capacitors of values C'_{ox} and C'_{SC}, connected in series as in Fig. B-8a.

It is interesting to consider the individual contributions of the depletion-region and inversion-layer charges to C'_{SC}. Thus it can be written that

$$C'_{SC} = -\frac{dQ'_B}{d\phi_s} + \left(-\frac{dQ'_I}{d\phi_s}\right) \tag{B-39}$$

The total space-charge capacitance C'_{SC} is then divided into two components: one due to the depletion-region charge and one due to the inversion-layer charge.

Analogously to Eq. (B-37), a depletion-region incremental capacitance per unit area can be defined as

$$C'_B = -\frac{dQ'_B}{d\phi_s} \tag{B-40}$$

This capacitance relates changes of the potential across the depletion region to the resulting changes of the charge in it.

Finally, a capacitance per unit area associated with the inversion layer can be defined. This capacitance should relate changes in the charge of that layer to the potential changes that cause them. In analogy with Eq. (B-40) it can be written that

$$C_I' = -\frac{dQ_I'}{d\phi_s} \tag{B-41}$$

Using Eqs. (B-40) and (B-41) in Eq. (B-39), it follows that

$$C_{SC}' = C_B' + C_I' \tag{B-42}$$

and Eq. (B-38) becomes

$$\frac{1}{C_{GB}'} = \frac{1}{C_{ox}'} + \frac{1}{C_B' + C_I'} \tag{B-43}$$

which can be represented by the circuit of Fig. B-8b.

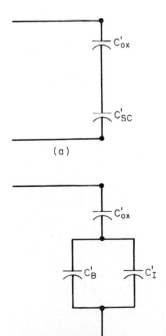

(a)

(b)

Figure B-8 Incremental capacitances for a two-terminal MOS structure [1].

REFERENCES

1. Y. P. Tsividis, *Operation and Modeling of the MOS Transistor,* McGraw-Hill, New York, 1987.
2. A. S. Grove, *Physics and Technology of Semiconductor Devices,* Wiley, New York, 1967.
3. S. M. Sze, *Physics of Semiconductor Devices,* Wiley, New York, 1969.
4. R. S. Muller and T. I. Kamins, *Device Electronics for Integrated Circuits,* Wiley, New York, 1977.
5. A. B. Glaser and G. Subak-Sharpe, *Integrated Circuit Engineering,* Addison-Wesley, Reading, Mass., 1979.
6. A. G. Milnes, *Seminconductor Devices and Integrated Electronics,* Van Nostrand Reinhold, New York, 1980.
7. DeWitt, G. Ong, *Modern MOS Technology: Processes, Devices, and Design,* McGraw-Hill, New York, 1984.

Index

ABOUT THE EDITORS

PAOLO ANTOGNETTI is a professor on the faculty of the University of Genoa, Italy. Dr. Antognetti's extensive background in semiconductor modeling and simulation includes teaching and research positions as a post-doctoral fellow at Stanford University and as visiting professor at MIT. He is the author or coauthor of more than thirty papers in the field, and the editor of five books on VLSI aspects of computer-aided design and process modeling, including *Power Integrated Circuits: Physics, Design, and Applications* from McGraw-Hill. Dr. Antognetti has also translated several books on electronic engineering into Italian.

GIUSEPPE MASSOBRIO is a research associate in the Department of Electronics (DIBE) of the University of Genoa, Italy, where he received his degree in 1976. He has done research in the areas of semiconductor device modeling, circuit simulation, and optimization. Mr. Massobrio has published several papers on modeling BJT devices for CAD applications. His current research interests include the development of models of ISFET-based biosensors for biomedical applications.